Sympathetic Nerves and Cardiovascular Diseases

Sympathetic Nerves and Cardiovascular Diseases

Editors

Yutang Wang
Kate Denton

MDPI

Basel • Beijing • Wuhan • Barcelona • Belgrade • Novi Sad • Cluj • Manchester

Editors
Yutang Wang
Life Science
Federation University
Australia
Ballarat
Australia

Kate Denton
Department of Physiology
Monash University
Melbourne
Australia

Editorial Office
MDPI
St. Alban-Anlage 66
4052 Basel, Switzerland

This is a reprint of articles from the Special Issue published online in the open access journal *International Journal of Molecular Sciences* (ISSN 1422-0067) (available at: https://www.mdpi.com/journal/ijms/special_issues/Nerves_Cardiovascular_Diseases).

For citation purposes, cite each article independently as indicated on the article page online and as indicated below:

Lastname, A.A.; Lastname, B.B. Article Title. *Journal Name* **Year**, *Volume Number*, Page Range.

ISBN 978-3-7258-0543-3 (Hbk)
ISBN 978-3-7258-0544-0 (PDF)
doi.org/10.3390/books978-3-7258-0544-0

Contents

Yutang Wang and Kate M. Denton
Special Issue "Sympathetic Nerves and Cardiovascular Diseases"
Reprinted from: *Int. J. Mol. Sci.* **2024**, *25*, 2633, doi:10.3390/ijms25052633 . . . . **1**

Lu Qin, Jian Cui and Jianhua Li
Sympathetic Nerve Activity and Blood Pressure Response to Exercise in Peripheral Artery Disease: From Molecular Mechanisms, Human Studies, to Intervention Strategy Development
Reprinted from: *Int. J. Mol. Sci.* **2022**, *23*, 10622, doi:10.3390/ijms231810622 . . . . **4**

Yu-Long Li
Stellate Ganglia and Cardiac Sympathetic Overactivation in Heart Failure
Reprinted from: *Int. J. Mol. Sci.* **2022**, *23*, 13311, doi:10.3390/ijms232113311 . . . . **31**

Evgenii Ivanov, Marina Akhmetshina, Aleksei Erdiakov and Svetlana Gavrilova
Sympathetic System in Wound Healing: Multistage Control in Normal and Diabetic Skin
Reprinted from: *Int. J. Mol. Sci.* **2023**, *24*, 2045, doi:10.3390/ijms24032045 . . . . **49**

Davide Martelli and Virginia L. Brooks
Leptin Increases: Physiological Roles in the Control of Sympathetic Nerve Activity, Energy Balance, and the Hypothalamic–Pituitary–Thyroid Axis
Reprinted from: *Int. J. Mol. Sci.* **2023**, *24*, 2684, doi:10.3390/ijms24032684 . . . . **69**

Christian A. Reynolds and Zeljka Minic
Chronic Pain-Associated Cardiovascular Disease: The Role of Sympathetic Nerve Activity
Reprinted from: *Int. J. Mol. Sci.* **2023**, *24*, 5378, doi:10.3390/ijms24065378 . . . . **92**

Niujin Shi, Jingbo Xia, Chaoge Wang, Jie Zhou, Junhao Huang, Min Hu and Jingwen Liao
Aerobic Exercise Prevents Arterial Stiffness and Attenuates Hyperexcitation of Sympathetic Nerves in Perivascular Adipose Tissue of Mice after Transverse Aortic Constriction
Reprinted from: *Int. J. Mol. Sci.* **2022**, *23*, 11189, doi:10.3390/ijms231911189 . . . . **110**

Xiao-Li Wang, Jing-Xiao Wang, Jun-Liu Chen, Wen-Yuan Hao, Wen-Zhou Xu, Zhi-Qin Xu, et al.
Asprosin in the Paraventricular Nucleus Induces Sympathetic Activation and Pressor Responses via cAMP-Dependent ROS Production
Reprinted from: *Int. J. Mol. Sci.* **2022**, *23*, 12595, doi:10.3390/ijms232012595 . . . . **121**

Roberto B. Pontes, Erika E. Nishi, Renato O. Crajoinas, Maycon I. O. Milanez, Adriana C. C. Girardi, Ruy R Campos and Cassia T Bergamaschi
Relative Contribution of Blood Pressure and Renal Sympathetic Nerve Activity to Proximal Tubular Sodium Reabsorption via NHE3 Activity
Reprinted from: *Int. J. Mol. Sci.* **2023**, *24*, 349, doi:10.3390/ijms24010349 . . . . **135**

Jan Naar, Mikulas Mlcek, Andreas Kruger, Dagmar Vondrakova, Marek Janotka, Michaela Popkova, et al.
Acute Severe Heart Failure Reduces Heart Rate Variability: An Experimental Study in a Porcine Model
Reprinted from: *Int. J. Mol. Sci.* **2023**, *24*, 493, doi:10.3390/ijms24010493 . . . . **148**

Marta Kantauskaite, Oliver Vonend, Mina Yakoub, Philipp Heilmann, Andras Maifeld, Peter Minko, et al.
The Effect of Renal Denervation on T Cells in Patients with Resistant Hypertension
Reprinted from: *Int. J. Mol. Sci.* **2023**, *24*, 2493, doi:10.3390/ijms24032493 . . . . **158**

Yutang Wang, Dinh Tam Nguyen, Jack Anesi, Ahmed Alramahi, Paul K. Witting, Zhonglin Chai, et al.
Moxonidine Increases Uptake of Oxidised Low-Density Lipoprotein in Cultured Vascular Smooth Muscle Cells and Inhibits Atherosclerosis in Apolipoprotein E-Deficient Mice
Reprinted from: *Int. J. Mol. Sci.* **2023**, *24*, 3857, doi:10.3390/ijms24043857 . . . . **170**

International Journal of
Molecular Sciences

Editorial

Special Issue "Sympathetic Nerves and Cardiovascular Diseases"

Yutang Wang [1,*] and Kate M. Denton [2,3]

1 Discipline of Life Science, Institute of Innovation, Science and Sustainability, Federation University Australia, Ballarat, VIC 3350, Australia
2 Department of Physiology, Monash University, Melbourne, VIC 3800, Australia; kate.denton@monash.edu
3 Cardiovascular Disease Program, Monash Biomedicine Discovery Institute, Monash University, Melbourne, VIC 3800, Australia
* Correspondence: yutang.wang@federation.edu.au

Cardiovascular diseases (CVDs) constitute a spectrum of disorders affecting the heart and blood vessels, which include coronary heart disease, cerebrovascular disease, and peripheral artery disease [1]. The major underlying cause of CVD is atherosclerosis, which can be exacerbated by risk factors including hypertension, hypercholesterolemia, and diabetes [2].

Despite significant advancements in our understanding of disease pathogenesis and the availability of numerous therapeutic drugs [3], the burden of CVD continues to increase globally [4]. CVDs remain the leading cause of death worldwide, responsible for 17.9 million deaths each year [1]. This suggests that our understanding of CVD pathogenesis remains limited and incomplete. Indeed, a growing body of evidence indicates the significant involvement of the sympathetic nervous system in CVD pathogenesis [5], a notion that has been largely overlooked until recently.

This Special Issue aims to assemble the latest perspectives and research findings to illuminate the role of altered sympathetic nerve activity (SNA) in CVD pathogenesis. It comprises five review articles and six original research articles, collectively emphasizing the significance of SNA in the pathogenesis of various CVDs, including hypertension, atherosclerosis, heart failure, peripheral artery disease, diabetic wound healing, and chronic pain-associated CVDs.

Review 1, titled "Sympathetic Nerve Activity and Blood Pressure Response to Exercise in Peripheral Artery Disease: From Molecular Mechanisms, Human Studies, to Intervention Strategy Development", explored the abnormal SNA-mediated blood pressure response during exercise in peripheral artery disease. This study elucidates the molecular mechanisms underlying such abnormal responses and intervention strategies to improve them, thus enhancing tolerance and performance during exercise in patients with peripheral artery disease.

Review 2, titled "Stellate Ganglia and Cardiac Sympathetic Overactivation in Heart Failure", delved into cardiac sympathetic remodeling in stellate ganglia in heart failure. It elucidates potential underlying mechanisms and emphasizes the therapeutic potential of targeting cardiac sympathetic remodeling against malignant cardiac arrhythmias in heart failure.

Review 3, titled "Sympathetic System in Wound Healing: Multistage Control in Normal and Diabetic Skin", explored sympathetic regulation and signaling pathways in normal and diabetic wound healing. The review highlights the potential utility of β2-adrenoceptor blockers and nicotinic acetylcholine receptor activators in treating diabetic ulcers with neuropathy.

Review 4, titled "Leptin Increases: Physiological Roles in the Control of Sympathetic Nerve Activity, Energy Balance, and the Hypothalamic–Pituitary–Thyroid Axis", explored

Citation: Wang, Y.; Denton, K.M. Special Issue "Sympathetic Nerves and Cardiovascular Diseases". *Int. J. Mol. Sci.* **2024**, *25*, 2633. https://doi.org/10.3390/ijms25052633

Received: 5 February 2024
Accepted: 20 February 2024
Published: 23 February 2024

the role of leptin in regulating sympathetic nerve activity and energy balance. It also discussed obesity-induced inflammation and its impact on leptin's actions during obesity.

Review 5, titled "Chronic Pain-Associated Cardiovascular Disease: The Role of Sympathetic Nerve Activity", explored the link between sympathetic nervous system dysfunction and chronic pain, highlighting its contribution to CVD in chronic pain settings.

The original research articles delve into various aspects related to SNA and its impact on cardiovascular health.

Original Research 1, titled "Aerobic Exercise Prevents Arterial Stiffness and Attenuates Hyperexcitation of Sympathetic Nerves in Perivascular Adipose Tissue of Mice after Transverse Aortic Constriction", investigated the effect of exercise on arterial stiffness and the potential role of sympathetic nerves within perivascular adipose tissue using the pressure overload-induced heart failure model in mice. This study found that regular aerobic exercise prevented arterial stiffness during heart failure, and sympathetic innervation and adiponectin within PVAT played a role in this process.

Original Research 2, titled "Asprosin in the Paraventricular Nucleus Induces Sympathetic Activation and Pressor Responses via cAMP-Dependent ROS Production", investigated the roles and underlying mechanisms of asprosin (an adipokine involved in metabolism) in the paraventricular nucleus in regulating sympathetic outflow and blood pressure. The study found that asprosin in the PVN increased sympathetic outflow, blood pressure, and heart rate via NADPH oxidase activation and subsequent superoxide production.

Original Research 3, titled "Relative Contribution of Blood Pressure and Renal Sympathetic Nerve Activity to Proximal Tubular Sodium Reabsorption via NHE3 Activity", investigated the effects of blood pressure and renal SNA on renal salt and water balance. This study found that an acute increase in blood pressure and renal SNA inhibited hydrogen exchanger 3 (NHE3) activity, thus leading to salt and water excretion.

Original Research 4, titled "Acute Severe Heart Failure Reduces Heart Rate Variability: An Experimental Study in a Porcine Model", investigated the effect of experimental acute heart failure on heart rate variability. This study found that acute severe cardiac failure, induced by global myocardial hypoxia, was associated with a significant reduction in heart rate variability.

Original Research 5, titled "The Effect of Renal Denervation on T Cells in Patients with Resistant Hypertension", investigated the effect of Renal Denervation on T-cell signatures in Resistant Hypertension. This study found that patients with resistant hypertension had higher levels of certain types of T cells in the blood, and the extent of the decrease in blood pressure by renal denervation was associated with T cell frequencies at baseline. Therefore, a detailed analysis of T cells might be useful in selecting patients for renal denervation.

Original Research 6, titled "Moxonidine Increases Uptake of Oxidised Low-Density Lipoprotein in Cultured Vascular Smooth Muscle Cells and Inhibits Atherosclerosis in Apolipoprotein E-Deficient Mice", investigated the effect of the sympatholytic drug moxonidine on atherosclerosis. This study found that moxonidine inhibited atherosclerosis in apolipoprotein E-deficient mice.

Overall, this Special Issue provides a comprehensive overview of the diverse effects of SNA on CVD, from hypertension and atherosclerosis [6] to peripheral artery disease and heart failure [7]. The insights derived from these reviews and original research studies enhance our understanding of the importance of the sympathetic nervous system in CVD pathogenesis and will facilitate the development of new therapies to treat CVDs.

Author Contributions: Y.W. prepared the manuscript. Y.W. and K.M.D. revised the manuscript. All authors have read and agreed to the published version of the manuscript.

Conflicts of Interest: The authors declare no conflicts of interest.

References

1. World Health Organization. Cardiovascular Diseases. Available online: https://www.who.int/health-topics/cardiovascular-diseases#tab=tab_1 (accessed on 4 January 2024).
2. Scott, J. Pathophysiology and biochemistry of cardiovascular disease. *Curr. Opin. Genet. Dev.* **2004**, *14*, 271–279. [CrossRef] [PubMed]
3. Hong, C.C. The grand challenge of discovering new cardiovascular drugs. *Front. Drug Discov.* **2022**, *2*, 1027401. [CrossRef]
4. Roth, G.A.; Mensah, G.A.; Johnson, C.O.; Addolorato, G.; Ammirati, E.; Baddour, L.M.; Barengo, N.C.; Beaton, A.Z.; Benjamin, E.J.; Benziger, C.P.; et al. Global Burden of Cardiovascular Diseases and Risk Factors, 1990–2019: Update from the GBD 2019 Study. *J. Am. Coll. Cardiol.* **2020**, *76*, 2982–3021. [CrossRef] [PubMed]
5. Hadaya, J.; Ardell, J.L. Autonomic Modulation for Cardiovascular Disease. *Front. Physiol.* **2020**, *11*, 617459. [CrossRef] [PubMed]
6. Wang, Y.; Nguyen, D.T.; Anesi, J.; Alramahi, A.; Witting, P.K.; Chai, Z.; Khan, A.W.; Kelly, J.; Denton, K.M.; Golledge, J. Moxonidine Increases Uptake of Oxidised Low-Density Lipoprotein in Cultured Vascular Smooth Muscle Cells and Inhibits Atherosclerosis in Apolipoprotein E-Deficient Mice. *Int. J. Mol. Sci.* **2023**, *24*, 3857. [CrossRef] [PubMed]
7. Zhang, D.Y.; Anderson, A.S. The sympathetic nervous system and heart failure. *Cardiol. Clin.* **2014**, *32*, 33–45. [CrossRef] [PubMed]

International Journal of
Molecular Sciences

MDPI

Review

Sympathetic Nerve Activity and Blood Pressure Response to Exercise in Peripheral Artery Disease: From Molecular Mechanisms, Human Studies, to Intervention Strategy Development

Lu Qin, Jian Cui and Jianhua Li *

Heart & Vascular Institute, The Penn State University College of Medicine, Hershey, PA 17033, USA
* Correspondence: jianhuali@pennstatehealth.psu.edu

Abstract: Sympathetic nerve activity (SNA) regulates the contraction of vascular smooth muscle and leads to a change in arterial blood pressure (BP). It was observed that SNA, vascular contractility, and BP are heightened in patients with peripheral artery disease (PAD) during exercise. The exercise pressor reflex (EPR), a neural mechanism responsible for BP response to activation of muscle afferent nerve, is a determinant of the exaggerated exercise-induced BP rise in PAD. Based on recent results obtained from a series of studies in PAD patients and a rat model of PAD, this review will shed light on SNA-driven BP response and the underlying mechanisms by which receptors and molecular mediators in muscle afferent nerves mediate the abnormalities in autonomic activities of PAD. Intervention strategies, particularly non-pharmacological strategies, improving the deleterious exercise-induced SNA and BP in PAD, and enhancing tolerance and performance during exercise will also be discussed.

Keywords: sympathetic nerve activity; arterial blood pressure; peripheral artery disease; static exercise; muscle afferent nerve; heat treatment

Citation: Qin, L.; Cui, J.; Li, J. Sympathetic Nerve Activity and Blood Pressure Response to Exercise in Peripheral Artery Disease: From Molecular Mechanisms, Human Studies, to Intervention Strategy Development. *Int. J. Mol. Sci.* **2022**, *23*, 10622. https://doi.org/10.3390/ijms231810622

Academic Editors: Yutang Wang and Kate Denton

Received: 19 August 2022
Accepted: 9 September 2022
Published: 13 September 2022

1. Introduction

Peripheral artery disease (PAD) is a common and disabling cardiovascular disease that affects over 200 million worldwide and ~20% of Americans over age 60 [1–4]. Patients with PAD are at a high risk of myocardial infarctions, cerebral vascular accidents, and all-cause mortality with a death rate like that in patients with coronary or cerebral vascular disease [5–7]. The atherosclerotic alternation in the affected vessel results in progressive narrowing of the lower extremity conduit vasculature and eventually leads to severe limb ischemia. As one of the consequences of limb ischemia, the syndrome of "intermittent claudication" in PAD patients which is characterized by pain in the lower limbs that occurs with walking and is relieved by rest limits their tolerance and performance in daily physical activities.

Compared with the major advances seen in the management of other cardiovascular diseases such as coronary artery disease and systolic heart failure, therapeutic options other than surgery for PAD remain extremely limited [8]. Several pharmacological interventions have been evaluated for use in patients with claudication symptoms, but efficacy has only been reported for cilostazol and anti-platelet agents [9,10]. In fact, exercise training (advice to walk more often) is commonly recommended for PAD patients. It has been supported by studies [11–13] that supervised treadmill exercise is effective in attenuating pain perception and improving the exercise performance of PAD patients. However, the implementation of exercise into the daily lives of PAD patients is met with significant challenges. During exercise, sympathetic nerve activity (SNA) and blood pressure (BP) responses are amplified in PAD patients [14–17], which is associated with a higher risk and incidence of cardiovascular events [18,19].

In this regard, experimental animal models are necessary for studying the underlying molecular mechanisms leading to the exaggerated SNA and BP responses in the pathological conditions of PAD; human clinical studies in both the healthy population and PAD patients are essential to examine the clinical conditions and validate the results of the mechanism studies with approved medications. More importantly, by incorporating the animal study and human study, intervention studies on the treatment targets are vital to develop strategies aiming to be effective and low-cost and to attenuate the above adverse conditions in PAD patients.

2. Sympathetic Nerves and BP Regulation during Exercise in PAD

2.1. Exercise Pressor Reflex (EPR)

During the muscle movements of exercise, the sympathetic nervous activity (SNA) increases, resulting in increased arterial BP and heart rate (HR), myocardial contractility, and peripheral vasoconstriction [20,21]. Two mechanisms: central command and exercise pressor reflex (EPR) [21–26] are considered involved in this regulatory process. Specifically, "Central Command" [27] is initiated by a volitional signal emanating from central motor units and then induces the enhancement of SNA; and the "Exercise Pressor Reflex" [26,28] originates from the signal inputs from the afferents of the contracting skeletal muscle and then induces a subsequent autonomic reflex. For the specific types of signal input, the EPR responds to metabolic stimulation (i.e., "metaboreceptor" stimulation in Group IV afferents) and mechanical deformation (i.e., "mechanoreceptor stimulation" in Group III afferents) in the muscle afferents receptive field [29]. Thin fiber muscle afferent nerves are engaged following the stimulation of the receptors during the exercise, therefore inducing the activation of cardiovascular nuclei in the brainstem [28]. Figure 1 illustrates the activation of the EPR and its neural pathways.

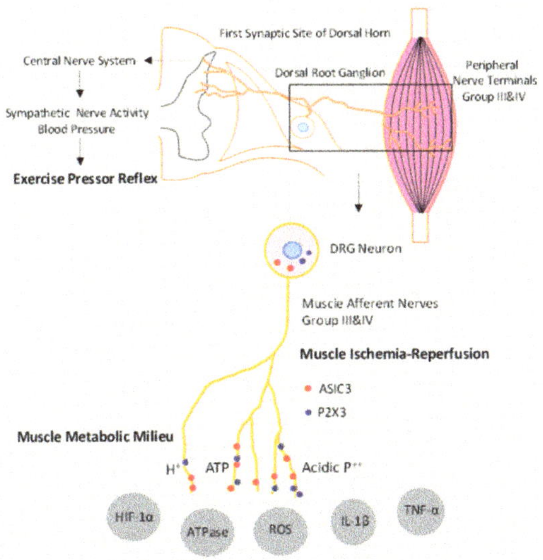

Figure 1. Diagram indicates the potential signaling pathways leading to the exaggerated EPR in IR rats via enhancing acidic metabolites, ATP, and proinflammatory cytokines (PICs) in skeletal muscle and thereby stimulating ASIC3 and P2X3 receptors in muscle sensory nerves. This proposal will examine the integrated signals in both skeletal muscle and primary sensory neurons involved in the EPR of rats with femoral artery occlusion followed by reperfusion. ASIC3: acid-sensing ion channel 3; P2X3: purinergic P2X subtype 3; HIF-1α: hypoxia-induced factor-1α; ROS: reactive oxygen species; IL-1β: interleukin-1β; TNF-α: tumor necrosis factor-α.

2.2. EPR in PAD Patients

In PAD patients, the pressor response to walking is significantly greater than that in healthy control subjects [14–17]. Human studies further indicate that the responses in BP, renal vasoconstriction, and total peripheral resistance (TPR) during plantar flexion exercise are accentuated in PAD [30,31]. In these studies, the increase in BP occurred before the subjects reported pain. Thus, it is believed that an exaggerated EPR is a major determinant of why BP rises with exercise in PAD [32]. The muscle SNA (MSNA) increases in PAD occurred early and were much greater than those at the same exercise time/workload in matched healthy control subjects [33]. Figure 2 shows increased MSNA in PAD patients. It is believed that an exaggerated MSNA response contributes to the accentuated TPR and BP responses to leg exercise in PAD patients and any mechanisms for intervention/therapy decreasing the MSNA response to exercise would alleviate the exaggerated EPR in PAD.

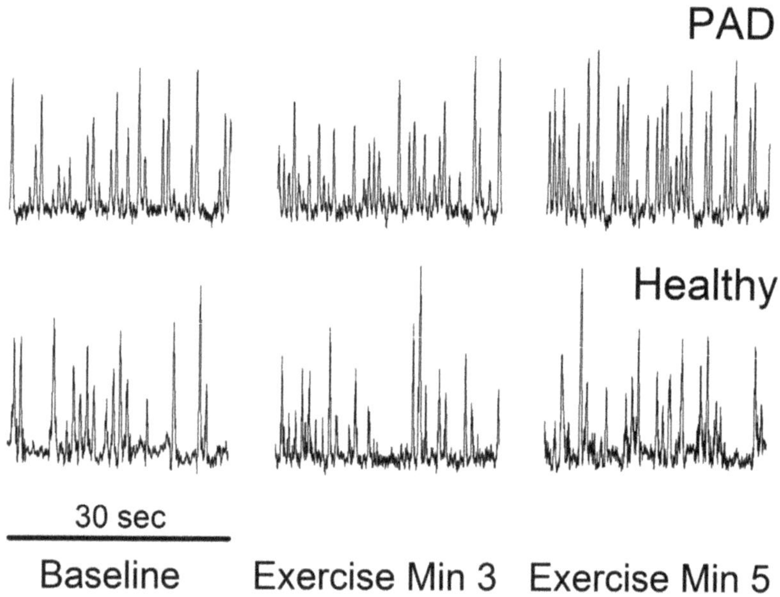

Figure 2. MSNA during plantar flexion with incremental loading (2 kg, +1 kg/min) in a PAD patient. The exercise was ended after min 5 due to the tolerance level of this patient. In the matched healthy control subject, increases in MSNA and BP occurred after min 7 with a 9 kg workload. (Abstract presented at EB 2021; unpublished figure).

3. Experimental Models to Study the Pathological Status in Human PAD

3.1. Blood Flow Restriction (BFR) and Ischemia-Reperfusion (IR) in Healthy Subject

Other than involving PAD patients, there are two popularized experimental models to simulate PAD, namely BFR and IR, in healthy humans. They provide low-risk and feasible ways to mimic the pathological status (e.g., ischemia and ischemia-reperfusion) in PAD.

BFR: Regarding the pathophysiology of PAD, the consequences of limb ischemia have been emphasized [34–38]. It is known that the EPR is amplified as oxygen delivery to skeletal muscle is reduced [39]. Acute flow limitation during exercise also raises BP in a canine hindlimb occlusion model [40]. In humans, BFR is achieved by placing a pressure cuff proximal to the working muscle and inflating it to achieve a pressure-limiting flow to the muscles. BFR has been used for augmenting the peripheral adaptations to resistance training [41–46]. Importantly, a recent report [47] and data showed that the BP response to exercise is accentuated under BFR conditions, even when the blood flow is not fully occluded. Thus, there is technical and ethical feasibility of using the BFR model in healthy humans to simulate the flow limitation in PAD.

IR: The IR injury is a main feature of various cardiovascular diseases, including myocardial infarction, stroke, and PAD [48,49]. Figure 3 illustrates the potential effect of IR on the metabolic milieu in the skeletal muscle tissue. The tissue damage associated with ischemic events occurs due to a combination of ischemia and paradoxical reperfusion following the restoration of blood flow to ischemic tissue, commonly referred to as IR injury. Following ischemia, the muscles may be salvaged by reperfusion. However, the re-introduction of oxygen to hypoxic muscles can also lead to damage by oxygen-derived free radicals. In PAD patients, IR injury was observed after limb revascularization [50,51]. Moreover, walking in PAD patients can induce ischemia (indicated with pain), while reperfusion can occur after stopping walking. Thus, intermittent claudication has been linked to ischemia followed by reperfusion leading to repeated IR injuries in their daily life [35,48,49,52].

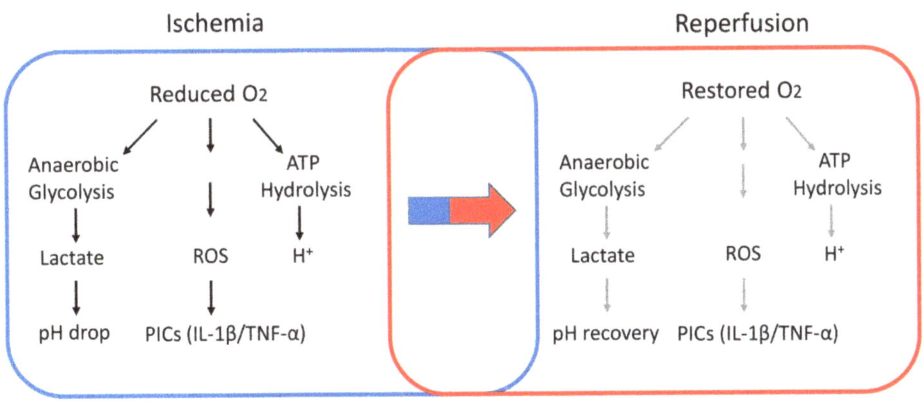

Figure 3. A diagram illustrates the potential effects of IR on muscle metabolic milieu in the hindlimb. We will study integration of the signaling pathways between skeletal muscle metabolites and primary muscle sensory neurons in regulation of the exercise pressor reflex in IR rats. ROS: reactive oxygen species; PIC: proinflammatory cytokine. Dark arrow indicates "increase" and light arrow indicates "alleviate". Note that the diagram is simplified to show the mechanisms more related to our proposed studies, but not all the molecular mediators responsible for IR are shown.

Prior studies [53–55] have employed a 20-min period of ischemia followed by a 20-min period of reperfusion (i.e., 20–20 min) to induce IR stress in a limb of healthy subjects to examine the effects of IR stress on vascular function [56,57]. This model has been used to simulate the IR stress in PAD and examine the effects of IR on the EPR. In our prior study [58], subjects performed fatiguing handgrip exercise before and ~20 min after a 20-min period of muscle ischemia (i.e., the limb circulation was totally occluded) in the control trial. The results showed that the MSNA responses to handgrip were accentuated after the 20-min period of muscle ischemia. Thus, the IR stress in healthy subjects and PAD patients with leg revascularization can be used to examine the role played by the IR in regulating the EPR.

3.2. Animal Models of Studying Human PAD

By effectively restricting and eliminating the blood flow in the affected limbs, the animal model of PAD proves an essential tool for studying the underlying molecular mechanism under the ischemic-related etiological and pathological conditions in PAD patients. In a recent review of the literature, we summarized the features of representative animal models of PAD [59]. Based on different feasibilities and approaches, the experimental animal options are rodents (e.g., mice, rats, and rabbits) and large animals such as swine. Methods of inducing blood flow restriction and elimination include single/double occlusion with or without blocking the branches in the femoral artery [60–65], ultra-sound assisted endovascular occlusion [66,67], and chemical-induced thrombus ischemia [68], etc. In the following sections, we will focus on discussing two animal models that mimic the blood flow restriction and ischemia-reperfusion status in human PAD: femoral artery occlusion/ligation and hindlimb ischemia-reperfusion.

3.2.1. Femoral Artery Occlusion/Ligation

Femoral artery occlusion in rats has been widely used to study human PAD [69] as it mimics one of the critical characteristics seen in PAD patients, namely intermittent claudication manifested by insufficient blood flow to the legs during exercise or slightly decreased blood flow to the legs under resting conditions. Notably, 24 h or 72 h femoral occlusion exaggerates BP response to muscle contraction (Figure 4 and muscle metabolites (e.g., acidic products and ATP) in the occluded limb (Figure 1), but not in the opposing control limb of the same rats [70,71]. Meanwhile, it has also been reported that the BP response during exercise is still exaggerated ~1 month and ~2 months after the femoral artery occlusion [72]. These findings parallel those reported in humans, showing that the BP response to walking is enhanced in PAD and the BP response during the exercise with the "diseased" limb is greater than that during the exercise with the "non-diseased" limb [16]. Therefore, it is indicated that a rat model of the femoral artery ligation is suitable for studying exercise-induced ischemia that occurs in PAD. Moreover, PAD in human subjects is not solely a disease of large vessel obstruction, but it is a disease of large vessel obstruction in the setting of a chronic disease process (atherosclerosis) that is influenced by oxidative stress and inflammation [73]. Femoral artery occlusion increases products of oxidative stress in the hindlimb muscles of rats and activates inflammatory signals (i.e., IL-6 and TNF-α) [74–76] as shown in Figure 1. This also makes the femoral occlusion model reflective of human conditions.

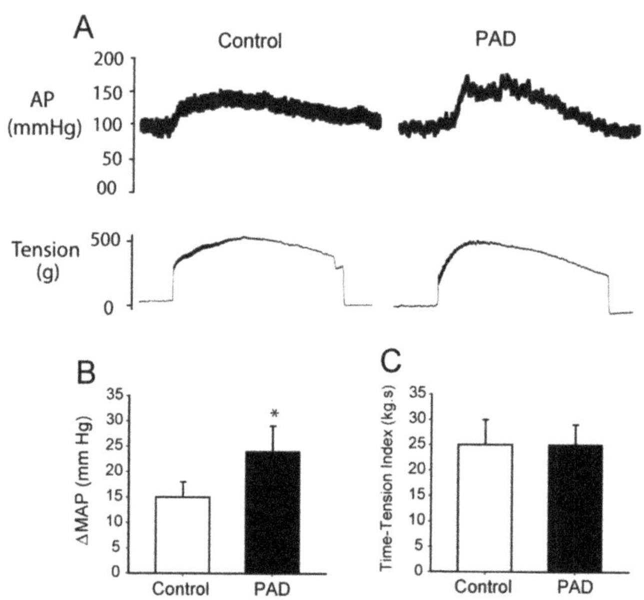

Figure 4. Mean arterial pressure (AP) response to static muscle contraction (30 s) induced by electrical stimulation of the L4&5 ventral roots in controls rat and PAD rats. (**A**) Typical recordings of arterial pressure (AP) response to static muscle contraction in control and PAD rats. (**B**,**C**) A greater MAP response was seen in PAD rats than in control rats, without different muscle tension between two groups * $p < 0.05$ between control and PAD.

3.2.2. Hindlimb Ischemia-Reperfusion

The hindlimb IR is induced by femoral artery ligation, followed by re-opening the ligation and is used to study PAD. To date, little is known about the engagement of a reperfusion component of IR injury in the pathophysiological processes of BP response in PAD. In the previous studies with forelimb IR models with mice, 18 h of blood flow reperfusion following 6 h of ischemia lowered the pain threshold of the affected limbs and increased the BP response during the dynamic global exercise. Meanwhile, the mRNA levels of primary sensory receptors (e.g., acid-sensing ion channel 3, ASIC3 and purinergic P2X3, P2X3) and the receptors for cytokines (e.g., interlukin-1β receptor, IL-1 βr) in the DRG were also increased [77]. With further development of the hindlimb IR model, we characterized a rat model of IR, showing that BP response to muscle contraction in different time courses following IR (e.g., 18, 66, and 114 h) were exaggerated. Notably, the increment of BP response 18 h following reperfusion was the most profound. It should also be noted that the BP response to muscle contraction was evaluated in decerebrated animals, which excluded the effect of central command. Further underlying mechanism studies with this IR model have shown that the intra-arterial injection of lactic acid (activator of ASIC3 receptor) and α,β me-ATP (activator of P2X3 receptor) were amplified in rats (IR18h) who experienced 6 h of femoral artery ligation followed by 18 h of reperfusion [78]. The increasing levels in BP response were similar in IR18h rats and rats with 24 h of femoral artery occlusion (24 h-occlusion rats). Moreover, the protein levels of ASIC3 and P2X3 expression in dorsal root ganglion (DRG) were increased to a similar degree in IR18h rats and 24 h-occlusion rats. These data suggest that reperfusion following 6 h ischemia is likely a factor leading to the remaining exaggeration of the BP response in IR rats and it is rational to utilize a rat model of IR18h for studying IR injury in PAD.

4. Molecular Mechanisms Leading to Exaggerated SNA and BP Responses in PAD

4.1. Effects of Muscle Metabolic Products and Their Responsive Receptors (Figures 5 and 6)

Using this PAD rat model with femoral artery ligation/occlusion, the previous studies have demonstrated that the SNA and pressor responses to muscle contraction and stimulation of muscle metabolite receptors i.e., acid-sensing ion channel 3 (ASIC3), purinergic P2X (subtype P2X3), transient receptor potential vanilloid 1 and ankyrin 1 (TRPV1 and TRPA1) are amplified in PAD rats as compared with control rats [71,79–83]. In addition, other receptors in muscle afferent nerves including μ-opioid/δ-opioid, bradykinin (BK) B2, prostaglandin (PGE2) EP4, and thromboxane (TP) receptors are engaged in the reflex responses in processing chronic ischemia of the hindlimb muscles [71,84–87].

In addition to the previous works on the roles of ASICs, TRPV1, and P2X receptors [70], we will focus on the updated research of ASIC3 and the interaction effect of ASIC3 and P2X3 on muscle sensory nerves in mediating the exaggerated sympathetic response in hindlimb muscle ischemia seen in PAD patients. Those receptors to be studied are expressed at both the peripheral terminals and the cell body of the sensory afferent neurons-DRG. With greater feasibility, receptor activity of DRG cell bodies has been used frequently as a surrogate for the nerve-ending receptor activity and physiology [88,89]. In particular, the whole cell patch-clamp methods are used to characterize the precise mechanisms by which those receptors mediate responses in the DRG neurons [88,89].

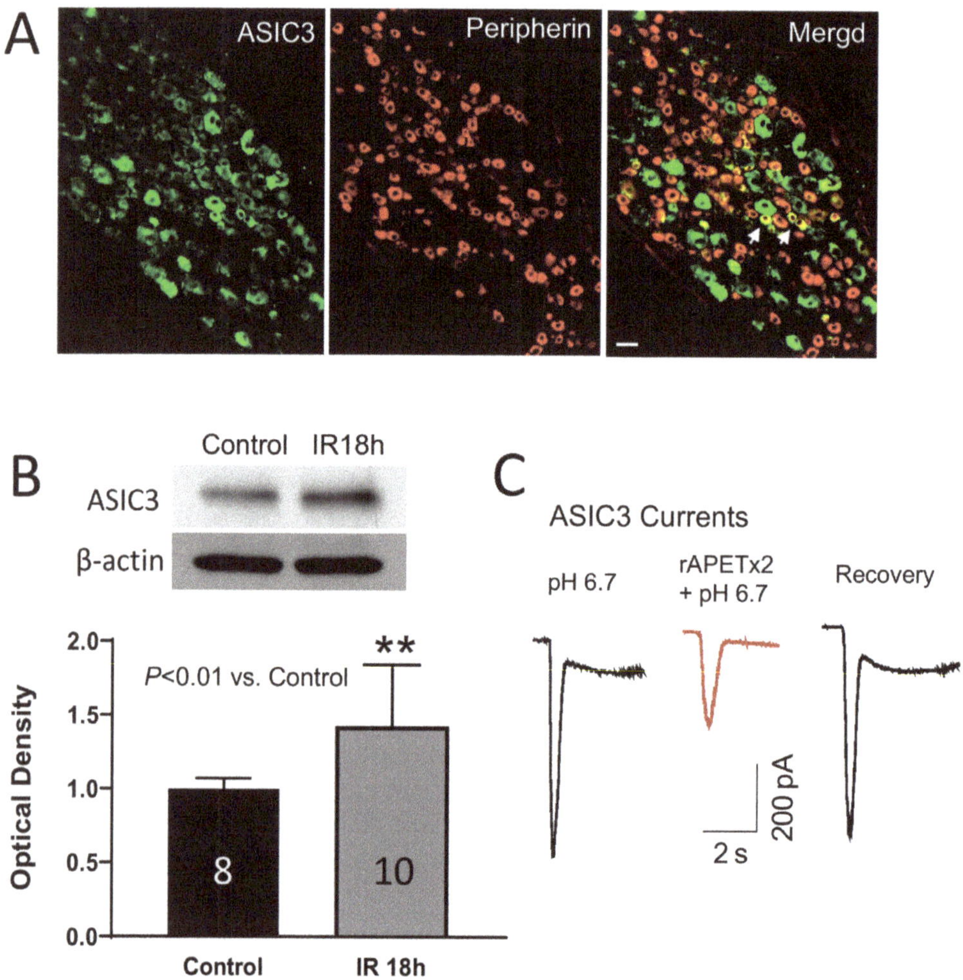

Figure 5. (**A**) The dual immunofluorescence method showing co-existence of ASIC3 and peripherin staining in DRG neurons. Arrows indicate representative cells positive for both ASIC3 and peripherin after they were merged. Scale bar = 50 μm. Peripherin was used to label C-fiber of DRG neurons. (**B**) Bands and averaged data (mean ± SD) showing that IR increased the protein levels of ASIC3 in DRGs. ** $p < 0.01$ between control and IR 18h rats. (**C**) Original traces of patch clamp showing that the amplitudes of ASIC current (elicited by pH 6.7 solution) in were largely decreased after application of ASIC3 antagonist rAPETx2 (1 μM).

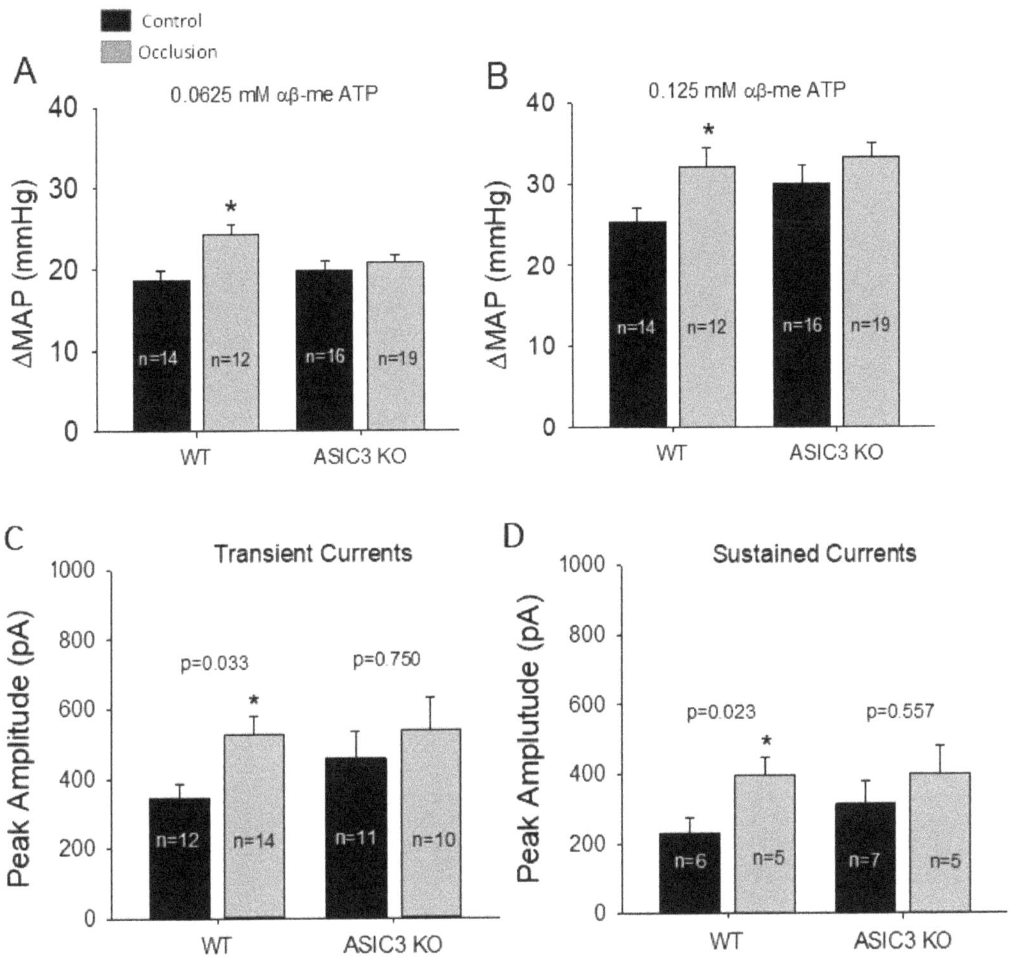

Figure 6. Effect of ASIC3 KO on BP response and DRG currents to activation of P2X3. (**A**,**B**): Two doses of αβ-me ATP were given intra-arterially into the hindlimb muscles to stimulate P2X3 in muscle afferents; and activation of P2X3 receptors amplified MAP response to a greater degree in WT rats with femoral occlusion, but not in ASIC3 KO rats. (**C**,**D**) Patch-clamp method shows averaged amplitude of P2X3 currents in Dil-labelled DRG neurons of WT rats and ASIC3 KO rats. (**C**): transient and (**D**): sustained P2X3 currents. An amount of 10 μM of αβ-me ATP was applied to induce current response. * $p < 0.05$ between control and occlusion in WT rats, but no significant difference in current response was seen between control and occlusion in ASIC3 KO rats. Data presented as mean ± SD.

4.1.1. ASIC3 KO Suppresses the Exercise Pressor Response under Ischemic Situation

ASICs are members of a family of amiloride-sensitive sodium channels and are considered as molecular sensors in afferent neurons [90–93]. They are almost ubiquitous in the mammalian nervous system and are activated as pH drops below 7.0. Among six different proteins of ASICs (ASIC1a, 1b, 2a, 2b, 3, and 4), encoded by four genes (ASIC1, 2, 3, and 4), the ASIC3 protein, however, is mostly found in DRG where it forms functional channels that are responsive to proton concentration fluctuation [90–93]. The pH range required to activate ASIC3 is approximately 6.5–7.0 [94,95], which is close to what is observed in exercising muscle and/or moderately ischemic tissues [96–99].

Prior to utilizing the genetic approach of ASIC3 knockout, several studies were performed to assess the role played by ASIC3 in evoking the exercise pressor reflex in PAD by comparing with sham control rats on the protein expression currents [82] of ASIC3 in DRG [81], and the blood pressure responses before and after the application of pharmacological blockades (e.g., amiloride and APETx2) [100]. ASIC3 knockout provides a powerful tool to validate the role of ASIC3 on the exercise pressor reflex in PAD as it not only blocks the function of this receptor but also induces a significant reduction of the protein expression.

In the ASIC3 KO rats, the peak mean arterial pressure (MAP) and the blood pressure index (BPI) following the static muscle contraction were similar to the wide type (WT) rats when the blood flow was freely perfused. However, under the ischemia conditions induced by femoral artery occlusion, the peak MAP and BPI were significantly lower in ASIC3 KO rats than in the WT rats. This effect was not seen in the EPR response induced by the passive tendon stretch. Researchers also injected the solutions of diprotonated phosphate (86 mM; pH 6.0), lactic acid (12 mM; pH 2.85), and capsaicin (0.2 μg; pH 7.2) to evoke the EPR response. Compared with the WT rats, the EPR response induced by diprotonated phosphate and lactic acid was significantly attenuated in ASIC3 KO rats. Interestingly, the EPR response induced by capsaicin (0.2 μg; pH 7.2) was also attenuated. However, blocking the ASIC3 in ligated WT rats by APETx2 did not suppress this capsaicin induced EPR response. This suggests there may be a special interaction or coupling effect between ASIC3 and TRPV1 receptors during the activation of TRPV1.

4.1.2. ASIC3 KO Attenuates the Exercise Pressor Response and the Activities of P2X3 under Ischemic Situation

Apart from the potential interaction between ASIC3 and TRPV1, it has been reported that there is a functional interaction between ASIC3 and P2X3 receptors [93,101]. In a published work, we used ASIC3 KO rats to examine the underlying mechanisms by which ASIC3 receptors affect P2X3 functions in regulating the EPR following femoral artery occlusion. Figure 6 shows that compared with wild-type (WT), ASIC3 KO attenuated the exaggeration of the BP response to injections of α,β-me ATP, a P2X3 agonist, into the arterial blood supply of the hindlimb muscles of occluded rats. This result is consistent with the notion suggested by our previous work that blocking ASIC3 signaling pathways can attenuate amplification of the BP response to stimulation of P2X3 receptors under the acidic milieu of the hindlimb muscles [102].

We further determined if ASIC3 KO attenuates P2X3 currents in DRG neurons innervating ischemic muscles. Figure 6 shows that muscle DRG neurons from both WT rats and ASIC3 KO rats exhibited the typical transient and sustained current responses with activation of P2X3 receptors by applying α,β-me ATP. The data further show that femoral artery occlusion augmented the amplitude of P2X3 currents in response to α,β-me ATP in muscle DRG neurons of WT rats, but this effect appeared to be less in ASIC3 KO rats. This result further supports the notion that inhibition of ASIC3 has a regulatory role in P2X3 function, and this is likely to be involved in causing the exaggerated EPR in PAD rats following femoral artery occlusion.

4.2. *Other Ischemia-Induced Products*

In addition to ASIC3, TRPV1, and P2X3, it must be pointed out that other muscle afferents' receptors, including μ-opioid and thromboxane (TP) receptors, etc., are engaged in processing chronic ischemia of the hindlimb muscles [84,85]. In addition, studies showed that bradykinin B2 and peripheral δ-opioid receptors contribute to the exaggerated exercise pressor reflex via a mechanically sensitive group III muscle afferents in rats with femoral artery occlusion [87,103]. Blocking PGE2 EP4 also attenuates the augmented BP response to static exercise observed in PAD rats [86]. Meanwhile, the role of nerve growth factor

(NGF) in regulating the metabolic receptors in the ischemia-muscles of PAD has been extensively discussed in one of the previous reviews [104], highlighting its elevation under the ischemia condition and upregulating the protein expression and function of ASIC3, P2X3, and TRPV1 in the DRG neurons. In this review, we will focus on extending our discussion on the role of hypoxia-inducible factor 1α (HIF-1α) and the reactive oxygen species (ROS).

4.2.1. HIF-1α

HIF-1 is a heterodimeric protein composed of constitutively expressed HIF-1α and HIF-1β subunits [105]. In the two subunits, oxygen-sensitive HIF-1α accumulates rapidly under hypoxic conditions and modulates the expression of several target genes in protecting tissues against ischemia and infarction [106–109]. HIF-1α is considered a transcription factor that mediates adaptive responses to hypoxia and ischemia [106–109]. Thus, we have examined if arterial occlusion increases the levels of HIF-1α in sensory neurons and if engagement of HIF-1α is responsible for the enhancement in the reflex cardiovascular responses induced by activation of muscle afferent nerves [110].

The first insight we gained in this previous study by using western blot analysis showed that HIF-1α protein expression is significantly increased in DRG neurons 6–72 h after femoral artery ligation as compared with non-ligated controls [110]. This result suggests that femoral occlusion induces HIF-1α response in sensory nerves. In addition, DMOG, an inhibitor of prolyl hydroxylase, has been shown to stabilize or increase HIF-1α protein and enhance the expression of downstream target genes [111,112]. It was reported that inhibition of endogenous HIF inactivation by DMOG induces angiogenesis in the ischemic skeletal muscles of mice [112]. In this previous study, we further examined the expression of HIF-1α protein in DRG neurons induced by intramuscular injection of DMOG [110]. HIF-1α protein expression was significantly increased in lumbar DRG neurons 24 hrs after injection of DMOG into the hindlimb muscles as compared with sham controls. In this prior report, we also examined the effects of femoral occlusion on the reflex cardiovascular responses evoked by activation of muscle afferent nerves [110]. Our data have shown that 24 h of femoral artery occlusion significantly increased arterial BP response induced by static muscle contraction. To determine if HIF-1α has a potential effect on the exercise pressor reflex, we injected DMOG into the hindlimb muscles. Then, BP and HR responses induced by static muscle contraction were examined 24 hrs after DMOG injection. Our results showed that there were no significant differences in increases of the reflex MAP and HR responses after DMOG as compared with controls [110].

In contrast, BAY87, a synthesized compound with characteristics of highly potent and specific suppressive effects on expression and activity of HIF-1α, was given into the arterial blood supply of the ischemic hindlimb muscles three hours before the exercise pressor reflex was evoked by static muscle contraction. First, arterial injection of BAY87 inhibited expression of HIF-1α in the DRG of occluded limbs three hours following its injection. Second, muscle contraction evoked a greater increase in BP in occluded rats and BAY87 attenuated the enhanced BP response in occluded rats to a greater degree than in control rats. Taken together, these data suggest that inhibition of HIF-1α alleviates exaggeration of the exercise pressor reflex in rats under ischemic circumstances of the hindlimbs in PAD induced by femoral artery occlusion; however, an increase in HIF-1α of DRG neurons per se may not alter the muscle pressor reflex.

Nonetheless, it should be noted that the time courses are very similar in increased HIF-1α expression, and elevated NGF and amplitude of DRG response to stimulation of ASIC3, P2X3, and TRPV1receptors after ischemic insult induced by the femoral artery occlusion [83,113–115]. This similarity may indicate that there is a close relationship between NGF and HIF-1α responses in the DRG neurons in the processing of muscle ischemia. Interestingly, published work shows that increasing HIF-1α or inhibiting HIF-1α prolyl hydroxylases can attenuate NGF deprivation-induced effects on neurons, suggesting that HIF-1α plays a regulatory role in affecting effects of NGF [116–118]. Therefore, we postulate that HIF-1α likely contributes to the effects of NGF on augmented muscle metabolic responses in the DRG neurons after arterial occlusion.

4.2.2. Reactive Oxidative Species

Notably, a number of studies suggest that reactive oxidative species (ROS) contribute to the regulation of discharges of vagal lung thin afferent fiber nerves [119,120]. Additionally, it has been reported that an increase in muscle NADPH oxidase-derived ROS sensitizes the exercise pressor reflex in a decerebrate rat model [121]. Likewise, a decrease in ROS can attenuate the reflex [121]. Thus, it is speculated that ROS is engaged in augmented SNA and BP response during activation of the exercise pressor reflex in rats with femoral occlusion. Superoxide dismutases (SOD), are a class of enzymes that catalyze the dismutation of superoxide into oxygen and hydrogen peroxide as considered an important antioxidant. In a published work, tempol, a mimic of SOD, was arterially injected into the hindlimb muscles of rats and results demonstrated that tempol attenuates BP response evoked by contraction of occluded hindlimb muscles, but the attenuation was not seen when contraction was induced in freely perfused control legs [122]. A following study suggested that effects of tempol on the BP response during contraction are via ATP-dependent potassium channels [123]. However, a prior study suggested that ROS plays an important role in regulating discharges of vagal lung thin afferent fiber nerves via engagement of TRPV1 and P2X receptors [119,120]. In those experiments, the reflex pulmonary chemical response induced by a ROS stimulant hydrogen peroxide was attenuated by the prior application of i-RTX (TRPV1 antagonist) and PPADS (P2X antagonist) [119,120]. Thus, it is likely that ROS can alter the response of sensory nerves with activation of TRPV1 and P2X. Nevertheless, the augmented exercise pressor reflex is significantly attenuated after tempol is given to compensate SOD in occluded muscles of rats [122].

In addition, ROS activates the transient receptor potential channel A1 (TRPA1) [124–126]. TPRA1 is a member of branch A of the transient receptor potential (TRP) family of nonselective cation channels and expressed in the sensory (nerves) neurons and is involved in acute and inflammatory pain [124,127–132]. A published work has demonstrated that intra-arterial injection of AITC, a TRPA1 agonist, into the hindlimb muscle circulation of healthy rats led to increases in SNA and BP via a reflex mechanism [133]. Additionally, this study has suggested that TRPA1 plays a role in regulating the exercise pressor reflex and acid phosphate, bradykinin, and arachidonic acid, which are accumulated in exercising muscles are likely engaged in the role played by TRPA1 as endogenous stimuli. Interestingly, it was observed that femoral artery occlusion (1) upregulates the protein levels of TRPA1 in DRG tissues; (2) selectively increases expression of TRPA1 in DRG neurons supplying metabolically sensitive afferent nerves of C-fiber (group IV); (3) enhances renal SNA and BP responses to AITC (a TRPA1 agonist) injected into the hindlimb muscles, and (4) blocks TRPA1 attenuates SNA and BP responses during muscle contraction to a greater degree in ligated rats than those responses in control rats. Overall, the results of these studies indicate that alternations in muscle afferent nerves' TRPA1 likely contribute to the enhanced sympathetic and BP responses via the metabolic component of the muscle reflex under circumstances of chronic muscle ischemia in PAD, and the effects of oxidative stress are also likely associated with expression and activities of TRPA1 in sensory nerves of PAD.

4.2.3. Endothelin-1 (ET-1)

ET-1 is originally characterized as an endothelium-derived peptide which majorly functions as constricting factor in the vasculature [134,135]. It affects several tissues including the smooth muscle and the nervous system [136]. During inflammatory conditions, ET-1 has been found to be associated with an inflammatory response involving the expression of proinflammatory cytokines including TNF-α, IL-1 and IL-6 [137]. Meanwhile, ROS stimulates the production of ET-1 in both in vivo and in vitro situations [138,139]. In PAD patients, it has been reported that the ET-1 was elevated in the plasma [140,141]. In an animal model of PAD, the ET-1 concentration was also elevated in the gastrocnemius muscle [142]. Of note, ET-1 is an important vasoconstrictor for the restraint of blood flow in active skeletal muscle and the maintenance of arterial BP during exercise [143]. The underlying mechanism of the ET-1 on EPR response in PAD patients has not yet been fully investigated. However, a number of previous studies of the peripheral nerve establish a fundamental rationale for further mechanism and intervention studies on the ET-1-related cellular and molecular pathways during evoking the EPR response in PAD.

In the peripheral nervous system, two subtypes of ET-1 receptors, ETA and ETB, are expressed in nociceptive primary sensory neurons such as the DRG neurons, whereas ETB is mainly found in satellite glial cells [144]. Numerous studies have demonstrated that ET-1 plays an important role in modifying peripheral pain signaling. In humans, exogenous ET-1 causes tactile allodynia and severe pain. In rodents, an intra-plantar injection of ET-1 produces mechanical and thermal hyperalgesia and spontaneous pain-like behaviors [145,146]. Paw withdrawal thresholds to mechanical stimuli and heat are significantly altered in conditioned ET-1 knockout mice [147]. Peripheral ET-1 acts on nociceptors through its cognate receptors, which subsequently modify pain-related ion channels to amplify signal generation [148]. In DRG neurons, ET-1 increases neuronal excitation by hyperpolarizing tetrodotoxin-resistant (TTX-R) Na+ channels and by suppressing the delayed outward rectifier K+ current via ETA receptor [149,150]. In contrast, ET-1 can decrease the excitability of DRG neurons through the activation of ETB receptors [151]. It has been also reported that ETA and ETB receptors play a role in regulating sensitivity of the peripheral sensory nerve in neuropathic pain induced by spinal nerve ligation in rats [152].

4.3. *Pro-Inflammation Cytokines and Ion Channels in Muscle Sensory Neurons (Figures 7 and 8)*

4.3.1. TNF-α and Activities of Nav Channels in Muscle DRG Neurons

The augmented exercise pressor reflex might be due in part to inflammation, specifically pro-inflammatory cytokines (PICs) associated with PAD. Numerous cells (i.e., leukocytes, myocytes, microglia, astrocytes, and Schwann cells) produce and release PICs [153], which include interleukins, lymphokines, and cell signaling molecules. In particular, the roles of tumor necrosis factor-α (TNF-α), interleukin-6 (IL-6), and interleukin-1β (IL-1β) are significant in regulating immune and inflammatory reactions. These PICs modulate the activities of many cell types in various diseases. For example, during diseased states, PICs help to recruit cells to inflammatory sites, stimulating cell survival, division, and enhancing proliferation and differentiation [154]. Evidence indicates that PICs are involved in regulating physiological functions, with their levels increasing in the circulation and in the affected tissues [153,155,156]. Increased circulating and intramuscular levels of PICs (such as IL-6 and TNF-α) were also found in coronary and/or atherosclerotic vascular disorders such as PAD [157–159].

It was first observed that the levels of TNF-α and protein expression of TNF-α receptor type 1 (TNFR1) were increased in the DRG of the hindlimbs of PAD rats. Note that TNF-α was observed within DRG neurons of C-fiber afferent nerves. Capsaicin (TRPV1 agonist) and AITC (TRPA1 agonist) were injected into the arterial blood supply of the hindlimbs to stimulate metabolically sensitive thin-fiber muscle afferents. The effects of these injections on the SNA and pressor responses were attenuated in PAD rats after TNF-α synthesis suppressor pentoxifylline (PTX) was previously administered into the hindlimb with femoral artery occlusion. These data suggest that TNF-α plays a role in modulating exaggerated SNA via the metabolic component of the exercise pressor reflex in PAD.

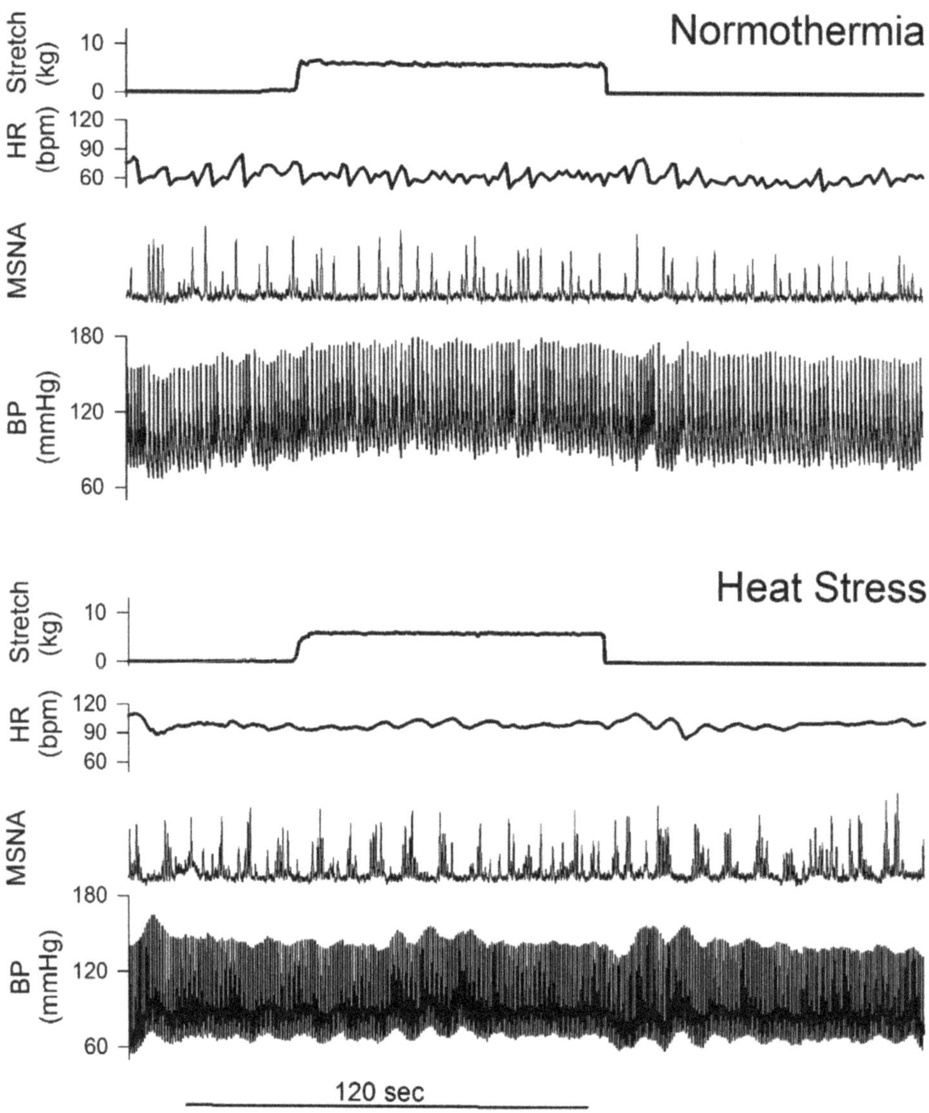

Figure 7. Passive stretch during post-exercise circulatory occlusion evoked MSNA and BP increases in normothermic conditions. These responses were attenuated under heat stress conditions. (Unpublished figure).

Tetrodotoxin (TTX)-resistant Na^+ (i.e., $Na_V1.8$) channels are highly expressed in group IV afferents [160]. The role played by $Na_V1.8$ in evoking the exercise pressor reflex was examined using whole animal preparations. A803467, a $Na_V1.8$ blocker, attenuates the pressor response evoked by arterial injection of lactic acid and capsaicin stimulating thin fiber afferents [161]. There is a linkage between TNF-α and the activity of Na^+ current in sensory nerves [162]. A prior study demonstrated the role of TNF-α in enhancing the current densities of Nav1.8 in DRG neurons [163]. In an additional work, the role played by TNF-α in regulating the activity of NaV1.8 currents in muscle DRG neurons of PAD rats was specifically examined. Results showed that peak amplitude of TTX-resistant (TTX-R) Nav and NaV1.8 currents in muscle DRG neurons were increased in PAD rats. Meanwhile, the amplification of TTX-R and NaV1.8 currents induced by TNF-α was attenuated in DRG neurons with pre-incubation with respective inhibitors of the intracellular signaling pathwaysp38-MAPK, JNK, and ERK. It was concluded that NaV1.8 is engaged in the role of TNF-α in amplifying muscle afferent inputs as the hindlimb muscles are ischemic in PAD. The pathways of p38-MAPK, JNK, and ERK are likely necessary to mediate the effects of TNF-α.

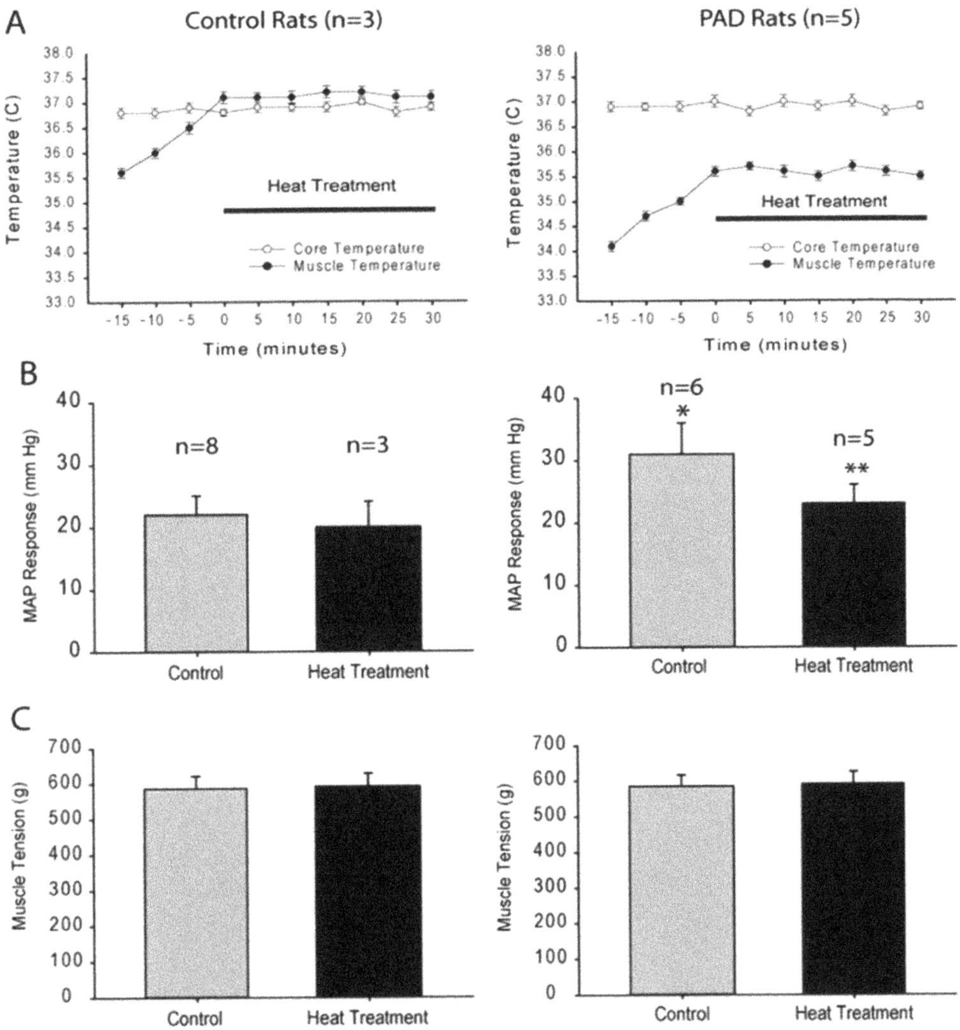

Figure 8. Tcore and Tm, MAP response and muscle tension in control rats (left panels) and PAD rats (right panels). (**A**): Baseline Tm was lower in PAD rats and Tcore was not altered during heat treatment. (**B**): MAP response to contraction was increased in PAD rats and amplification of pressor response was attenuated after heat treatment. * $p < 0.05$ vs. control rats; and ** $p < 0.05$ vs. control group without heat treatment. (**C**): No difference in muscle tension among groups ($p > 0.05$).

4.3.2. IL-6 and Activities of Kv4 Channels in Muscle DRG Neurons

Increased circulating and intramuscular levels of interleukin-6 (IL-6) are detected in PAD patients [164,165]. The activity of exercise induces a greater increase in the levels of IL-6 of the mixed venous blood in PAD patients than those levels in healthy age-matched subjects [166,167]. Consistently, during the exercise ischemic insult also enhances the circulating IL-6 levels compared with non-ischemic exercise [168].

Seventy-two hours of femoral artery occlusion increases products of oxidative stress in the hindlimb muscles of rats and activates inflammatory signaling pathways [74–76]. IL-6 also plays a role in regulating the exaggerated BP response to static exercise in PAD rats [169] likely via membrane-bound IL-6R or gp130 trans-signaling pathways assembled by soluble forms of IL-6R [163,170,171]. Thus, it was anticipated that the activity of IL-6 signaling would be increased in muscle afferent nerves involving the exercise pressor reflex in PAD rats.

We found that the protein levels of IL-6 and its receptor IL-6R expression were increased in the DRGs of PAD rats with 72 h of femoral artery occlusion. Inhibition of muscle afferents' IL-6 trans-signaling pathway (gp130) by intra-arterial administration of SC144, a gp130 inhibitor, into the hindlimb muscles of PAD rats alleviated BP to static muscle contraction. On the other hand, it was found that PAD decreased amplitude of Kv4 currents in rat muscle DRG neurons. The homo IL-6/IL-6Rα fusion protein (H. IL-6/6Rα) but not IL-6 alone significantly inhibited Kv4 currents in muscle DRG neurons; the effect of H. IL-6/6Rα was largely reverted by SC144. Consistent with the previous findings, these data suggest that via trans-signaling pathway upregulated IL-6 in muscle afferent nerves by ischemic hindlimb muscles inhibits the activity of Kv4 channels and therefore likely leads to adjustments of the exercise pressor reflex in PAD.

5. Heat Treatment and Nutrition Intervention on Improving Exercise-Induced Exaggerated SNA and BP Responses in PAD

Supervised exercise intervention is one of the most effective means of maintaining or restoring the exercise tolerance of PAD patients [12,172]. However, as above mentions, the challenge exists in the adherence to exercise training programs due to the symptom of intermittent claudication, which is partly attributed to the exaggerated EPR response. Therefore, in this review, we discuss three other non-pharmacological interventions that may be helpful to ameliorate the hyper-amplified EPR response in PAD patients. With the introduction of those promising economic strategies, we are aiming to incorporate them into the well-established exercise training protocols to enhance the adherence to exercise training, improve the efficacy of the intervention protocols, and benefit the overall well-being of PAD patients.

5.1. Heat Treatment (Figures 7 and 8)

In recent years, heat treatment has been obtaining significant attention in terms of its beneficial effect on cardiovascular patients including PAD. In a rat model of PAD, the heat treatment protocol of increasing the muscle temperature (Tm) by 1.5 °C (30 min period each heating protocol, 2 times/day for 3 days) attenuated ET-1 in both red and white portions of gastrocnemius muscle in PAD [142]. It has been reported that repeated heating exposure suppressed the production of plasma ET-1 in human participants with symptomatic PAD in both rest [173] and post-exercise situations [174]. Meanwhile, heating exposure increases the contraction force of the ischemia-induced damaged skeletal muscle [175]. In a mice obesity model, the heating exposure also decreased the percentage of fat and increase the ratio of muscle mass to body mass even if capillary density and collateral supply diameter was unchanged [176].

In terms of the blood flow dynamic response, heat exposure increases skin blood flow (SkBF), heart rate, cardiac output [177], ejection fraction, and systolic function in healthy individuals [178,179] and heart failure patients [180,181]. Based on these observations, thermal therapy has been suggested for patients with heart diseases [180–184]. Moreover, it has been shown that heat treatment (e.g., dry sauna) may improve chronic endothelial function in patients with heart diseases or in those with atherosclerotic risk factors [183,185–187]. A recent report also shows that hot water immersion raised the blood flow in lower limbs of PAD patients [188]. It should be noted that these prior studies [188–190] only focused on the effects of heating on blood flow and vascular function in PAD. The effects of heat exposure on EPR in PAD have not been examined.

Thus, the effects of heating on EPR were studied in PAD patients and PAD rats induced by femoral artery occlusion. First, to determine how whole-body heating alters muscle mechanoreflex and metaboreflex responses, we measured MSNA in healthy subjects during fatiguing isometric handgrip exercise, PECO, and passive muscle stretch (extension of wrist, EOW) during PECO. The protocol was performed under both normothermic and whole-body heating (ΔTcore ~0.6 °C via a heating suit) conditions. Under normothermic conditions, passive stretches during PECO evoked significant increases in mean arterial pressure (MAP) and MSNA. However, during heating, passive stretch did not significantly increase MSNA or MAP (Figure 7). These data show that sympathetic response to the mechanoreceptor stimulation [191] is attenuated by heat exposure when body temperature is elevated [192]. The attenuated MSNA response to stretch during heating should not be a "ceiling effect" because there was no significant difference in the MSNA burst incidence during stretches between thermal conditions.

Although there was no difference in MSNA response to PECO (i.e., non-specific metaboreceptor stimulation), the MAP response to PECO during heating was much less (by ~50%) than in normothermic conditions. Thus, the EPR (i.e., pressor response) is attenuated during heating. It is speculated that the BP response to sympathetic activation is also attenuated during heating [193]. It is known that MSNA response to metaboreceptor stimulation is attenuated in heart failure [194]. On the other hand, it is unclear if the MSNA response to metaboreceptor stimulation is altered in PAD. Therefore, it is necessary to examine both MSNA and BP responses to exercise in PAD.

A prior study [195] demonstrated that local heating of an isometric exercising forearm muscle group augmented the increase in MSNA during fatiguing exercise. They speculated that the elevated Tm might sensitize muscle mechanosensitive afferents. It should be noted that in those studies, local heating increased forearm Tm from ~34 to 39 °C [195–197]. Our pilot study shows that whole-body heating only raised forearm Tm of ~1.5 °C. Thus, the fewer increases in the Tm in our studies are likely to lead to the different effects on the muscle afferents at the receptor level, necessitating the study engagement of P2X in the EPR after heat exposure.

Group III and IV respond to changes in Tm [198,199]. In animals, we have shown that a higher Tm response is linked to a lower BP response and elevated Tm attenuates the P2X receptor-mediated reflex activation of muscle mechano- and metabo-receptors [200]. We have shown that arterial injection of α,β-me ATP into the hindlimb muscles evoked a dose-dependent response, and the peak pressor response evoked by α,β-me ATP was attenuated as Tm was increased by heat exposure. Additionally, α,β me ATP amplified the reflex BP response evoked by stretch and the effect was blunted with heat exposure.

The effects of heat exposure on EPR in PAD have not been well understood. Due to limb ischemia, the lower limb temperature is lower and revascularization therapy raised the limb temperature [201]. In addition, a decrease in the Tm induces a decelerated rate of ATP turnover [202], which likely leads to an elevation of ATP concentration in the extracellular space. It is, therefore, speculated that the lower Tm may contribute to the accentuated EPR in PAD. In turn, a suitable rise in Tm may attenuate the sympathetic response and decrease the exaggerated BP response to exercise in PAD. To obtain the same degree of Tm increase in animal models, the temperature in rat hindlimb muscle was monitored and increased by 1.5 °C (30 min period each heating protocol, two times/day for three days) and then BP response to static muscle contraction was examined (Figure 8). These data demonstrated that a raise in muscle temperature attenuated the exaggerated BP response to muscle contraction. This study further showed that a protocol with increasing muscle temperature by 1.5 °C decreased expression and current response of P2X3 in DRG neurons of PAD rats, suggesting P2X3 signaling is a part of the mechanisms leading to inhibition of BP response. Based on those published results in humans and animals, it is speculated that heat exposure and/or heat treatment would be beneficial to attenuate the sympathetic response and decrease the exaggerated BP response in PAD.

The effect of a short period heating intervention on relieving the symptoms of intermittent claudication is also intriguing to study. In one of the previous studies, the one-time acute effect on the EPR response in PAD was evaluated. The muscle temperature was increased by 1.5 °C and the length was 5 min. The EPR response induced by both static muscle contraction and α,β-Me-ATP injection was evaluated 20 min before, immediately after, and 20 min after the heat exposure. The results of this study were interesting as the static muscle contraction induced EPR was attenuated following the heat exposure, and the EPR response attenuation was recovered 20 min after the muscle temperature returned to the baseline. However, the α,β-Me-ATP-induced EPR response did not alter with heat exposure. This suggested the attenuation of static-muscle-contraction-induced EPR response following one-time heat exposure may not work through alternating the expression and function of P2X3 receptors. Instead, it may work through the alternation of the ATP metabolism enzyme activity, e.g., ATPase, which will be one of the further directions of the mechanism study on this topic. As intermittent claudication frequently occurs and interrupts the daily physical movement of the patients, this study provides a fundamental basis for the daily base intervention strategy for the PAD patients.

5.2. *Effects of Supplemental Nutrients*

5.2.1. Vitamin B6

A diet deficient in vitamin B6 leads to a decreased activity of cystathionine β-synthase and cystathionase in the liver. Dietary supplementation of vitamin B6 stimulates the activity of these enzymes and increases the endogenous synthesis of cysteine from methionine. In hypertensive animals and humans, increased production of cysteine would lead to more efficient excretion of excess metabolic aldehydes, normalizing vascular calcium channels and lowering blood pressure [203]. A regression study has also shown that an increase in the daily intake of vitamin B6 by one standard deviation (approximately 0.5 mg per day) would reduce the risk of PAD by 29% [204]. More importantly, once consumed vitamin B6 will be converted into a P2-purinoceptor antagonist called pyridoxal-5-phosphate (PLP) [205]. In animal studies, the intraperitoneal injection of the vitamin B complex (B1/B6/B12 = 100/100/2 mg/kg) attenuated the expression of P2X3 in DRG of diabetic rats [206]. By locally infusing the vitamin B6 into human participants' forearms, previous studies [207] suggested that the MSNA responses to fatiguing handgrip, post-exercise circulatory occlusion (PECO), and PECO + passive stretch were all significantly less than those before pyridoxine. The blood pressure responses were also significantly less than those before vitamin B6 infusion.

5.2.2. Vitamin C

Low levels of Vitamin C supplementation (assessed by dietary intake or plasma analysis) are associated with multiple conditions, including high blood pressure (BP), endothelial dysfunction, heart disease, atherosclerosis, and stroke [208]. For the mechanism work on the Vitamin C supplementation on blood pressure regulation, the information is lacking in terms of the efficacy of Vitamin C supplementation on the ERP response in cardiovascular patients, especially in PAD, and the IR injury of PAD patients following the revascularization surgery. A previously performed human study [17], investigated the efficacy of Vitamin C intravenous infusion in attenuating oxidative stress and therefore the subsequent EPR responses in PAD patients. In this study, the Vitamin C infusion elicited a lower MAP response to low-intensity rhythmic plantar flexion in the affected legs of PAD patients than that in the condition without Vitamin C infusion.

6. Conclusions

Studies using a rat model of femoral artery occlusion show that sympathetic responses of the exercise pressor reflex engagement are exaggerated as observed in PAD patients. As summarized in Figure 1, findings of the completed studies suggest that enhanced protein levels of ASIC3, P2X3, and TRPV1 in muscle afferent nerves and amplified responses of those receptors contribute to the exaggerated reflexive sympathetic and pressor responses to their individual receptor stimulus. The findings further suggest that NGF is likely responsible for enhanced ASIC3, P2X3, and TRPV1 and plays a role in modulating the metaboreceptor component of the exercise pressor reflex in hindlimb muscle ischemia. Lactic acid, ATP, and acid phosphate are the major muscle by-products in exercising muscles and ASIC3, P2X3, and TRPV1 receptors are sensitive to those individual metabolites and/or combined metabolites. Overall data presented here provide evidence that alteration in chemically sensitive receptors ASIC3, P2X3, and TRPV1 in primary afferent neurons innervating ischemic muscles plays an important role in the development of the exaggerated reflexive sympathetic responses, likely leading to worsening exercise capacity in patients with PAD. Moreover, NGF in sensory nerves plays a role in regulating abnormal responses of those metabolic receptors. Also, HIF-1α likely contributes to the effects of NGF on augmented muscle metabolic responses in the DRG neurons after arterial occlusion. A study limitation needs to be mentioned based on the fact that the studies included in this current review are varied from factors such as sample size, gender ratio, and the choices of different animal models as well as human populations. Therefore, additional studies are warranted to verify and confirm the underlying mechanism and clinical results. More animal models to exemplify different stages or pathological conditions of PAD are also necessary to explore. In combination with the fundamental work performed by the previous studies, those mechanistic foundations formed by studies in animal models will shed light on the targets of the translational intervention studies to alleviate the adverse effects that increase the cardiovascular event risk in PAD patients.

Author Contributions: L.Q., J.C. and J.L. performed literature search and prepared the figures. L.Q. and J.L. drafted and revised the manuscript. L.Q., J.C. and J.L. proof-read and approved the manuscript for submission. All authors have read and agreed to the published version of the manuscript.

Funding: The supported funding resources of this study are NIH R01 HL141198, NIH R01 HL164571 and AMERICAN HEART ASSOCIATION GRANT # 940567/Lu Qin/2022.

Conflicts of Interest: The authors declare no conflict of interest.

References

1. Criqui, M.H.; Aboyans, V. Epidemiology of peripheral artery disease. *Circ. Res.* **2015**, *116*, 1509–1526. [CrossRef] [PubMed]
2. Fowkes, F.G.; Aboyans, V.; Fowkes, F.J.; McDermott, M.M.; Sampson, U.K.; Criqui, M.H. Peripheral artery disease: Epidemiology and global perspectives. *Nat. Rev. Cardiol.* **2017**, *14*, 156–170. [CrossRef] [PubMed]
3. Criqui, M.H.; Langer, R.D.; Fronek, A.; Feigelson, H.S.; Klauber, M.R.; McCann, T.J.; Browner, D. Mortality over a period of 10 years in patients with peripheral arterial disease. *N. Engl. J. Med.* **1992**, *326*, 381–386. [CrossRef] [PubMed]
4. Mozaffarian, D.; Benjamin, E.J.; Go, A.S.; Arnett, D.K.; Blaha, M.J.; Cushman, M.; Das, S.R.; de Ferranti, S.; Despres, J.P.; Fullerton, H.J.; et al. Executive Summary: Heart Disease and Stroke Statistics–2016 Update: A Report From the American Heart Association. *Circulation* **2016**, *133*, 447–454. [CrossRef]
5. Ouriel, K. Peripheral arterial disease. *Lancet* **2001**, *358*, 1257–1264. [CrossRef]
6. Anand, S.S.; Caron, F.; Eikelboom, J.W.; Bosch, J.; Dyal, L.; Aboyans, V.; Abola, M.T.; Branch, K.R.H.; Keltai, K.; Bhatt, D.L.; et al. Major Adverse Limb Events and Mortality in Patients With Peripheral Artery Disease: The COMPASS Trial. *J. Am. Coll. Cardiol.* **2018**, *71*, 2306–2315. [CrossRef]
7. Bauersachs, R.; Zeymer, U.; Brière, J.B.; Marre, C.; Bowrin, K.; Huelsebeck, M. Burden of Coronary Artery Disease and Peripheral Artery Disease: A Literature Review. *Cardiovasc. Ther.* **2019**, *2019*, 8295054. [CrossRef]

8. Hirsch, A.T.; Haskal, Z.J.; Hertzer, N.R.; Bakal, C.W.; Creager, M.A.; Halperin, J.L.; Hiratzka, L.F.; Murphy, W.R.; Olin, J.W.; Puschett, J.B.; et al. ACC/AHA Guidelines for the Management of Patients with Peripheral Arterial Disease (lower extremity, renal, mesenteric, and abdominal aortic): A collaborative report from the American Associations for Vascular Surgery/Society for Vascular Surgery, Society for Cardiovascular Angiography and Interventions, Society for Vascular Medicine and Biology, Society of Interventional Radiology, and the ACC/AHA Task Force on Practice Guidelines (writing committee to develop guidelines for the management of patients with peripheral arterial disease)—Summary of recommendations. *J. Vasc. Interv. Radiol.* **2006**, *17*, 1383–1397, quiz 1398. [CrossRef]
9. Clagett, G.P.; Sobel, M.; Jackson, M.R.; Lip, G.Y.H.; Tangelder, M.; Verhaeghe, R. Antithrombotic Therapy in Peripheral Arterial Occlusive Disease. The Seventh ACCP Conference on Antithrombotic and Thrombolytic Therapy. *Chest* **2004**, *126*, 609S–626S. [CrossRef]
10. Kitrou, P.; Katsanos, K.; Karnabatidis, D.; Reppas, L.; Brountzos, E.; Spiliopoulos, S. Current Evidence and Future Perspectives on Anti-platelet and Statin Pharmacotherapy for Patients with Symptomatic Peripheral Arterial Disease. *Curr. Vasc. Pharmacol.* **2017**, *15*, 430–445. [CrossRef]
11. Brendle, D.C.; Joseph, L.J.; Corretti, M.C.; Gardner, A.W.; Katzel, L.I. Effects of exercise rehabilitation on endothelial reactivity in older patients with peripheral arterial disease. *Am. J. Cardiol.* **2001**, *87*, 324–329. [CrossRef]
12. Hamburg, N.M.; Balady, G.J. Exercise rehabilitation in peripheral artery disease: Functional impact and mechanisms of benefits. *Circulation* **2011**, *123*, 87–97. [CrossRef]
13. McDermott, M.M. Exercise Rehabilitation for Peripheral Artery Disease: A REVIEW. *J. Cardiopulm. Rehabil. Prev.* **2018**, *38*, 63–69. [CrossRef] [PubMed]
14. Baccelli, G.; Reggiani, P.; Mattioli, A.; Corbellini, E.; Garducci, S.; Catalano, M. The exercise pressor reflex and changes in radial arterial pressure and heart rate during walking in patients with arteriosclerosis obliterans. *Angiology* **1999**, *50*, 361–374. [CrossRef] [PubMed]
15. Bakke, E.F.; Hisdal, J.; Jorgensen, J.J.; Kroese, A.; Stranden, E. Blood pressure in patients with intermittent claudication increases continuously during walking. *Eur. J. Vasc. Endovasc. Surg.* **2007**, *33*, 20–25. [CrossRef] [PubMed]
16. Lorentsen, E. Systemic arterial blood pressure during exercise in patients with atherosclerosis obliterans of the lower limbs. *Circulation* **1972**, *46*, 257–263. [CrossRef]
17. Muller, M.D.; Drew, R.C.; Blaha, C.A.; Mast, J.L.; Cui, J.; Reed, A.B.; Sinoway, L.I. Oxidative stress contributes to the augmented exercise pressor reflex in peripheral arterial disease patients. *J. Physiol.* **2012**, *590*, 6237–6246. [CrossRef]
18. Lewis, G.D.; Gona, P.; Larson, M.G.; Plehn, J.F.; Benjamin, E.J.; O'Donnell, C.J.; Levy, D.; Vasan, R.S.; Wang, T.J. Exercise blood pressure and the risk of incident cardiovascular disease (from the Framingham Heart Study). *Am. J. Cardiol.* **2008**, *101*, 1614–1620. [CrossRef]
19. Thanassoulis, G.; Lyass, A.; Benjamin, E.J.; Larson, M.G.; Vita, J.A.; Levy, D.; Hamburg, N.M.; Widlansky, M.E.; O'Donnell, C.J.; Mitchell, G.F.; et al. Relations of exercise blood pressure response to cardiovascular risk factors and vascular function in the Framingham Heart Study. *Circulation* **2012**, *125*, 2836–2843. [CrossRef]
20. Sinoway, L.; Prophet, S.; Gorman, I.; Mosher, T.; Shenberger, J.; Dolecki, M.; Briggs, R.; Zelis, R. Muscle Acidosis during Static Exercise Is Associated with Calf Vasoconstriction. *J. Appl. Physiol.* **1989**, *66*, 429–436. [CrossRef]
21. Victor, R.G.; Bertocci, L.; Pryor, S.; Nunnally, R. Sympathetic nerve discharge is coupled to muscle cell pH during exercise in humans. *J. Clin. Investig.* **1988**, *82*, 1301–1305. [CrossRef] [PubMed]
22. Goodwin, G.M.; McCloskey, D.I.; Mitchell, J.H. Cardiovascular and respiratory responses to changes in central command during isometric exercise at constant muscle tension. *J. Physiol.* **1972**, *226*, 173–190. [CrossRef] [PubMed]
23. Waldrop, T.G.; Mullins, D.C.; Millhorn, D.E. Control of respiration by the hypothalamus and by feedback from contracting muscles in cats. *Respir. Physiol.* **1986**, *64*, 317–328. [CrossRef]
24. Victor, R.G.; Mark, A.L. Interaction of cardiopulmonary and carotid baroreflex control of vascular resistance in humans. *J. Clin. Investig.* **1985**, *76*, 1592–1598. [CrossRef]
25. Freyschuss, U. Cardiovascular adjustment to somatomotor activation. The elicitation of increments in heart rate, aortic pressure and venomotor tone with the initiation of muscle contraction. *Acta Physiol. Scand.* **1970**, *342*, 1–63.
26. Coote, J.H.; Hilton, S.M.; Pérez-González, J.F. The reflex nature of the pressor response to muscular exercise. *J. Physiol.* **1971**, *215*, 789–804. [CrossRef]
27. Waldrop, T.G.; Eldridge, F.L.; Iwamoto, G.A.; Mitchell, J.H. Central neural control of respiration and circulation during exercise. Chapter 9. In *Handbook of Physiology—Section 12, Exercise: Regulation and Integration of Multiple Systems*; Rowell, L.B., Shepherd, J.T., Eds.; Oxford University Press: New York, NY, USA, 1996; pp. 333–380.
28. Mitchell, J.H.; Kaufman, M.P.; Iwamoto, G.A. The exercise pressor reflex: Its cardiovascular effects, afferent mechanisms, and central pathways. *Annu. Rev. Physiol.* **1983**, *45*, 229–242. [CrossRef]
29. Kaufman, M.P.; Forster, H.V. Reflexes controlling circulatory, ventilatory and airway responses to exercise. Chapter 10. In *Handbook of Physiology—Section 12, Exercise: Regulation and Integration of Multiple Systems*; Rowell, L.B., Shepherd, J.T., Eds.; Oxford University Press: New York, NY, USA, 1996; pp. 381–447.

30. Drew, R.C.; Muller, M.D.; Blaha, C.A.; Mast, J.L.; Heffernan, M.J.; Estep, L.E.; Cui, J.; Reed, A.B.; Sinoway, L.I. Renal vasoconstriction is augmented during exercise in patients with peripheral arterial disease. *Physiol. Rep.* **2013**, *1*, e00154. [CrossRef]
31. Kim, D.J.; Kuroki, M.; Cui, J.; Gao, Z.; Luck, J.C.; Pai, S.; Miller, A.; Sinoway, L. Systemic and regional hemodynamic response to activation of the exercise pressor reflex in patients with peripheral artery disease. *Am. J. Physiol. Heart Circ. Physiol.* **2020**, *318*, H916–H924. [CrossRef]
32. Ritti-Dias, R.M.; Meneses, A.L.; Parker, D.E.; Montgomery, P.S.; Khurana, A.; Gardner, A.W. Cardiovascular responses to walking in patients with peripheral artery disease. *Med. Sci. Sports Exerc.* **2011**, *43*, 2017–2023. [CrossRef]
33. Cui, J.; Leuenberger, U.; Kim, D.; Luck, J.; Pai, S.; Blaha, C.; Cauffman, A.; Sinoway, L. Muscle Sympathetic Nerve Activity Responses to Exercise in Patients with Peripheral Artery Disease. *FASEB J.* **2021**, *35*, 35. [CrossRef]
34. Garcia, L.A. Epidemiology and pathophysiology of lower extremity peripheral arterial disease. *J. Endovasc. Ther.* **2006**, *13*, II3–II9. [CrossRef] [PubMed]
35. Hiatt, W.R.; Brass, E.P. Pathophysiology of Peripheral Artery Disease, Intermittent Claudication, and Critical Limb Ischemia. Chapter 17. In *Vascular Medicine: A Companion to Braunwald's Heart Disease*; Creager, M., Beckman, J.A., Loscalzo, J., Eds.; Saunders: Philadelphia, PA, USA, 2013; pp. 223–230.
36. Hirsch, A.T.; Criqui, M.H.; Treat-Jacobson, D.; Regensteiner, J.G.; Creager, M.A.; Olin, J.W.; Krook, S.H.; Hunninghake, D.B.; Comerota, A.J.; Walsh, M.E.; et al. Peripheral arterial disease detection, awareness, and treatment in primary care. *JAMA* **2001**, *286*, 1317–1324. [CrossRef] [PubMed]
37. Isbell, D.C.; Berr, S.S.; Toledano, A.Y.; Epstein, F.H.; Meyer, C.H.; Rogers, W.J.; Harthun, N.L.; Hagspiel, K.D.; Weltman, A.; Kramer, C.M. Delayed calf muscle phosphocreatine recovery after exercise identifies peripheral arterial disease. *J. Am. Coll. Cardiol.* **2006**, *47*, 2289–2295. [CrossRef]
38. Anderson, J.D.; Epstein, F.H.; Meyer, C.H.; Hagspiel, K.D.; Wang, H.; Berr, S.S.; Harthun, N.L.; Weltman, A.; Dimaria, J.M.; West, A.M.; et al. Multifactorial determinants of functional capacity in peripheral arterial disease: Uncoupling of calf muscle perfusion and metabolism. *J. Am. Coll. Cardiol.* **2009**, *54*, 628–635. [CrossRef]
39. Sheriff, D.; Wyss, C.; Rowell, L.; Scher, A. Does inadequate oxygen delivery trigger pressor response to muscle hypoperfusion during exercise? *Am. J. Physiol. Heart Circ. Physiol.* **1987**, *253*, H1199–H1207. [CrossRef]
40. O'Leary, D.S. Point: The muscle metaboreflex does restore blood flow to contracting muscles. *J. Appl. Physiol.* **2006**, *100*, 357–358, discussion 360–361. [CrossRef]
41. Crisafulli, A.; de Farias, R.R.; Farinatti, P.; Lopes, K.G.; Milia, R.; Sainas, G.; Pinna, V.; Palazzolo, G.; Doneddu, A.; Magnani, S.; et al. Blood Flow Restriction Training Reduces Blood Pressure During Exercise Without Affecting Metaboreflex Activity. *Front. Physiol.* **2018**, *9*, 1736. [CrossRef]
42. Jessee, M.B.; Buckner, S.L.; Mouser, J.G.; Mattocks, K.T.; Dankel, S.J.; Abe, T.; Bell, Z.W.; Bentley, J.P.; Loenneke, J.P. Muscle Adaptations to High-Load Training and Very Low-Load Training With and Without Blood Flow Restriction. *Front. Physiol.* **2018**, *9*, 1448. [CrossRef]
43. Neto, G.R.; Novaes, J.S.; Dias, I.; Brown, A.; Vianna, J.; Cirilo-Sousa, M.S. Effects of resistance training with blood flow restriction on haemodynamics: A systematic review. *Clin. Physiol. Funct. Imaging* **2017**, *37*, 567–574. [CrossRef]
44. Patterson, S.D.; Hughes, L.; Warmington, S.; Burr, J.; Scott, B.R.; Owens, J.; Abe, T.; Nielsen, J.L.; Libardi, C.A.; Laurentino, G.; et al. Blood Flow Restriction Exercise: Considerations of Methodology, Application, and Safety. *Front. Physiol.* **2019**, *10*, 533. [CrossRef] [PubMed]
45. Libardi, C.A.; Chacon-Mikahil, M.P.; Cavaglieri, C.R.; Tricoli, V.; Roschel, H.; Vechin, F.C.; Conceicao, M.S.; Ugrinowitsch, C. Effect of concurrent training with blood flow restriction in the elderly. *Int. J. Sports Med.* **2015**, *36*, 395–399. [CrossRef] [PubMed]
46. Stavres, J.; Singer, T.J.; Brochetti, A.; Kilbane, M.J.; Brose, S.W.; McDaniel, J. The Feasibility of Blood Flow Restriction Exercise in Patients With Incomplete Spinal Cord Injury. *PM R* **2018**, *10*, 1368–1379. [CrossRef] [PubMed]
47. Singer, T.J.; Stavres, J.; Elmer, S.J.; Kilgas, M.A.; Pollock, B.S.; Kearney, S.G.; McDaniel, J. Knee extension with blood flow restriction: Impact of cuff pressure on hemodynamics. *Eur. J. Appl. Physiol.* **2020**, *120*, 79–90. [CrossRef]
48. Cryer, H.G. Therapeutic approaches for clinical ischemia and reperfusion injury. *Shock* **1997**, *8*, 26–32. [CrossRef]
49. Maxwell, S.R.; Lip, G.Y. Reperfusion injury: A review of the pathophysiology, clinical manifestations and therapeutic options. *Int. J. Cardiol.* **1997**, *58*, 95–117. [CrossRef]
50. Haimovici, H. Arterial embolism with acute massive ischemic myopathy and myoglobinuria: Evaluation of a hitherto unreported syndrome with report of two cases. *Surgery* **1960**, *47*, 739–747.
51. Lau, C.S.; Scott, N.; Shaw, J.W.; Belch, J.J. Increased activity of oxygen free radicals during reperfusion in patients with peripheral arterial disease undergoing percutaneous peripheral artery balloon angioplasty. *Int. Angiol.* **1991**, *10*, 244–246.
52. Edwards, A.T.; Blann, A.D.; Suarez-Mendez, V.J.; Lardi, A.M.; McCollum, C.N. Systemic responses in patients with intermittent claudication after treadmill exercise. *Br. J. Surg.* **1994**, *81*, 1738–1741. [CrossRef]
53. Loukogeorgakis, S.P.; van den Berg, M.J.; Sofat, R.; Nitsch, D.; Charakida, M.; Haiyee, B.; de Groot, E.; MacAllister, R.J.; Kuijpers, T.W.; Deanfield, J.E. Role of NADPH oxidase in endothelial ischemia/reperfusion injury in humans. *Circulation* **2010**, *121*, 2310–2316. [CrossRef]

54. Seeger, J.P.; Lenting, C.J.; Schreuder, T.H.; Landman, T.R.; Cable, N.T.; Hopman, M.T.; Thijssen, D.H. Interval exercise, but not endurance exercise, prevents endothelial ischemia-reperfusion injury in healthy subjects. *Am. J. Physiol. Heart Circ. Physiol.* **2015**, *308*, H351–H357. [CrossRef] [PubMed]
55. Loukogeorgakis, S.P.; Panagiotidou, A.T.; Broadhead, M.W.; Donald, A.; Deanfield, J.E.; MacAllister, R.J. Remote ischemic preconditioning provides early and late protection against endothelial ischemia-reperfusion injury in humans: Role of the autonomic nervous system. *J. Am. Coll. Cardiol.* **2005**, *46*, 450–456. [CrossRef] [PubMed]
56. Brunt, V.E.; Jeckell, A.T.; Ely, B.R.; Howard, M.J.; Thijssen, D.H.; Minson, C.T. Acute hot water immersion is protective against impaired vascular function following forearm ischemia-reperfusion in young healthy humans. *Am. J. Physiol. Regul. Integr. Comp. Physiol.* **2016**, *311*, R1060–R1067. [CrossRef] [PubMed]
57. Ely, B.R.; Francisco, M.A.; Halliwill, J.R.; Bryan, S.D.; Comrada, L.N.; Larson, E.A.; Brunt, V.E.; Minson, C.T. Heat therapy reduces sympathetic activity and improves cardiovascular risk profile in women who are obese with polycystic ovary syndrome. *Am. J. Physiol. Regul. Integr. Comp. Physiol.* **2019**, *317*, R630–R640. [CrossRef] [PubMed]
58. Cui, J.; McQuillan, P.; Momen, A.; Blaha, C.; Moradkhan, R.; Mascarenhas, V.; Hogeman, C.S.; Krishnan, A.; Sinoway, L.I. The role of the cyclooxygenase products in evoking sympathetic activation in exercise. *Am. J. Physiol. Heart Circ. Physiol.* **2007**, *293*, H1861–H1868. [CrossRef]
59. Qin, L.; Li, J. Sympathetic Nerve Control of Blood Pressure Response during Exercise in Peripheral Artery Disease and Current Application of Experimental Disease Models. *Am. J. Biomed Sci. Res.* **2020**, *9*, 204–209. [CrossRef]
60. Limbourg, A.; Korff, T.; Napp, L.C.; Schaper, W.; Drexler, H.; Limbourg, F.P. Evaluation of postnatal arteriogenesis and angiogenesis in a mouse model of hind-limb ischemia. *Nat. Protoc.* **2009**, *4*, 1737–1746. [CrossRef]
61. Yu, J.; de Muinck, E.D.; Zhuang, Z.; Drinane, M.; Kauser, K.; Rubanyi, G.M.; Qian, H.S.; Murata, T.; Escalante, B.; Sessa, W.C. Endothelial nitric oxide synthase is critical for ischemic remodeling, mural cell recruitment, and blood flow reserve. *Proc. Natl. Acad. Sci. USA* **2005**, *102*, 10999–11004. [CrossRef]
62. Westvik, T.S.; Fitzgerald, T.N.; Muto, A.; Maloney, S.P.; Pimiento, J.M.; Fancher, T.T.; Magri, D.; Westvik, H.H.; Nishibe, T.; Velazquez, O.C.; et al. Limb ischemia after iliac ligation in aged mice stimulates angiogenesis without arteriogenesis. *J. Vasc. Surg.* **2009**, *49*, 464–473. [CrossRef]
63. Yu, J.; Dardik, A. A Murine Model of Hind Limb Ischemia to Study Angiogenesis and Arteriogenesis. *Methods Mol. Biol.* **2018**, *1717*, 135–143. [CrossRef]
64. Schuler, D.; Sansone, R.; Nicolaus, C.; Kelm, M.; Heiss, C. Repetitive remote occlusion (RRO) stimulates eNOS-dependent blood flow and collateral expansion in hindlimb ischemia. *Free. Radic. Biol. Med.* **2018**, *129*, 520–531. [CrossRef]
65. Goggi, J.L.; Haslop, A.; Boominathan, R.; Chan, K.; Soh, V.; Cheng, P.; Robins, E.G.; Bhakoo, K.K. Imaging the Proangiogenic Effects of Cardiovascular Drugs in a Diabetic Model of Limb Ischemia. *Contrast Media Mol. Imaging* **2019**, *2019*, 2538909. [CrossRef] [PubMed]
66. Liddell, R.P.; Patel, T.H.; Weiss, C.R.; Lee, D.S.; Matsuhashi, T.; Brown, P.R.; Gabrielson, K.L.; Rodriguez, E.R.; Eng, J.; Kimura, H.; et al. Endovascular model of rabbit hindlimb ischemia: A platform to evaluate therapeutic angiogenesis. *J. Vasc. Interv. Radiol. JVIR* **2005**, *16*, 991–998. [CrossRef] [PubMed]
67. Long, C.A.; Timmins, L.H.; Koutakis, P.; Goodchild, T.T.; Lefer, D.J.; Pipinos, I.I.; Casale, G.P.; Brewster, L.P. An endovascular model of ischemic myopathy from peripheral arterial disease. *J. Vasc. Surg.* **2017**, *66*, 891–901. [CrossRef] [PubMed]
68. Seo, H.S.; Kim, H.W.; Roh, D.H.; Yoon, S.Y.; Kwon, Y.B.; Han, H.J.; Chung, J.M.; Beitz, A.J.; Lee, J.H. A new rat model for thrombus-induced ischemic pain (TIIP); development of bilateral mechanical allodynia. *Pain* **2008**, *139*, 520–532. [CrossRef]
69. Waters, R.E.; Terjung, R.L.; Peters, K.G.; Annex, B.H. Preclinical models of human peripheral arterial occlusive disease: Implications for investigation of therapeutic agents. *J. Appl. Physiol.* **2004**, *97*, 773–780. [CrossRef]
70. Li, J.; Xing, J. Muscle afferent receptors engaged in augmented sympathetic responsiveness in peripheral artery disease. *Front. Physiol.* **2012**, *3*, 247. [CrossRef]
71. Tsuchimochi, H.; McCord, J.L.; Hayes, S.G.; Koba, S.; Kaufman, M.P. Chronic femoral artery occlusion augments exercise pressor reflex in decerebrated rats. *Am. J. Physiol. Heart Circ. Physiol.* **2010**, *299*, H106–H113. [CrossRef]
72. Kuczmarski, J.M.; Unrath, K.; Thomas, G.D. Exaggerated cardiovascular responses to treadmill running in rats with peripheral arterial insufficiency. *Am. J. Physiol. Heart Circ. Physiol.* **2018**, *314*, H114–H121. [CrossRef]
73. Signorelli, S.S.; Katsiki, N. Oxidative Stress and Inflammation: Their Role in the Pathogenesis of Peripheral Artery Disease with or Without Type 2 Diabetes Mellitus. *Curr. Vasc. Pharmacol.* **2018**, *16*, 547–554. [CrossRef]
74. Harms, J.E.; Kuczmarski, J.M.; Kim, J.S.; Thomas, G.D.; Kaufman, M.P. The role played by oxidative stress in evoking the exercise pressor reflex in health and simulated peripheral artery disease. *J. Physiol.* **2017**, *595*, 4365–4378. [CrossRef] [PubMed]
75. Xing, J.; Lu, J.; Liu, J.; Li, J. Local Injections of Superoxide Dismutase Attenuate the Exercise Pressor Reflex in Rats with Femoral Artery Occlusion. *Front. Physiol.* **2018**, *9*, 39. [CrossRef] [PubMed]
76. Xing, J.; Lu, J.; Li, J. Role of TNF-α in Regulating the Exercise Pressor Reflex in Rats With Femoral Artery Occlusion. *Front. Physiol.* **2018**, *9*, 1461. [CrossRef] [PubMed]

77. Ross, J.L.; Queme, L.F.; Shank, A.T.; Hudgins, R.C.; Jankowski, M.P. Sensitization of group III and IV muscle afferents in the mouse after ischemia and reperfusion injury. *J. Pain* **2014**, *15*, 1257–1270. [CrossRef]
78. Qin, L.; Li, Q.; Li, J. Exaggerated Activities of P2X3 and ASIC3 Signaling Pathway in Muscle Afferent Following Himblimb Muscle Ischemia-Reperfusion. *FASEB J.* **2022**, *36*, 1257–1270. [CrossRef]
79. Liu, J.; Li, J.D.; Lu, J.; Xing, J.; Li, J. Contribution of nerve growth factor to upregulation of P2X(3) expression in DRG neurons of rats with femoral artery occlusion. *Am. J. Physiol. Heart Circ. Physiol.* **2011**, *301*, H1070–H1079. [CrossRef]
80. Xing, J.; Lu, J.; Li, J. TRPA1 mediates amplified sympathetic responsiveness to activation of metabolically sensitive muscle afferents in rats with femoral artery occlusion. *Front. Physiol.* **2015**, *6*, 249. [CrossRef]
81. Liu, J.; Gao, Z.; Li, J. Femoral artery occlusion increases expression of ASIC3 in dorsal root ganglion neurons. *Am. J. Physiol. Heart Circ. Physiol.* **2010**, *299*, H1357–H1364. [CrossRef]
82. Xing, J.; Lu, J.; Li, J. Acid-sensing ion channel subtype 3 function and immunolabelling increases in skeletal muscle sensory neurons following femoral artery occlusion. *J. Physiol.* **2012**, *590*, 1261–1272. [CrossRef]
83. Xing, J.; Lu, J.; Li, J. Augmented P2X response and immunolabeling in dorsal root ganglion neurons innervating skeletal muscle following femoral artery occlusion. *J. Neurophysiol.* **2013**, *109*, 2161–2168. [CrossRef]
84. Leal, A.K.; Yamauchi, K.; Kim, J.; Ruiz-Velasco, V.; Kaufman, M.P. Peripheral delta-opioid receptors attenuate the exercise pressor reflex. *Am. J. Physiol. Heart Circ. Physiol.* **2013**, *305*, H1246–H1255. [CrossRef] [PubMed]
85. Leal, A.K.; McCord, J.L.; Tsuchimochi, H.; Kaufman, M.P. Blockade of the TP receptor attenuates the exercise pressor reflex in decerebrated rats with chronic femoral artery occlusion. *Am. J. Physiol. Heart Circ. Physiol.* **2011**, *301*, H2140–H2146. [CrossRef] [PubMed]
86. Tsuchimochi, H.; McCord, J.L.; Kaufman, M.P. Peripheral mu-opioid receptors attenuate the augmented exercise pressor reflex in rats with chronic femoral artery occlusion. *Am. J. Physiol. Heart Circ. Physiol.* **2010**, *299*, H557–H565. [CrossRef]
87. Yamauchi, K.; Kim, J.S.; Stone, A.J.; Ruiz-Velasco, V.; Kaufman, M.P. Endoperoxide 4 receptors play a role in evoking the exercise pressor reflex in rats with simulated peripheral artery disease. *J. Physiol.* **2013**, *591*, 2949–2962. [CrossRef]
88. Puntambekar, P.; Van Buren, J.; Raisinghani, M.; Premkumar, L.S.; Ramkumar, V. Direct interaction of adenosine with the TRPV1 channel protein. *J. Neurosci.* **2004**, *24*, 3663–3671. [CrossRef] [PubMed]
89. Tsuzuki, K.; Ase, A.; Seguela, P.; Nakatsuka, T.; Wang, C.Y.; She, J.X.; Gu, J.G. TNPATP-resistant P2X ionic current on the central terminals and somata of rat primary sensory neurons. *J. Neurophysiol.* **2003**, *89*, 3235–3242. [CrossRef] [PubMed]
90. Waldmann, R.; Bassilana, F.; de Weille, J.; Champigny, G.; Heurteaux, C.; Lazdunski, M. Molecular cloning of a non-inactivating proton-gated Na+ channel specific for sensory neurons. *J. Biol. Chem.* **1997**, *272*, 20975–20978. [CrossRef]
91. Waldmann, R.; Champigny, G.; Bassilana, F.; Heurteaux, C.; Lazdunski, M. A proton-gated cation channel involved in acid-sensing. *Nature* **1997**, *386*, 173–177. [CrossRef]
92. Waldmann, R.; Champigny, G.; Lingueglia, E.; De Weille, J.R.; Heurteaux, C.; Lazdunski, M. H(+)-gated cation channels. *Ann. N. Y. Acad. Sci.* **1999**, *868*, 67–76. [CrossRef]
93. Light, A.R.; Hughen, R.W.; Zhang, J.; Rainier, J.; Liu, Z.; Lee, J. Dorsal root ganglion neurons innervating skeletal muscle respond to physiological combinations of protons, ATP, and lactate mediated by ASIC, P2X, and TRPV1. *J. Neurophysiol.* **2008**, *100*, 1184–1201. [CrossRef]
94. Deval, E.; Noël, J.; Gasull, X.; Delaunay, A.; Alloui, A.; Friend, V.; Eschalier, A.; Lazdunski, M.; Lingueglia, E. Acid-sensing ion channels in postoperative pain. *J. Neurosci.* **2011**, *31*, 6059–6066. [CrossRef] [PubMed]
95. Deval, E.; Noël, J.; Lay, N.; Alloui, A.; Diochot, S.; Friend, V.; Jodar, M.; Lazdunski, M.; Lingueglia, E. ASIC3, a sensor of acidic and primary inflammatory pain. *EMBO J.* **2008**, *19*, 3047–3055. [CrossRef] [PubMed]
96. MacLean, D.A.; LaNoue, K.F.; Gray, K.S.; Sinoway, L.I. Effects of hindlimb contraction on pressor and muscle interstitial metabolite responses in the cat. *J. Appl. Physiol.* **1998**, *85*, 1583–1592. [CrossRef]
97. Rotto, D.M.; Stebbins, C.L.; Kaufman, M.P. Reflex cardiovascular and ventilatory responses to increasing H+ activity in cat hindlimb muscle. *J. Appl. Physiol.* **1989**, *67*, 256–263. [CrossRef] [PubMed]
98. Yagi, J.; Wenk, H.N.; Naves, L.A.; McCleskey, E.W. Sustained currents through ASIC3 ion channels at the modest pH changes that occur during myocardial ischemia. *Circ. Res.* **2006**, *99*, 501–509. [CrossRef]
99. MacLean, D.A.; Imadojemu, V.A.; Sinoway, L.I. Interstitial pH, K+, lactate and phosphate determined with MSNA during exercise in humans. *Am. J. Physiol. Regul. Integr. Comp. Physiol.* **2000**, *278*, R563–R571. [CrossRef] [PubMed]
100. Tsuchimochi, H.; Yamauchi, K.; McCord, J.L.; Kaufman, M.P. Blockade of acid sensing ion channels attenuates the augmented exercise pressor reflex in rats with chronic femoral artery occlusion (Abstract). *FASEB J.* **2011**, *589*, 6173–6189.
101. Stephan, G.; Huang, L.; Tang, Y.; Vilotti, S.; Fabbretti, E.; Yu, Y.; Nörenberg, W.; Franke, H.; Gölöncsér, F.; Sperlágh, B.; et al. The ASIC3/P2X3 cognate receptor is a pain-relevant and ligand-gated cationic channel. *Nat. Commun.* **2018**, *9*, 1354. [CrossRef] [PubMed]
102. Gao, Z.; Li, J.D.; Sinoway, L.I.; Li, J. Effect of muscle interstitial pH on P2X and TRPV1 receptor-mediated pressor response. *J. Appl. Physiol.* **2007**, *102*, 2288–2293. [CrossRef]

103. Lu, J.; Xing, J.; Li, J. Bradykinin B2 receptor contributes to the exaggerated muscle mechanoreflex in rats with femoral artery occlusion. *Am. J. Physiol. Heart Circ. Physiol.* **2013**, *304*, H1166–H1174. [CrossRef]
104. Qin, L.; Li, J. Nerve growth factor in muscle afferent neurons of peripheral artery disease and autonomic function. *Neural. Regen. Res.* **2021**, *16*, 694–699. [CrossRef] [PubMed]
105. Wang, G.L.; Jiang, B.H.; Rue, E.A.; Semenza, G.L. Hypoxia-inducible factor 1 is a basic-helix-loop-helix-PAS heterodimer regulated by cellular O2 tension. *Proc. Natl. Acad. Sci. USA* **1995**, *92*, 5510–5514. [CrossRef] [PubMed]
106. Ceradini, D.J.; Kulkarni, A.R.; Callaghan, M.J.; Tepper, O.M.; Bastidas, N.; Kleinman, M.E.; Capla, J.M.; Galiano, R.D.; Levine, J.P.; Gurtner, G.C. Progenitor cell trafficking is regulated by hypoxic gradients through HIF-1 induction of SDF-1. *Nat. Med.* **2004**, *10*, 858–864. [CrossRef] [PubMed]
107. Iyer, N.V.; Kotch, L.E.; Agani, F.; Leung, S.W.; Laughner, E.; Wenger, R.H.; Gassmann, M.; Gearhart, J.D.; Lawler, A.M.; Yu, A.Y.; et al. Cellular and developmental control of O2 homeostasis by hypoxia-inducible factor 1 alpha. *Genes. Dev.* **1998**, *12*, 149–162. [CrossRef]
108. Kelly, B.D.; Hackett, S.F.; Hirota, K.; Oshima, Y.; Cai, Z.; Berg-Dixon, S.; Rowan, A.; Yan, Z.; Campochiaro, P.A.; Semenza, G.L. Cell type-specific regulation of angiogenic growth factor gene expression and induction of angiogenesis in nonischemic tissue by a constitutively active form of hypoxia-inducible factor 1. *Circ. Res.* **2003**, *93*, 1074–1081. [CrossRef]
109. Manalo, D.J.; Rowan, A.; Lavoie, T.; Natarajan, L.; Kelly, B.D.; Ye, S.Q.; Garcia, J.G.; Semenza, G.L. Transcriptional regulation of vascular endothelial cell responses to hypoxia by HIF-1. *Blood* **2005**, *105*, 659–669. [CrossRef]
110. Gao, W.; Li, J. Femoral artery occlusion increases muscle pressor reflex and expression of hypoxia-inducible factor-1α in sensory neurons. *J. Cardiovasc. Dis.* **2013**, *1*, 34–40.
111. Jaakkola, P.; Mole, D.R.; Tian, Y.M.; Wilson, M.I.; Gielbert, J.; Gaskell, S.J.; Kriegsheim, A.; Hebestreit, H.F.; Mukherji, M.; Schofield, C.J.; et al. Targeting of HIF-alpha to the von Hippel-Lindau ubiquitylation complex by O2-regulated prolyl hydroxylation. *Science* **2001**, *292*, 468–472. [CrossRef]
112. Milkiewicz, M.; Pugh, C.W.; Egginton, S. Inhibition of endogenous HIF inactivation induces angiogenesis in ischaemic skeletal muscles of mice. *J. Physiol.* **2004**, *560*, 21–26. [CrossRef]
113. Xing, J.; Gao, Z.; Lu, J.; Sinoway, L.I.; Li, J. Femoral artery occlusion augments TRPV1-mediated sympathetic responsiveness. *Am. J. Physiol. Heart Circ. Physiol.* **2008**, *295*, H1262–H1269. [CrossRef]
114. Xing, J.; Lu, J.; Li, J. Contribution of nerve growth factor to augmented TRPV1 responses of muscle sensory neurons by femoral artery occlusion. *Am. J. Physiol. Heart Circ. Physiol.* **2009**, *296*, H1380–H1387. [CrossRef] [PubMed]
115. Xing, J.; Lu, J.; Li, J. ASIC3 Function and Immunolabeling Increases in Skeletal Muscle Sensory Neurons Following Femoral Artery Occlusion. *J. Physiol.* **2012**, *590*, 1261–1272. [CrossRef] [PubMed]
116. Farinelli, S.E.; Greene, L.A. Cell cycle blockers mimosine, ciclopirox, and deferoxamine prevent the death of PC12 cells and postmitotic sympathetic neurons after removal of trophic support. *J. Neurosci.* **1996**, *16*, 1150–1162. [CrossRef]
117. Lomb, D.J.; Straub, J.A.; Freeman, R.S. Prolyl hydroxylase inhibitors delay neuronal cell death caused by trophic factor deprivation. *J. Neurochem.* **2007**, *103*, 1897–1906. [CrossRef] [PubMed]
118. Xie, L.; Johnson, R.S.; Freeman, R.S. Inhibition of NGF deprivation-induced death by low oxygen involves suppression of BIMEL and activation of HIF-1. *J. Cell Biol.* **2005**, *168*, 911–920. [CrossRef]
119. Ruan, T.; Lin, Y.S.; Lin, K.S.; Kou, Y.R. Sensory transduction of pulmonary reactive oxygen species by capsaicin-sensitive vagal lung afferent fibres in rats. *J. Physiol.* **2005**, *565*, 563–578. [CrossRef] [PubMed]
120. Ruan, T.; Lin, Y.S.; Lin, K.S.; Kou, Y.R. Mediator mechanisms involved in TRPV1 and P2X receptor-mediated, ROS-evoked bradypneic reflex in anesthetized rats. *J. Appl. Physiol.* **2006**, *101*, 644–654. [CrossRef]
121. Wang, H.J.; Pan, Y.X.; Wang, W.Z.; Zucker, I.H.; Wang, W. NADPH oxidase-derived reactive oxygen species in skeletal muscle modulates the exercise pressor reflex. *J. Appl. Physiol.* **2009**, *107*, 450–459. [CrossRef]
122. McCord, J.L.; Tsuchimochi, H.; Yamauchi, K.; Leal, A.; Kaufman, M.P. Tempol attenuates the exercise pressor reflex independently of neutralizing reactive oxygen species in femoral artery ligated rats. *J. Appl. Physiol.* **2011**, *111*, 971–979. [CrossRef]
123. Yamauchi, K.; Stone, A.J.; Stocker, S.D.; Kaufman, M.P. Blockade of ATP-sensitive potassium channels prevents the attenuation of the exercise pressor reflex by tempol in rats with ligated femoral arteries. *Am. J. Physiol. Heart Circ. Physiol.* **2012**, *303*, H332–H340. [CrossRef]
124. Bandell, M.; Story, G.M.; Hwang, S.W.; Viswanath, V.; Eid, S.R.; Petrus, M.J.; Earley, T.J.; Patapoutian, A. Noxious cold ion channel TRPA1 is activated by pungent compounds and bradykinin. *Neuron* **2004**, *41*, 849–857. [CrossRef]
125. Bessac, B.F.; Sivula, M.; von Hehn, C.A.; Escalera, J.; Cohn, L.; Jordt, S.-E. TRPA1 is a major oxidant sensor in murine airway sensory neurons. *J. Clin. Investig.* **2008**, *118*, 1899–1910. [CrossRef] [PubMed]
126. Trevisani, M.; Siemens, J.; Materazzi, S.; Bautista, D.M.; Nassini, R.; Campi, B.; Imamachi, N.; Andre, E.; Patacchini, R.; Cottrell, G.S.; et al. 4-Hydroxynonenal, an endogenous aldehyde, causes pain and neurogenic inflammation through activation of the irritant receptor TRPA1. *Proc. Natl. Acad. Sci. USA* **2007**, *104*, 13519–13524. [CrossRef]
127. Bautista, D.M.; Jordt, S.-E.; Nikai, T.; Tsuruda, P.R.; Read, A.J.; Poblete, J.; Yamoah, E.N.; Basbaum, A.I.; Julius, D. TRPA1 mediates the inflammatory actions of environmental irritants and proalgesic agents. *Cell* **2006**, *124*, 1269–1282. [CrossRef]

128. Katsura, H.; Obata, K.; Mizushima, T.; Yamanaka, H.; Kobayashi, K.; Dai, Y.; Fukuoka, T.; Tokunaga, A.; Sakagami, M.; Noguchi, K. Antisense knock down of TRPA1, but not TRPM8, alleviates cold hyperalgesia after spinal nerve ligation in rats. *Exp. Neurol.* **2006**, *200*, 112–123. [CrossRef] [PubMed]
129. Kwan, K.Y.; Allchorne, A.J.; Vollrath, M.A.; Christensen, A.P.; Zhang, D.-S.; Woolf, C.J.; Corey, D.P. TRPA1 contributes to cold, mechanical, and chemical nociception but is not essential for hair-cell transduction. *Neuron* **2006**, *50*, 277–289. [CrossRef]
130. Macpherson, L.J.; Dubin, A.E.; Evans, M.J.; Marr, F.; Schultz, P.G.; Cravatt, B.F.; Patapoutian, A. Noxious compounds activate TRPA1 ion channels through covalent modification of cysteines. *Nature* **2007**, *445*, 541–545. [CrossRef]
131. Obata, K.; Katsura, H.; Mizushima, T.; Yamanaka, H.; Kobayashi, K.; Dai, Y.; Fukuoka, T.; Tokunaga, A.; Tominaga, M.; Noguchi, K. TRPA1 induced in sensory neurons contributes to cold hyperalgesia after inflammation and nerve injury. [Erratum appears in *J. Clin. Investig.* 2010 Jan;120(1):394]. *J. Clin. Investig.* **2005**, *115*, 2393–2401. [CrossRef]
132. Story, G.M.; Peier, A.M.; Reeve, A.J.; Eid, S.R.; Mosbacher, J.; Hricik, T.R.; Earley, T.J.; Hergarden, A.C.; Andersson, D.A.; Hwang, S.W.; et al. ANKTM1, a TRP-like channel expressed in nociceptive neurons, is activated by cold temperatures. *Cell* **2003**, *112*, 819–829. [CrossRef]
133. Koba, S.; Hayes, S.G.; Sinoway, L.I. Transient receptor potential A1 channel contributes to activation of the muscle reflex. *Am. J. Physiol. Heart Circ. Physiol.* **2011**, *300*, H201–H213. [CrossRef]
134. Böhm, F.; Pernow, J. The importance of endothelin-1 for vascular dysfunction in cardiovascular disease. *Cardiovasc. Res.* **2007**, *76*, 8–18. [CrossRef] [PubMed]
135. Yanagisawa, M.; Kurihara, H.; Kimura, S.; Tomobe, Y.; Kobayashi, M.; Mitsui, Y.; Yazaki, Y.; Goto, K.; Masaki, T. A novel potent vasoconstrictor peptide produced by vascular endothelial cells. *Nature* **1988**, *332*, 411–415. [CrossRef] [PubMed]
136. Sokolovsky, M. Endothelin receptor subtypes and their role in transmembrane signaling mechanisms. *Pharmacol. Ther.* **1995**, *68*, 435–471. [CrossRef]
137. Yeager, M.E.; Belchenko, D.D.; Nguyen, C.M.; Colvin, K.L.; Ivy, D.D.; Stenmark, K.R. Endothelin-1, the unfolded protein response, and persistent inflammation: Role of pulmonary artery smooth muscle cells. *Am. J. Respir Cell Mol. Biol.* **2012**, *46*, 14–22. [CrossRef]
138. Chen, H.C.; Guh, J.Y.; Shin, S.J.; Tsai, J.H.; Lai, Y.H. Reactive oxygen species enhances endothelin-1 production of diabetic rat glomeruli in vitro and in vivo. *J. Lab. Clin. Med.* **2000**, *135*, 309–315. [CrossRef]
139. Ruef, J.; Moser, M.; Kubler, W.; Bode, C. Induction of endothelin-1 expression by oxidative stress in vascular smooth muscle cells. *Cardiovasc. Pathol.* **2001**, *10*, 311–315. [CrossRef]
140. De Haro Miralles, J.; Gonzalez, A.F.; Varela Casariego, C.; Garcia, F.A. Onset of peripheral arterial disease: Role of endothelin in endothelial dysfunction. *Interact. Cardiovasc. Thorac. Surg.* **2010**, *10*, 760–765. [CrossRef]
141. Mangiafico, R.A.; Malatino, L.S.; Santonocito, M.; Sarnataro, F.; Dell'Arte, S.; Messina, R.; Santangelo, B. Plasma endothelin-1 levels in patients with peripheral arterial occlusive disease at different Fontaine's stages. *Panminerva Med.* **1999**, *41*, 22–26.
142. Qin, L.; Li, J. Heat treatment attenuates activities of ET-1 signaling pathway and enhances the protective HSP72-SOD2 in ischemic skeletal muscle of rats with peripheral artery disease. *FASEB J.* **2021**, *35*. [CrossRef]
143. Holwerda, S.W.; Restaino, R.M.; Fadel, P.J. Adrenergic and non-adrenergic control of active skeletal muscle blood flow: Implications for blood pressure regulation during exercise. *Auton. Neurosci. Basic Clin.* **2015**, *188*, 24–31. [CrossRef]
144. Pomonis, J.D.; Rogers, S.D.; Peters, C.M.; Ghilardi, J.R.; Mantyh, P.W. Expression and localization of endothelin receptors: Implications for the involvement of peripheral glia in nociception. *J. Neurosci. Off. J. Soc. Neurosci.* **2001**, *21*, 999–1006. [CrossRef]
145. Motta, E.M.; Calixto, J.B.; Rae, G.A. Mechanical hyperalgesia induced by endothelin-1 in rats is mediated via phospholipase C, protein kinase C, and MAP kinases. *Exp. Biol. Med.* **2006**, *231*, 1141–1145.
146. Motta, E.M.; Chichorro, J.G.; D'Orléans-Juste, P.; Rae, G.A. Roles of endothelin ETA and ETB receptors in nociception and chemical, thermal and mechanical hyperalgesia induced by endothelin-1 in the rat hindpaw. *Peptides* **2009**, *30*, 918–925. [CrossRef] [PubMed]
147. Hasue, F.; Kuwaki, T.; Kisanuki, Y.Y.; Yanagisawa, M.; Moriya, H.; Fukuda, Y.; Shimoyama, M. Increased sensitivity to acute and persistent pain in neuron-specific endothelin-1 knockout mice. *Neuroscience* **2005**, *130*, 349–358. [CrossRef] [PubMed]
148. Barr, T.P.; Kam, S.; Khodorova, A.; Montmayeur, J.P.; Strichartz, G.R. New perspectives on the endothelin axis in pain. *Pharmacol. Res.* **2011**, *63*, 532–540. [CrossRef]
149. Feng, B.; Strichartz, G. Endothelin-1 raises excitability and reduces potassium currents in sensory neurons. *Brain Res. Bull.* **2009**, *79*, 345–350. [CrossRef]
150. Zhou, Z.; Davar, G.; Strichartz, G. Endothelin-1 (ET-1) selectively enhances the activation gating of slowly inactivating tetrodotoxin-resistant sodium currents in rat sensory neurons: A mechanism for the pain-inducing actions of ET-1. *J. Neurosci. Off. J. Soc. Neurosci.* **2002**, *22*, 6325–6330. [CrossRef]
151. Mule, N.K.; Singh, J.N.; Shah, K.U.; Gulati, A.; Sharma, S.S. Endothelin-1 Decreases Excitability of the Dorsal Root Ganglion Neurons via ET(B) Receptor. *Mol. Neurobiol.* **2018**, *55*, 4297–4310. [CrossRef]
152. Werner, M.F.; Trevisani, M.; Campi, B.; André, E.; Geppetti, P.; Rae, G.A. Contribution of peripheral endothelin ETA and ETB receptors in neuropathic pain induced by spinal nerve ligation in rats. *Eur. J. Pain* **2010**, *14*, 911–917. [CrossRef]

153. Miller, R.J.; Jung, H.; Bhangoo, S.; White, F.A. Cytokine and chemokine regulation of sensory neuron function. In *Handbook of Experimental Pharmacology*; Canning, B.J., Spina, D., Eds.; Springer: Berlin/Heidelberg, Germany, 2009; Volume 194, pp. 417–449.
154. Oppenheim, J.J. Cytokines: Past, present, and future. *Int. J. Hematol.* **2001**, *74*, 3–8. [CrossRef]
155. Jonsdottir, I.H.; Schjerling, P.; Ostrowski, K.; Asp, S.; Richter, E.A.; Pedersen, B.K. Muscle contractions induce interleukin-6 mRNA production in rat skeletal muscles. *J. Physiol.* **2000**, *528*, 157–163. [CrossRef] [PubMed]
156. Steensberg, A.; van Hall, G.; Osada, T.; Sacchetti, M.; Saltin, B.; Pedersen, B.K. Production of interleukin-6 in contracting human skeletal muscles can account for the exercise-induced increase in plasma interleukin-6. *J. Physiol.* **2000**, *529*, 237–242. [CrossRef] [PubMed]
157. Mann, D.L. Inflammatory mediators and the failing heart: Past, present, and the foreseeable future. *Circ. Res.* **2002**, *91*, 988–998. [CrossRef] [PubMed]
158. El-Menyar, A.A. Cytokines and myocardial dysfunction: State of the art. *J. Card Fail.* **2008**, *14*, 61–74. [CrossRef]
159. Yndestad, A.; Damas, J.K.; Oie, E.; Ueland, T.; Gullestad, L.; Aukrust, P. Systemic inflammation in heart failure–the whys and wherefores. *Heart Fail. Rev.* **2006**, *11*, 83–92. [CrossRef]
160. Lai, J.; Porreca, F.; Hunter, J.C.; Gold, M.S. Voltage-gated sodium channels and hyperalgesia. *Annu. Rev. Pharmacol. Toxicol.* **2004**, *44*, 371–397. [CrossRef]
161. Stone, A.J.; Kim, J.S.; Yamauchi, K.; Ruiz-Velasco, V.; Kaufman, M.P. Attenuation of autonomic reflexes by A803467 may not be solely caused by blockade of NaV 1.8 channels. *Neurosci. Lett.* **2013**, *543*, 177–182. [CrossRef]
162. Chen, X.; Pang, R.P.; Shen, K.F.; Zimmermann, M.; Xin, W.J.; Li, Y.Y.; Liu, X.G. TNF-alpha enhances the currents of voltage gated sodium channels in uninjured dorsal root ganglion neurons following motor nerve injury. *Exp. Neurol.* **2011**, *227*, 279–286. [CrossRef] [PubMed]
163. Scheller, J.; Garbers, C.; Rose-John, S. Interleukin-6: From basic biology to selective blockade of pro-inflammatory activities. *Semin. Immunol.* **2014**, *26*, 2–12. [CrossRef]
164. Chaparala, R.P.; Orsi, N.M.; Lindsey, N.J.; Girn, R.S.; Homer-Vanniasinkam, S. Inflammatory profiling of peripheral arterial disease. *Ann. Vasc. Surg.* **2009**, *23*, 172–178. [CrossRef]
165. Girn, H.R.; Orsi, N.M.; Homer-Vanniasinkam, S. An overview of cytokine interactions in atherosclerosis and implications for peripheral arterial disease. *Vasc. Med.* **2007**, *12*, 299–309. [CrossRef] [PubMed]
166. Palmer-Kazen, U.; Religa, P.; Wahlberg, E. Exercise in patients with intermittent claudication elicits signs of inflammation and angiogenesis. *Eur. J. Vasc. Endovasc. Surg.* **2009**, *38*, 689–696. [CrossRef] [PubMed]
167. Signorelli, S.S.; Mazzarino, M.C.; Di Pino, L.; Malaponte, G.; Porto, C.; Pennisi, G.; Marchese, G.; Costa, M.P.; Digrandi, D.; Celotta, G.; et al. High circulating levels of cytokines (IL-6 and TNFalpha), adhesion molecules (VCAM-1 and ICAM-1) and selectins in patients with peripheral arterial disease at rest and after a treadmill test. *Vasc. Med.* **2003**, *8*, 15–19. [CrossRef]
168. Shill, D.D.; Polley, K.R.; Willingham, T.B.; Call, J.A.; Murrow, J.R.; McCully, K.K.; Jenkins, N.T. Experimental intermittent ischemia augments exercise-induced inflammatory cytokine production. *J. Appl. Physiol. (1985)* **2017**, *123*, 434–441. [CrossRef]
169. Copp, S.W.; Stone, A.J.; Li, J.; Kaufman, M.P. Role played by interleukin-6 in evoking the exercise pressor reflex in decerebrate rats: Effect of femoral artery ligation. *Am. J. Physiol. Heart Circ. Physiol.* **2015**, *309*, H166–H173. [CrossRef]
170. Jostock, T.; Müllberg, J.; Ozbek, S.; Atreya, R.; Blinn, G.; Voltz, N.; Fischer, M.; Neurath, M.F.; Rose-John, S. Soluble gp130 is the natural inhibitor of soluble interleukin-6 receptor transsignaling responses. *Eur. J. Biochem.* **2001**, *268*, 160–167. [CrossRef]
171. Wolf, J.; Rose-John, S.; Garbers, C. Interleukin-6 and its receptors: A highly regulated and dynamic system. *Cytokine* **2014**, *70*, 11–20. [CrossRef] [PubMed]
172. Fakhry, F.; Spronk, S.; van der Laan, L.; Wever, J.J.; Teijink, J.A.; Hoffmann, W.H.; Smits, T.M.; van Brussel, J.P.; Stultiens, G.N.; Derom, A.; et al. Endovascular Revascularization and Supervised Exercise for Peripheral Artery Disease and Intermittent Claudication: A Randomized Clinical Trial. *JAMA* **2015**, *314*, 1936–1944. [CrossRef]
173. Monroe, J.C.; Song, Q.; Emery, M.S.; Hirai, D.M.; Motaganahalli, R.L.; Roseguini, B.T. Acute effects of leg heat therapy on walking performance and cardiovascular and inflammatory responses to exercise in patients with peripheral artery disease. *Physiol. Rep.* **2021**, *8*, e14650. [CrossRef]
174. Neff, D.; Kuhlenhoelter, A.M.; Lin, C.; Wong, B.J.; Motaganahalli, R.L.; Roseguini, B.T. Thermotherapy reduces blood pressure and circulating endothelin-1 concentration and enhances leg blood flow in patients with symptomatic peripheral artery disease. *Am. J. Physiol. Regul. Integr. Comp. Physiol.* **2016**, *311*, R392–R400. [CrossRef]
175. Kim, K.; Reid, B.A.; Ro, B.; Casey, C.A.; Song, Q.; Kuang, S.; Roseguini, B.T. Heat therapy improves soleus muscle force in a model of ischemia-induced muscle damage. *J. Appl. Physiol. (1985)* **2019**, *127*, 215–228. [CrossRef] [PubMed]
176. Kim, K.; Ro, B.; Damen, F.W.; Gramling, D.P.; Lehr, T.D.; Song, Q.F.; Goergen, C.J.; Roseguini, B.T. Heat therapy improves body composition and muscle function but does not affect capillary or collateral growth in a model of obesity and hindlimb ischemia. *J. Appl. Physiol.* **2021**, *130*, 355–368. [CrossRef] [PubMed]
177. Johnson, J.M.; Proppe, D.W. Cardiovascular adjustments to heat stress. In *Handbook of Physiology—Environmental Physiology*; Fregly, M.J., Blatteis, C.M., Eds.; Oxford University Press: New York, NY, USA, 1996; pp. 215–243.
178. Brothers, R.M.; Bhella, P.S.; Shibata, S.; Wingo, J.E.; Levine, B.D.; Crandall, C.G. Cardiac systolic and diastolic function during whole body heat stress. *Am. J. Physiol. Heart Circ. Physiol.* **2009**, *296*, H1150–H1156. [CrossRef]

179. Crandall, C.G.; Wilson, T.E.; Marving, J.; Vogelsang, T.W.; Kjaer, A.; Hesse, B.; Secher, N.H. Effects of passive heating on central blood volume and ventricular dimensions in humans. *J. Physiol.* **2008**, *586*, 293–301. [CrossRef]
180. Tei, C.; Horikiri, Y.; Park, J.C.; Jeong, J.W.; Chang, K.S.; Toyama, Y.; Tanaka, N. Acute hemodynamic improvement by thermal vasodilation in congestive heart failure. *Circulation* **1995**, *91*, 2582–2590. [CrossRef]
181. Kisanuki, A.; Daitoku, S.; Kihara, T.; Otsuji, Y.; Tei, C. Thermal therapy improves left ventricular diastolic function in patients with congestive heart failure: A tissue doppler echocardiographic study. *J. Cardiol.* **2007**, *49*, 187–191. [PubMed]
182. Michalsen, A.; Ludtke, R.; Buhring, M.; Spahn, G.; Langhorst, J.; Dobos, G.J. Thermal hydrotherapy improves quality of life and hemodynamic function in patients with chronic heart failure. *Am. Heart J.* **2003**, *146*, E11. [CrossRef]
183. Biro, S.; Masuda, A.; Kihara, T.; Tei, C. Clinical implications of thermal therapy in lifestyle-related diseases. *Exp. Biol. Med.* **2003**, *228*, 1245–1249. [CrossRef]
184. Weber, A.A.; Silver, M.A. Heat therapy in the management of heart failure. *Congest. Heart Fail.* **2007**, *13*, 81–83. [CrossRef]
185. Ikeda, Y.; Biro, S.; Kamogawa, Y.; Yoshifuku, S.; Eto, H.; Orihara, K.; Yu, B.; Kihara, T.; Miyata, M.; Hamasaki, S.; et al. Repeated sauna therapy increases arterial endothelial nitric oxide synthase expression and nitric oxide production in cardiomyopathic hamsters. *Circ. J.* **2005**, *69*, 722–729. [CrossRef]
186. Imamura, M.; Biro, S.; Kihara, T.; Yoshifuku, S.; Takasaki, K.; Otsuji, Y.; Minagoe, S.; Toyama, Y.; Tei, C. Repeated thermal therapy improves impaired vascular endothelial function in patients with coronary risk factors. *J. Am. Coll. Cardiol.* **2001**, *38*, 1083–1088. [CrossRef]
187. Kihara, T.; Biro, S.; Imamura, M.; Yoshifuku, S.; Takasaki, K.; Ikeda, Y.; Otuji, Y.; Minagoe, S.; Toyama, Y.; Tei, C. Repeated sauna treatment improves vascular endothelial and cardiac function in patients with chronic heart failure. *J. Am. Coll. Cardiol.* **2002**, *39*, 754–759. [CrossRef]
188. Thomas, K.N.; van Rij, A.M.; Lucas, S.J.; Cotter, J.D. Lower-limb hot-water immersion acutely induces beneficial hemodynamic and cardiovascular responses in peripheral arterial disease and healthy, elderly controls. *Am. J. Physiol. Regul. Integr. Comp. Physiol.* **2017**, *312*, R281–R291. [CrossRef] [PubMed]
189. Tei, C.; Shinsato, T.; Miyata, M.; Kihara, T.; Hamasaki, S. Waon therapy improves peripheral arterial disease. *J. Am. Coll. Cardiol.* **2007**, *50*, 2169–2171. [CrossRef]
190. Shinsato, T.; Miyata, M.; Kubozono, T.; Ikeda, Y.; Fujita, S.; Kuwahata, S.; Akasaki, Y.; Hamasaki, S.; Fujiwara, H.; Tei, C. Waon therapy mobilizes CD34+ cells and improves peripheral arterial disease. *J. Cardiol.* **2010**, *56*, 361–366. [CrossRef]
191. Cui, J.; Mascarenhas, V.; Moradkhan, R.; Blaha, C.; Sinoway, L.I. Effects of muscle metabolites on responses of muscle sympathetic nerve activity to mechanoreceptor(s) stimulation in healthy humans. *Am. J. Physiol. Regul. Integr. Comp. Physiol.* **2008**, *294*, R458–R466. [CrossRef]
192. Cui, J.; Blaha, C.; Sinoway, L.I. Whole body heat stress attenuates the pressure response to muscle metaboreceptor stimulation in humans. *J. Appl. Physiol. (1985)* **2016**, *121*, 1178–1186. [CrossRef]
193. Cui, J.; Shibasaki, M.; Low, D.A.; Keller, D.M.; Davis, S.L.; Crandall, C.G. Heat stress attenuates the increase in arterial blood pressure during the cold pressor test. *J. Appl. Physiol.* **2010**, *109*, 1354–1359. [CrossRef]
194. Sterns, D.A.; Ettinger, S.M.; Gray, K.S.; Whisler, S.K.; Mosher, T.J.; Smith, M.B.; Sinoway, L.I. Skeletal muscle metaboreceptor exercise responses are attenuated in heart failure. *Circulation* **1991**, *84*, 2034–2039. [CrossRef]
195. Ray, C.A.; Gracey, K.H. Augmentation of exercise-induced muscle sympathetic nerve activity during muscle heating. *J. Appl. Physiol.* **1997**, *82*, 1719–1725. [CrossRef]
196. Kuipers, N.T.; Sauder, C.L.; Kearney, M.L.; Ray, C.A. Interactive effect of aging and local muscle heating on renal vasoconstriction during isometric handgrip. *Am. J. Physiol. Renal. Physiol.* **2009**, *297*, F327–F332. [CrossRef] [PubMed]
197. Kuipers, N.T.; Sauder, C.L.; Kearney, M.L.; Ray, C.A. Changes in forearm muscle temperature alter renal vascular responses to isometric handgrip. *Am. J. Physiol. Heart Circ. Physiol.* **2007**, *293*, H3432–H3439. [CrossRef] [PubMed]
198. Hertel, H.C.; Howaldt, B.; Mense, S. Responses of group IV and group III muscle afferents to thermal stimuli. *Brain Res.* **1976**, *113*, 201–205. [CrossRef]
199. Kumazawa, T.; Mizumura, K. Thin-fibre receptors responding to mechanical, chemical, and thermal stimulation in the skeletal muscle of the dog. *J. Physiol.* **1977**, *273*, 179–194. [CrossRef] [PubMed]
200. Gao, Z.; Kehoe, V.; Xing, J.; Sinoway, L.; Li, J. Temperature modulates P2X receptor-mediated cardiovascular responses to muscle afferent activation. *Am. J. Physiol. Heart Circ. Physiol.* **2006**, *291*, H1255–H1261. [CrossRef]
201. Bombor, I.; Wissgott, C.; Andresen, R. Lumbar sympathicolysis in patients with severe peripheral artery disease: Hemodynamics of the lower limbs determined by near-infrared spectroscopy, color coded duplex sonography, and temperature measurement. *Clin. Med. Insights Cardiol.* **2014**, *8*, 29–36. [CrossRef]
202. Gray, S.R.; De Vito, G.; Nimmo, M.A.; Farina, D.; Ferguson, R.A. Skeletal muscle ATP turnover and muscle fiber conduction velocity are elevated at higher muscle temperatures during maximal power output development in humans. *Am. J. Physiol. Regul. Integr. Comp. Physiol.* **2006**, *290*, R376–R382. [CrossRef]
203. Dakshinamurti, K.; Dakshinamurti, S. Blood pressure regulation and micronutrients. *Nutr. Res. Rev.* **2001**, *14*, 3–44. [CrossRef]
204. Wilmink, A.B.; Welch, A.A.; Quick, C.R.; Burns, P.J.; Hubbard, C.S.; Bradbury, A.W.; Day, N.E. Dietary folate and vitamin B6 are independent predictors of peripheral arterial occlusive disease. *J. Vasc. Surg.* **2004**, *39*, 513–516. [CrossRef]

205. Khakh, B.S.; Humphrey, P.P.; Surprenant, A. Electrophysiological properties of P2X-purinoceptors in rat superior cervical, nodose and guinea-pig coeliac neurones. *J. Physiol.* **1995**, *484*, 385–395. [CrossRef]
206. He, D.D.; Gao, Y.; Wang, S.; Xie, Z.; Song, X.J. Systematic Administration of B Vitamins Alleviates Diabetic Pain and Inhibits Associated Expression of P2X3 and TRPV1 in Dorsal Root Ganglion Neurons and Proinflammatory Cytokines in Spinal Cord in Rats. *Pain. Res. Manag.* **2020**, *2020*, 3740162. [CrossRef] [PubMed]
207. Cui, J.; Leuenberger, U.A.; Blaha, C.; King, N.C.; Sinoway, L.I. Effect of P2 receptor blockade with pyridoxine on sympathetic response to exercise pressor reflex in humans. *J. Physiol.* **2011**, *589*, 685–695. [CrossRef] [PubMed]
208. Morelli, M.B.; Gambardella, J.; Castellanos, V.; Trimarco, V.; Santulli, G. Vitamin C and Cardiovascular Disease: An Update. *Antioxidants* **2020**, *9*, 1227. [CrossRef] [PubMed]

International Journal of *Molecular Sciences*

MDPI

Review

Stellate Ganglia and Cardiac Sympathetic Overactivation in Heart Failure

Yu-Long Li [1,2]

[1] Department of Emergency Medicine, University of Nebraska Medical Center, Omaha, NE 68198, USA; yulongli@unmc.edu; Tel.: +1-402-559-3016; Fax: +1-402-559-9659
[2] Department of Cellular & Integrative Physiology, University of Nebraska Medical Center, Omaha, NE 68198, USA

Abstract: Heart failure (HF) is a major public health problem worldwide, especially coronary heart disease (myocardial infarction)-induced HF with reduced ejection fraction (HFrEF), which accounts for over 50% of all HF cases. An estimated 6 million American adults have HF. As a major feature of HF, cardiac sympathetic overactivation triggers arrhythmias and sudden cardiac death, which accounts for nearly 50–60% of mortality in HF patients. Regulation of cardiac sympathetic activation is highly integrated by the regulatory circuitry at multiple levels, including afferent, central, and efferent components of the sympathetic nervous system. Much evidence, from other investigators and us, has confirmed the afferent and central neural mechanisms causing sympathoexcitation in HF. The stellate ganglion is a peripheral sympathetic ganglion formed by the fusion of the 7th cervical and 1st thoracic sympathetic ganglion. As the efferent component of the sympathetic nervous system, cardiac postganglionic sympathetic neurons located in stellate ganglia provide local neural coordination independent of higher brain centers. Structural and functional impairments of cardiac postganglionic sympathetic neurons can be involved in cardiac sympathetic overactivation in HF because normally, many effects of the cardiac sympathetic nervous system on cardiac function are mediated via neurotransmitters (e.g., norepinephrine) released from cardiac postganglionic sympathetic neurons innervating the heart. This review provides an overview of cardiac sympathetic remodeling in stellate ganglia and potential mechanisms and the role of cardiac sympathetic remodeling in cardiac sympathetic overactivation and arrhythmias in HF. Targeting cardiac sympathetic remodeling in stellate ganglia could be a therapeutic strategy against malignant cardiac arrhythmias in HF.

Keywords: arrhythmia; autonomic nervous system; cardiac sympathetic activation; heart failure; inflammation; stellate ganglion

Citation: Li, Y.-L. Stellate Ganglia and Cardiac Sympathetic Overactivation in Heart Failure. *Int. J. Mol. Sci.* **2022**, *23*, 13311. https://doi.org/10.3390/ijms232113311

Academic Editors: Yutang Wang and Kate Denton

Received: 16 October 2022
Accepted: 29 October 2022
Published: 1 November 2022

1. Introduction

Heart failure (HF) is a major public health problem worldwide, presented by an inability of the heart to provide metabolic demands and perfusion of organs/tissues and characterized by symptoms and signs, including shortness of breath, fatigue, rapid to irregular heart rate, lung crepitations, elevated jugular venous pressure, and peripheral tissue edema [1,2]. HF affects more than 26 million adults worldwide and an estimated 6 million American adults have HF [2,3], in which coronary heart disease (myocardial infarction, MI)-induced HF with reduced ejection fraction (HFrEF) accounts for about 50% of all HF cases [4–7]. Considering the stable incidence of HF with an annual increase, the actual burden of treatment and diagnosis in patients with HF has obviously exceeded the projected burden in the United States and worldwide, especially accounting for other factors, including an increased comorbidity burden and advancing age of the population [2,8,9]. Despite advances in diagnosis and therapeutic management of HF, HF still has a high morbidity and mortality rate. As a major feature of HF, cardiac sympathetic overactivation [10–14] triggers malignant arrhythmias and sudden cardiac death [15–23], which accounts for nearly 50–60% of the mortality in HF patients [20,24–35]. The role of cardiac sympathetic hyperactivation in HF is highlighted by the use of β-blockers and cardiac sympathetic denervation as the key approach to the current therapy of HF [36–43]. However, such pharmacological treatment may not be ideal because some studies have demonstrated that β-blockers do not provide satisfactory protection against sudden cardiac death, and some patients are either intolerant or refractory to this therapy [44–50]. Additionally, despite being an alternative in managing refractory ventricular arrhythmias [38,43,51,52], cardiac sympathetic denervation has adverse complications (including Horner's syndrome, hyperhidrosis, paresthesia, and sympathetic fight/fight response loss) that severely limit the use of procedures in HF patients [53,54]. These drawbacks have increased the focus on exploring the mechanisms responsible for HF-increased cardiac sympathetic activation and on identifying effective therapeutic interventions, which are crucial for improving prognosis of HF and reducing its mortality.

The regulation of cardiac sympathetic activation is highly integrated by the regulatory circuitry at multiple levels, including afferent, central, and efferent components of the sympathetic nervous system [55,56]. Much evidence, from other investigators and us, has confirmed the afferent and central neural mechanisms causing sympathoexcitation in HF [57–67]. The stellate ganglion is a peripheral sympathetic ganglion formed by the fusion of the 7th cervical and 1st thoracic sympathetic ganglion. As the efferent component of the sympathetic nervous system, cardiac postganglionic sympathetic neurons located in stellate ganglia provide local neural coordination independent of higher brain centers [56,68]. These neurons innervate the heart to regulate cardiac function through neurotransmitters (e.g., norepinephrine, NE) released from cardiac sympathetic nerve terminals [69]. Much evidence from clinical studies and animal experiments has demonstrated that the remodeling of cardiac postganglionic sympathetic neurons in stellate ganglia could contribute to cardiac sympathetic overactivation and malignant ventricular arrhythmias in HF. In this review, therefore, we discuss cardiac sympathetic remodeling in stellate ganglia and potential mechanisms and the role of cardiac sympathetic remodeling in cardiac sympathetic overactivation and arrhythmias in HF.

2. Anatomy and Physiology of Stellate Ganglia (Figure 1)

The sympathetic nervous system is one of the two divisions of the autonomic nervous system, the other being the parasympathetic nervous system. The sympathetic nervous system is composed of preganglionic and postganglionic neurons that are involved in signal transmission to regulate a variety of functions in all peripheral organs/tissues. The cardiac preganglionic sympathetic neurons originate in the intermediolateral column of the spinal cord in the thoracic region with the somata located in the gray rami communicantes bilaterally and symmetrically. Their axons are very short and pass through the white rami communicantes to form the synapses with cardiac postganglionic sympathetic neurons located in the lower cervical and upper thoracic paravertebral ganglia, releasing a neurotransmitter, acetylcholine, from cardiac preganglionic nerve terminals [70,71]. Usually, the 7th cervical and 1st thoracic paravertebral sympathetic ganglia fuse into stellate ganglia, and the latter play a key role in a substantial amount of cardiac neurotransmission [71–73]. When acetylcholine released from cardiac preganglionic sympathetic nerve endings activates nicotinic acetylcholine receptors on cardiac postganglionic sympathetic neurons in stellate ganglia, the longer cardiac postganglionic sympathetic nerve terminals innervated the heart release some neurotransmitters (such as norepinephrine, neuropeptide Y, and galanin) to regulate the functions of the heart through the activation of adrenergic receptors and other peptide receptors [72,74].

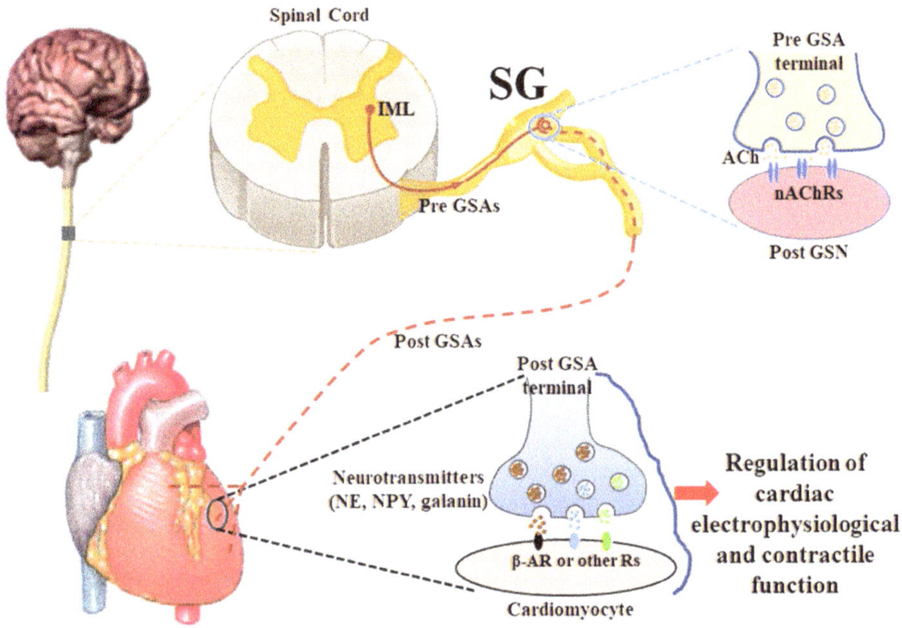

Figure 1. A schematic diagram illustrating the anatomy and physiology of stellate ganglia. ACh: acetylcholine; β-AR: β-adrenergic receptor; IML: intermediolateral nucleus; nAChR: nicotinic acetylcholine receptor; NE: norepinephrine; NPY: neuropeptide Y; Pre GSA: preganglionic sympathetic axon; Post GSA: postganglionic sympathetic axon; Post GSN: postganglionic sympathetic neuron; SG: stellate ganglion.

In the physiological condition, the sympathetic nervous system is responsible for up/downregulating various homeostatic mechanisms in many organs/tissues, especially mediating the fight-or-flight response in situations in which survival is threatened [75,76]. Norepinephrine, which is released from cardiac postganglionic sympathetic neurons in stellate ganglia with their nerve terminals, binds with beta-adrenergic receptors to affect cardiac electrophysiological and contractile functions, including heart rate, cardiac conduction, and myocardial contraction, which finally regulates cardiac output to supply the whole body with oxygenated blood and nutrients [14,75]. It is widely recognized that cardiac postganglionic sympathetic nerve terminals innervate the sinoatrial node, atrioventricular node, His bundle, and contractile myocardium [77]. However, the innervation of the heart with cardiac postganglionic sympathetic neurons in the stellate ganglia presents lateral and regional variations (such as anterior/posterior and left/right divisions of the heart) [78]. In particular, there is an obvious variation and overlap in the innervation of the cardiac tissues from the left and right stellate ganglia [78,79]. The sinoatrial node is primarily innervated by cardiac postganglionic sympathetic nerve terminals from the right stellate ganglion [78]. The conduction system, including the sinoatrial node, atrioventricular node, and His bundle, is more densely innervated than the contractile myocardium [80,81]. Compared to the endocardium of the heart, there is a high density of postganglionic sympathetic innervation on the epicardium of the heart [78,81]. The posterior surface of the heart is mostly innervated by cardiac postganglionic sympathetic nerve terminals from the left stellate ganglion, whereas the anterior surface of the heart is principally innervated by sympathetic nerve terminals from the right stellate ganglion, measured by activation recovery internal (ARI) shortening as a probe of functional innervation [78]. Additionally, it is possible that certain areas of the heart are not innervated by cardiac postganglionic sympathetic nerve terminals from both sides of the stellate ganglia.

Although the amount of neurotransmitters, including norepinephrine, is primarily determined by the intensity of the cardiac postganglionic sympathetic nerve activity, the number of norepinephrine molecules that bind to cardiac adrenergic receptors and induce biological effects on the heart is not only determined by the release of norepinephrine from cardiac postganglionic sympathetic nerve terminals but also by its elimination from the synaptic cleft [82]. Normally, more than 90% of the norepinephrine released into the synaptic cleft is removed by the norepinephrine transporter (NET) [83]. NET (also named the noradrenaline transporter, NAT), a 617 amino acid protein, comprises 12 transmembrane domains at cardiac postganglionic sympathetic nerve terminals [84]. As a member of the sodium/chloride-dependent family of neurotransmitter transporters, NET can take up norepinephrine from the interstitial space to sympathetic nerve terminals with the stochiometric exchange of sodium and chloride against their electrochemical gradient [84]. Therefore, the expression and activity of NET are key factors affecting the level of norepinephrine molecules binding with cardiac adrenergic receptors and maintenance of the intrinsic myocardial electrophysiology and contractility.

In addition to cardiac postganglionic sympathetic neurons, satellite glial cells are also located in stellate ganglia. Satellite glial cells exist ubiquitously in peripheral ganglia, including sympathetic, parasympathetic, and sensory ganglia, which almost envelop peripheral ganglionic neuronal cell somata [85]. Although astrocytes, a counterpart of satellite glial cells in the central nervous system, have been widely studied, a few studies reported the morphology and function of satellite glial cells in peripheral ganglia, including stellate ganglia. Satellite glial cells are derived from the neural crest and have a relatively small volume with a thinner sheath and flattened processes [86]. Normally, satellite glial cells around a given neural cell body are in close contact with each other, which forms a neuron–glial unit to almost separate the connection between neuronal cells [85–87]. The distance between satellite glial cells and the membrane of peripheral ganglionic neuronal cells is about 20 nm, which is similar to that of the synaptic cleft [85]. This close synapse-like structural pattern could provide a structural basis for the neuron–satellite glial cell interaction, although little is known about the function of satellite glial cells in peripheral ganglia, especially stellate ganglia.

Although satellite glial cells are electrically non-excitable without voltage-gated sodium and calcium channels, the inwardly rectifying potassium channels (Kir4.1) are expressed in satellite glial cells [88]. Satellite glial cells also express gap junction channels (such as connexin 43, Cx43) and purinergic 2 (P2) receptors (such as P2X and P2Y receptors) on the cell membrane [85,86]. Modulation of Kir4.1 permeability and activation of Cx43 and P2 receptors might depolarize the membrane of satellite glial cells and increase the intracellular calcium concentration to induce the release of excitatory mediators (such as ATP and some cytokines) from these cells for further activation of adjacent neurons [85,87,89]. Additionally, the release of nerve growth factor (NGF) from satellite glial cells may be involved in the maintenance and restoration of adjacent neurons [90]. Moreover, using single-cell RNA sequencing, one recent study demonstrated that the mature satellite glial cells in stellate ganglia are classed into five subpopulations of satellite glial cells with different functions of the subclusters [91]. Therefore, satellite glial cells in stellate ganglia could play an important role in the regulation of sympathetic neuronal function and maintenance of these adjacent neurons.

3. Remodeling of Cardiac Postganglionic Sympathetic Neurons and Its Role in Cardiac Sympathetic Overactivation, Malignant Arrhythmias, and Cardiac Sudden Death in HF

Although cardiac sympathetic remodeling can contribute to cardiac sympathetic overactivation and has not been systematically explored during HF progression, scattered information about HF-triggered cardiac sympathetic remodeling is demonstrated by most previous studies, including our work. These include structural and functional changes in cardiac postganglionic sympathetic cell somata and their nerve terminals.

3.1. Structural Remodeling in Cardiac Postganglionic Sympathetic Neurons Located in Stellate Ganglia

The majority of sympathetic nerves projecting to the heart originate in cardiac sympathetic postganglionic neurons located in stellate ganglia. There are limited data on the actual remodeling of cardiac sympathetic neuronal structures in stellate ganglia. One previous study demonstrated that stellate ganglionic nerve sprouting and density are elevated at one to four weeks after coronary artery ligation-induced rabbit myocardial infarction, which is mediated by nerve growth factor [92]. In a porcine chronic myocardial infarction model (six weeks after left anterior coronary descending artery occlusion-induced myocardial infarction), chronic myocardial infarction significantly increased the size of neuronal somata in the left stellate ganglion [93]. Using growth-associated protein 43, synaptophysin, and tyrosine hydroxylase as immunohistochemistry markers of synapses and sympathetic neurons in stellate ganglia, Han et al. found that the synaptic density and size of sympathetic neurons in the left stellate ganglion increased in dogs two months post-myocardial infarction [94]. Tan et al. also reported that the sympathetic nerve density (immunoreactivity of tyrosine hydroxylase) in stellate ganglia was markedly increased in canines 12 weeks after premature ventricular contraction-induced cardiomyopathy [95]. Similarly, Ajijola et al. demonstrated that the size of stellate ganglionic neurons increased with an increase in the neuronal adrenergic phenotype and neuropeptide Y-positive neurons in pigs at 6 weeks post left circumflex or right coronary artery occlusion-induced myocardial infarction [96]. In humans with cardiomyopathy, the size of the stellate ganglionic neurons is significantly increased without ganglionic fibrosis and changes in the neuronal density (cell number/tissue area) and synaptic density [97]. Although these studies in myocardial infarction-induced animal HF models and humans with cardiomyopathy are not totally consistent and it is unclear how structural alterations of cardiac postganglionic sympathetic neurons affect the progression and prognosis of HF, the morphological changes in these neurons could be associated with increased stellate ganglionic nerve activities and further related to cardiac sympathetic overactivation and malignant arrhythmias. Certainly, investigation of the structures of subcellular organelles (including nucleus, mitochondria, lysosome, and secretory vesicles) by electron microscopy is necessary to explore the cellular and molecular mechanisms underlying the structural remodeling of cardiac postganglionic sympathetic neuronal somata in HF.

3.2. Functional Remodeling in Cardiac Postganglionic Sympathetic Neurons Located in Stellate Ganglia

The function of neurons is to transmit electrical signals over long distances through the generation of action potentials. The left stellate ganglionic nerve activity is increased in ambulatory dogs with pacing- or coronary artery occlusion-induced HF [94,98]. Tu et al. demonstrated that the cell excitability in cardiac postganglionic sympathetic neurons located in stellate ganglia increases in coronary artery ligation-induced HF rats [99]. Although various types of ion channels (such as voltage-gated sodium, calcium, and potassium channels) can contribute to the generation of action potentials in cardiac postganglionic sympathetic neurons, voltage-gated calcium channels should be considered as the mechanism governing the increased cell excitability of these sympathetic neurons in HF because calcium influx through voltage-gated calcium channels is a key trigger for the release of neurotransmitters from neuronal nerve terminals [100–102]. There are five subtypes of voltage-gated calcium channels (T, L, N, P/Q, and R) functionally characterized in central and peripheral neurons [103,104]. A pore-forming α-subunit in all subtypes of calcium channels determines the biophysical and pharmacological properties of calcium channels [105]. There are three major families of α-subunits: (1) Cav1 (Cav1.1, Cav1.2, and Cav1.3) encodes L-type calcium channels; (2) Cav2 encodes P/Q (Cav2.1), N (Cav2.2), and R (Cav2.3) types of calcium channels; and (3) Cav3 encodes T-type calcium channels [105,106]. In fact, Tu et al. reported that the L, N, P/Q, and R types of calcium channels are expressed in cardiac postganglionic sympathetic neurons [99]. However, HF only increases N-type calcium currents and does not affect the mRNA and protein expression of all calcium channel subtypes in these sympathetic neurons [99]. Some previous studies demonstrated that N-type calcium channels, predominantly expressed in the nervous system, play an important role in modulating neurotransmitter release at sympathetic neve terminals [107,108]. More importantly, increased N-type calcium currents in cardiac postganglionic sympathetic neurons contribute to the elevated cell excitability of these neurons, cardiac sympathetic overactivation, and malignant arrhythmias in HF [109,110]. Until now, there has been no report about the involvement of other ion channels in HF-increased cell excitability of cardiac postganglionic sympathetic neurons.

3.3. Structural Remodeling in Cardiac Postganglionic Sympathetic Nerve Terminals

Cardiac sympathetic nerve terminals are directly embedded in the myocardium with a heterogeneous distribution. Most previous studies reported the information about structural remodeling in cardiac postganglionic sympathetic nerve terminals during HF progression from acute myocardial infarction to chronic HF, with no consistent conclusion. Acute myocardial infarction could result in sympathetic nerve terminal denervation in the scar and viable myocardium beyond the infarcted area [111–113]. Then, the regeneration of sympathetic nerve terminals in the heart has been characterized by nerve spouting and a high density of nerve fibers in the periphery of the necrotic myocardium of failed hearts [114–116]. Additionally, some previous studies also reported cardiac sympathetic nerve terminal denervation in HF [117,118]. Regions of cardiac sympathetic nerve terminal denervation and hyperinnervation are present in the same failed heart to form the heterogeneity of the cardiac sympathetic nerve distribution [119,120]. Iodine-123 meta-iodobenzylguanidine (^{123}I-MIBG) or other radiolabeled neurotransmitter analogs (including the recently used F-18 meta-fluorobenzylguanidine) and cardiac neurotransmission imaging with single-photon emission computed tomography (SPECT) and positron emission tomography (PET) have been employed to noninvasively assess the integrity of human NET and further evaluate cardiac sympathetic nerve innervation [121–123]. However, poor imaging quality, difficulty in distinguishing different cardiac structures, and high cost limit this technique's application in animal studies, especially small animal studies. Additionally, previous studies used immunohistochemical staining in several myocardial slices to evaluate the structural remodeling of cardiac sympathetic nerve terminals, which cannot represent the distribution of cardiac sympathetic nerve terminals in the whole heart with a neglected heterogeneous distribution of nerve terminals. Although these structural alterations of cardiac postganglionic sympathetic nerve terminals are considered to create a high-yield substrate for malignant arrhythmias in HF [124], the conclusion from these previous studies should be questioned. Using three-dimensional assessment of the cardiac sympathetic network in cleared transparent murine hearts, one recent study demonstrated both cardiac sympathetic nerve terminal hyperinnervation and denervation in the whole heart at 2 weeks post myocardial infarction [125]. It is not clear whether the same phenomenon (sympathetic nerve terminal hyperinnervation and denervation) is also present in the whole heart with HF using three-dimensional assessment of the cardiac sympathetic network. Therefore, the timing and patterns of the cardiac sympathetic nerve terminal remodeling in HF should be re-evaluated in future studies. Indeed, structural remodeling and norepinephrine release in cardiac postganglionic sympathetic nerve terminals in HF should be combined to assess the association of cardiac sympathetic activation and malignant arrhythmias because it is unclear whether reinnervated cardiac sympathetic nerve terminals can release norepinephrine like mature sympathetic nerve terminals in the heart.

3.4. Functional Remodeling in Cardiac Postganglionic Sympathetic Nerve Terminals

The function of cardiac sympathetic nerve terminals is to release neurotransmitters, including norepinephrine, which bind to adrenergic receptors to regulate cardiac function in physiological and pathophysiological conditions. Cardiac norepinephrine spillover is measured by calculating the amount of plasma norepinephrine entering the heart and the amount of norepinephrine exiting the heart. An elevation in cardiac norepinephrine spillover occurs in HF [126–131], which primarily results from the increase in cardiac norepinephrine synthesis and release, and the decrease in norepinephrine reuptake [84,132]. By ^{123}I-MIBG with SPECT and PET images, some clinical studies measured the heart-to-mediastinum (H/M) ratio and ^{123}I-MIBG washout (WO) rate to demonstrate an increased level of sympathetic neurotransmitter in HF [133–137]. Although the above studies indirectly tested norepinephrine release and demonstrated that these measured scientific parameters are strong predictors of cardiac sympathetic overactivation, heart failure progression, life-threatening arrhythmias, and cardiac sudden death in HF, there is limited information available on the direct measurement of norepinephrine release from cardiac sympathetic nerve terminals during HF progression. In vivo cardiac microdialysis with HPLC can directly test norepinephrine release from cardiac sympathetic terminals [138–140], but it is hard to obtain stable data of norepinephrine release due to the heterogeneous distribution of cardiac sympathetic nerve terminals as described above. Zhang et al. recently reported an electrochemistry recording, patch-clamp technique with a carbon fiber electrode for the catecholamine release from adrenal chromaffin cells [141]. The development of this recording in in vivo cardiac slices could be an innovative technique for the direct measurement of norepinephrine release from cardiac sympathetic nerve terminals, which possibly avoids the interference of the heterogeneous distribution of cardiac sympathetic nerve terminals and also analyzes norepinephrine release kinetics (including the maximal amplitudes of norepinephrine release and reuptake).

Cardiac sympathetic activation is dependent on two major components, namely circulating catecholamines from the adrenal medulla and local norepinephrine release from cardiac postganglionic sympathetic nerve terminals. Stellate ganglion stimulation, including left, right, or bilateral stellate ganglion stimulation, produces distinct patterns of cardiac myocyte repolarization in the normal porcine heart, evaluated by the analysis of epicardial and endocardial electrograms, whereas marked dispersion of cardiac myocyte repolarization does not occur when exogenous norepinephrine is infused (circulating norepinephrine) [142]. From these data, it has been demonstrated that stellate-ganglion-stimulated dispersion of cardiac myocyte repolarization is highly arrhythmogenic, compared to the more uniform changes in cardiac myocyte repolarization triggered by circulating norepinephrine [142,143]. As described above, cardiac sympathetic denervation, a key approach to the current therapy of HF, highlights the role of cardiac sympathetic remodeling in cardiac sympathetic overactivation, malignant arrhythmias, and cardiac sudden death in HF [38,42,43]. Using in vivo shRNA transfection into stellate ganglia, Zhang et al. demonstrated that ion channel remodeling in cardiac postganglionic sympathetic neurons is involved in cardiac sympathetic overactivation and ventricular arrhythmogenesis in coronary artery ligation-induced HF [110].

Recent studies have reported that elevated neuropeptide Y and other sympathetic co-transmitters released from cardiac sympathetic neurons act on neuropeptide Y or other related receptors on the membrane of cardiac myocytes to cause the development of HF and ventricular arrhythmias [144–148]. As a result, cardiac postganglionic sympathetic remodeling, including alterations of the sympathetic co-neurotransmitter release (such as norepinephrine, neuropeptide Y, and galanin), could be associated with cardiac sympathetic overactivation, malignant arrhythmias, and cardiac sudden death in HF.

4. Mechanisms Underlying the Remodeling of Cardiac Postganglionic Sympathetic Neurons in HF

The mechanisms responsible for the remodeling of cardiac postganglionic sympathetic neurons in HF are not well understood and could be multifactorial. Additionally, cardiac postganglionic sympathetic nerve terminals are embedded in the myocardium. Therefore, the microenvironment surrounding cardiac postganglionic sympathetic nerve terminals in the myocardium should be the key factor for the structural and functional remodeling of these nerve terminals, although the factors that modulate cardiac postganglionic sympathetic cell somata might also affect their nerve terminals.

4.1. Mechanisms Associated with the Remodeling of Cardiac Postganglionic Sympathetic Cell Somata

Nerve growth factor (NGF), a prototypical member of the neurotrophin family, is normally involved in the maintenance, proliferation, and survival of neurons. The action of NGF in targeted cells is initiated by its high-affinity binding to the tropomyosin receptor kinase A (TrkA, also named neurotrophic tyrosine kinase receptor 1) receptors in mature sympathetic neurons [149,150]. NGF is usually produced by sympathetic innervated organ/tissues. A high level of NGF is present in stellate ganglia, in which NGF is accumulated by retrograde axonal transport to affect the function of neurons [151,152]. The local production of NGF by satellite glial cells in stellate ganglia could be another source because NGF mRNA is expressed in neurons and satellite glial cells from trigeminal ganglia [90]. Although a high level of NGF is probably involved in the structural and functional remodeling of cardiac postganglionic sympathetic cell somata, including an increased cell size of sympathetic neurons, synaptic density, and neuronal excitability in stellate ganglia in HF, further studies are needed to provide direct evidence.

As mentioned above, satellite glial cells exist ubiquitously in peripheral ganglia, including sympathetic, parasympathetic, and sensory ganglia. In addition to the local production of NGF by satellite glial cells in stellate ganglia, satellite glial cell–macrophage communication could be another mechanism responsible for the remodeling of cardiac postganglionic sympathetic neurons in HF. One recent study demonstrated that acute or chronic intestinal inflammation activates enteric glial cells to release macrophage colony-stimulating factor (M-CSF) through the connexin-43 hemichannel-cytosolic PKC-MAPK-cell membrane TNFα-converting enzyme (TACE) signaling pathway [89]. M-CSF is a key factor for the regulation of macrophage survival, proliferation, migration, and activation through binding with M-CSF receptors on the macrophage membrane [153,154]. Chronic inflammation, with activation of both cytokines and immune cells (such as macrophages), is a major feature of HF [155–157]. Macrophages play a key role in mediating inflammatory responses in the post-myocardial infarction heart [158,159]. Our recent study already found that elevation of cytokines and macrophages in stellate ganglia is involved in cardiac sympathetic overactivation and ventricular arrhythmogenesis in HF [160]. Growing evidence suggests that inflammation-raised cytokines increase the expression and activation of cyclin-dependent kinase 5 (CDK5, a proline-directed serine/threonine kinase) in some tissues and cell lines [161–163]. CDK5 can phosphorylate the N-type calcium channels and the latter induce cardiac sympathetic overactivation and ventricular arrhythmias in HF [110,164,165]. Therefore, further understanding of the relations among satellite glial cells, macrophages, and cardiac postganglionic sympathetic neurons in stellate ganglia can provide therapeutic targets against cardiac sympathetic overactivation and malignant ventricular arrhythmias in HF.

Oxidative stress is also considered to be another factor for the remodeling in cardiac sympathetic neurons in HF. Ajijola et al. found that stellate ganglia from patients with cardiomyopathy and arrhythmias exhibit oxidative stress [166]. In small animal models of myocardial infarction-induced HF, oxidative signaling is increased in stellate ganglia [167]. Therefore, the effect of oxidative stress on the remodeling in cardiac sympathetic neurons and interaction between oxidative stress and inflammation-raised cytokines in HF should be explored in future studies.

4.2. Mechanisms Associated with the Remodeling in Cardiac Postganglionic Sympathetic Never Terminals

As described above, NGF is mainly produced in the sympathetic innervated organs/tissues. Western bolt analysis in the left ventricle demonstrated an elevation of NGF in explanted, failing human hearts compared to normal, donor hearts [168]. NGF overexpression in the sympathetic targeted organs/tissues causes sympathetic nerve hyperinnervation, which could be involved in nerve spouting and a high density of sympathetic nerve terminals in the periphery of the necrotic myocardium of failed hearts [116,149,152,169]. When NGF binds with TrkA receptors on sympathetic nerve terminals, activated TrkA receptors regulate cytoskeletal dynamics through successive activation of the MAPK and PI3K-Akt pathways, and endocytosed TrkA receptors promote sympathetic nerve growth and hyperinnervation through a calcineurin-dynamin 1 signaling pathway [170–172]. Although NGF could be considered as the key factor modulating sympathetic nerve innervation in targeted organs/tissues in physiological and pathophysiological conditions, some other factors also contribute to cardiac sympathetic hyperinnervation in either an independent style or by association with NGF [149,152]. These endogenous factors include growth differentiation factor 5 (GDF5), TNF receptor 1, leukemia inhibitory factor, cardiotrophin-1, and leptin [173–176]. Additionally, pro-NGF, pro-brain-derived neurotrophic factor (pro-BDNF), and brain-derived neurotrophic factor (BDNF) activate the p75 neurotrophin receptors (p75NTRs) and death receptor 6 (DR6), two members of the TNF super-family, to stimulate sympathetic nerve denervation [177–183], which occurs in the cardiac infarcted area and myocardium adjacent to the scar in myocardial infarction-induced HF.

Using a mouse model of cardiac ischemia-reperfusion, one recent study demonstrates that therapeutics-restored sympathetic reinnervation of the infarcted area decreases M1-like macrophages and elevates the numbers of dendritic cells, M2-like macrophages, and Treg cells [184]. There are different contributions of M1-like and M2-like macrophages to cardiac sympathetic remodeling. Therefore, future studies are needed to assess the interaction between cardiac sympathetic remodeling and the different types of macrophages in HF

Functional remodeling in cardiac postganglionic sympathetic nerve terminals includes the change in norepinephrine synthesis, release, and reuptake in HF. For norepinephrine synthesis, normally, tyrosine is converted to 3,4-dihydrooxyphenyl alanine (DOPA) by the rate-limiting enzyme, tyrosine hydroxylase (TH). DOPA is then converted to dopamine by L-aromatic acid decarboxylase. After that, the vesicular monoamine transporter translocates dopamine into storage vesicles, in which dopamine is converted to norepinephrine by dopamine β-hydroxylase. Although it has been shown that cardiac norepinephrine synthesis is increased in HF [132], so far, it is unclear which enzyme and potential mechanism(s) are responsible for the increase in cardiac norepinephrine synthesis in HF. For norepinephrine release from sympathetic nerve terminals, Huang et al. demonstrated that the action potential modulates calcium-dependent and -independent (voltage-dependent) quantal norepinephrine release in the mammalian sympathetic nervous system [185]. Our recent studies found that the activated macrophage-triggered increase in the N-type calcium currents results in neuronal overexcitation of cardiac postganglionic sympathetic neurons in a myocardial infarction-induced HF animal model [110,160], which could contribute to the increase in the norepinephrine release from cardiac sympathetic nerve terminals in the HF state. For norepinephrine reuptake, it is responsible for the rapid removal of interstitial norepinephrine through the norepinephrine transporter (NET) after norepinephrine release from cardiac sympathetic nerve terminals. Much evidence confirms a reduction in the neuronal NET density and activity at cardiac sympathetic nerve terminals in the failing heart [186–188]. In cultured rat neuroblastoma cells (PC12 cell line), exogenous norepinephrine significantly reduces the expression of NET protein but not NET mRNA [189,190], which suggests post-transcriptional downregulation for NET in HF [82]. Normally, NET located in the cell membrane is regulated by glycosylation [191]. Therefore, the effect of norepinephrine on the reduction in the neuronal NET density in HF could be associated with endoplasmic reticulum stress-reduced glycosylation, causing the trafficking of NET to the cell membrane [190]. Additionally, endothelin binding with endothelin receptors also inhibits NET activity in HF by affecting NET phosphorylation because the NET activity is, in the short and long term, modulated by protein kinase A, C, and G and calcium-calmodulin-dependent protein kinase [191,192]. It is unclear whether NET activity and expression are modulated by other endogenous factors in HF, such as the renin-angiotensin-aldosterone system, bradykinin, nitric oxide, natriuretic peptides, etc.

5. Conclusions

In this review, we updated the information about the cardiac sympathetic remodeling in stellate ganglia and potential mechanisms and the role of cardiac sympathetic remodeling in cardiac sympathetic overactivation, arrhythmias, and cardiac dysfunction in HF (Figure 2). Using droplet-based high-throughput single-cell RNA sequencing, one recent study separated eight large clusters in the superior cervical ganglion (one type of peripheral sympathetic ganglia) from young adult mice based on the expression of canonical marker genes [193]. These clusters include satellite glial cells (Plp1, Fabp7), Schwann cells (Plp1, Ncmap), sympathetic neurons (Snap25, Th), vascular endothelial cells (Ly6c1), macrophages (C1qb), T cells (Trb2), fibroblasts (Dcn), and mural cells (Rgs5) [193]. Following the development of innovative techniques, including epigenetics, transcriptomics, proteomics, optogenetics, single-cell RNA sequence analysis, etc., more details on stellate ganglionic cell remodeling with related mechanisms in HF will be explored. Stellate ganglia could be a therapeutic target against cardiac sympathetic overactivation and myocardial electrophysiological and contractile dysfunction in HF.

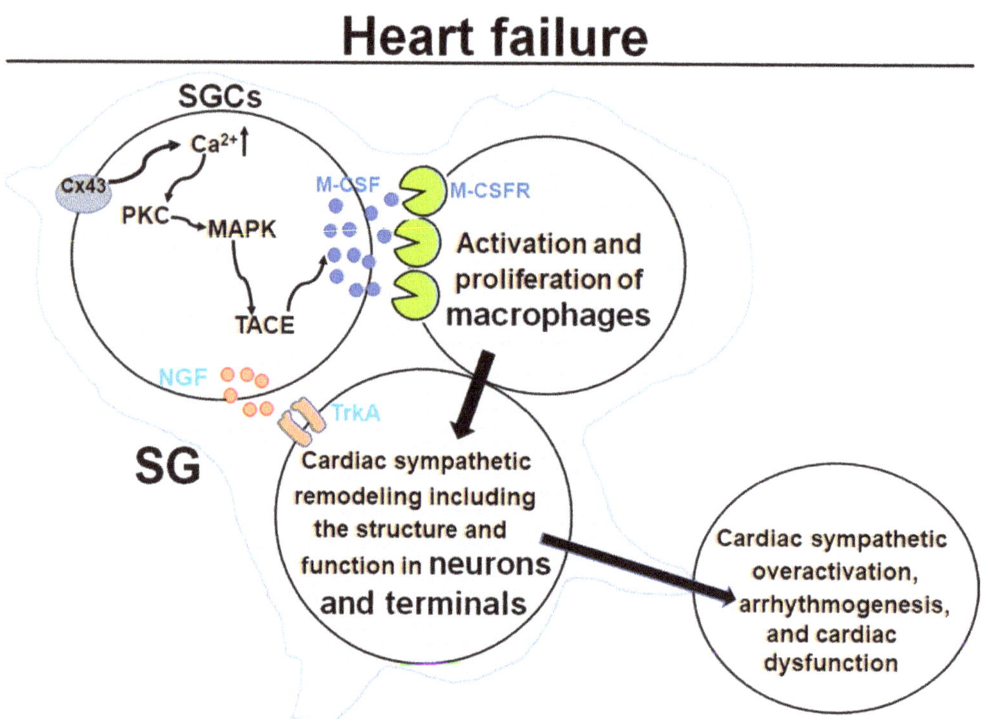

Figure 2. Mechanisms underlying the remodeling of cardiac postganglionic sympathetic neurons in HF. SGC: satellite glial cell; Cx43: connexin 43; PKC: protein kinase C; MAPK: mitogen-activated protein kinase; TACE: tumor necrosis-converting enzyme; M-CSF; macrophage colony-stimulating factor; M-CSFR: macrophage colony-stimulating factor receptor; NGF: nerve growth factor; TrkA: tropomyosin receptor kinase A; SG: stellate ganglion.

Funding: This study was supported by the National Institute of Health's National Heart, Lung, and Blood Institute, USA (R01HL-137832 and R01HL-144146 to Y-LL).

Conflicts of Interest: The author declares no conflict of interest.

References

1. Sweitzer, N.K. Looking ahead: Circulation: Heart failure in 2022. *Circ. Heart Fail.* **2022**, *15*, e009405. [CrossRef] [PubMed]
2. Tsao, C.W.; Aday, A.W.; Almarzooq, Z.I.; Alonso, A.; Beaton, A.Z.; Bittencourt, M.S.; Boehme, A.K.; Buxton, A.E.; Carson, A.P.; Commodore-Mensah, Y.; et al. Heart disease and stroke statistics-2022 update: A report from the American Heart Association. *Circulation* **2022**, *145*, e153–e639. [CrossRef] [PubMed]
3. Savarese, G.; D'Amario, D. Sex differences in heart failure. *Adv. Exp. Med. Biol.* **2018**, *1065*, 529–544. [PubMed]
4. Elgendy, I.Y.; Mahtta, D.; Pepine, C.J. Medical therapy for heart failure caused by ischemic heart disease. *Circ. Res.* **2019**, *124*, 1520–1535. [CrossRef]
5. He, J.; Ogden, L.G.; Bazzano, L.A.; Vupputuri, S.; Loria, C.; Whelton, P.K. Risk factors for congestive heart failure in US men and women: NHANESIepidemiologic follow-up study. *Arch. Intern. Med.* **2001**, *161*, 996–1002. [CrossRef]
6. Velagaleti, R.S.; Vasan, R.S. Heart failure in the twenty-first century: Is it a coronary artery disease or hypertension problem? *Cardiol. Clin.* **2007**, *25*, 487–495. [CrossRef]
7. Virani, S.S.; Alonso, A.; Aparicio, H.J.; Benjamin, E.J.; Bittencourt, M.S.; Callaway, C.W.; Carson, A.P.; Chamberlain, A.M.; Cheng, S.; Delling, F.N.; et al. Heart disease and stroke statistics-2021 update: A report from the American Heart Association. *Circulation* **2021**, *143*, e254–e743. [CrossRef]
8. Emmons-Bell, S.; Johnson, C.; Roth, G. Prevalence, incidence and survival of heart failure: A systematic review. *Heart* **2022**, *108*, 1351–1360. [CrossRef]
9. Savarese, G.; Becher, P.M.; Lund, L.H.; Seferovic, P.; Rosano, G.M.C.; Coats, A. Global burden of heart failure: A comprehensive and updated review of epidemiology. *Cardiovasc. Res.* 2022, cvac013, *online ahead of print*.
10. Creager, M.A.; Faxon, D.P.; Cutler, S.S.; Kohlmann, O.; Ryan, T.J.; Gavras, H. Contribution of vasopressin to vasoconstriction in patients with congestive heart failure: Comparison with the renin-angiotensin system and the sympathetic nervous system. *J. Am. Coll. Cardiol.* **1986**, *7*, 758–765. [CrossRef]
11. Floras, J.S. Sympathetic nervous system activation in human heart failure: Clinical implications of an updated model. *J. Am. Coll. Cardiol.* **2009**, *54*, 375–385. [CrossRef]
12. Saul, J.P.; Arai, Y.; Berger, R.D.; Lilly, L.S.; Colucci, W.S.; Cohen, R.J. Assessment of autonomic regulation in chronic congestive heart failure by heart rate spectral analysis. *Am. J. Cardiol.* **1988**, *61*, 1292–1299. [CrossRef]
13. Schwartz, P.J.; De Ferrari, G.M. Sympathetic-parasympathetic interaction in health and disease: Abnormalities and relevance in heart failure. *Heart Fail. Rev.* **2011**, *16*, 101–107. [CrossRef] [PubMed]
14. Triposkiadis, F.; Karayannis, G.; Giamouzis, G.; Skoularigis, J.; Louridas, G.; Butler, J. The sympathetic nervous system in heart failure physiology, pathophysiology, and clinical implications. *J. Am. Coll. Cardiol.* **2009**, *54*, 1747–1762. [CrossRef] [PubMed]
15. Du, X.J.; Cox, H.S.; Dart, A.M.; Esler, M.D. Sympathetic activation triggers ventricular arrhythmias in rat heart with chronic infarction and failure. *Cardiovasc. Res.* **1999**, *43*, 919–929. [CrossRef]
16. Gilmour, R.F. Life out of balance: The sympathetic nervous system and cardiac arrhythmias. *Cardiovasc. Res.* **2001**, *51*, 625–626. [CrossRef]
17. Kalla, M.; Herring, N.; Paterson, D.J. Cardiac sympatho-vagal balance and ventricular arrhythmia. *Auton. Neurosci.* **2016**, *199*, 29–37. [CrossRef]
18. Podrid, P.J.; Fuchs, T.; Candinas, R. Role of the sympathetic nervous system in the genesis of ventricular arrhythmia. *Circulation* **1990**, *82*, I103–I113.
19. Schwartz, P.J. Cardiac sympathetic denervation to prevent life-threatening arrhythmias. *Nat. Rev. Cardiol.* **2014**, *11*, 346–353. [CrossRef]
20. Thompson, B.S. Sudden cardiac death and heart failure. *AACN Adv. Crit. Care* **2009**, *20*, 356–365.
21. Tomaselli, G.F.; Zipes, D.P. What causes sudden death in heart failure? *Circ. Res.* **2004**, *95*, 754–763. [CrossRef]
22. Zhou, S.; Jung, B.C.; Tan, A.Y.; Trang, V.Q.; Gholmieh, G.; Han, S.W.; Lin, S.F.; Fishbein, M.C.; Chen, P.S.; Chen, L.S. Spontaneous stellate ganglion nerve activity and ventricular arrhythmia in a canine model of sudden death. *Heart Rhythm* **2008**, *5*, 131–139. [CrossRef] [PubMed]
23. Zipes, D.P. Heart-brain interactions in cardiac arrhythmias: Role of the autonomic nervous system. *Cleve. Clin. J. Med.* **2008**, *75*, S94–S96. [CrossRef] [PubMed]
24. Carson, P.; Anand, I.; O'Connor, C.; Jaski, B.; Steinberg, J.; Lwin, A.; Lindenfeld, J.; Ghali, J.; Barnet, J.H.; Feldman, A.M.; et al. Mode of death in advanced heart failure: The Comparison of Medical, Pacing, and Defibrillation Therapies in Heart Failure (COMPANION) trial. *J. Am. Coll. Cardiol.* **2005**, *46*, 2329–2334. [CrossRef] [PubMed]
25. Cygankiewicz, I.; Zareba, W.; Vazquez, R.; Vallverdu, M.; Gonzalez-Juanatey, J.R.; Valdes, M.; Almendral, J.; Cinca, J.; Caminal, P.; de Luna, A.B. Heart rate turbulence predicts all-cause mortality and sudden death in congestive heart failure patients. *Heart Rhythm* **2008**, *5*, 1095–1102. [CrossRef] [PubMed]
26. Doval, H.C.; Nul, D.R.; Grancelli, H.O.; Varini, S.D.; Soifer, S.; Corrado, G.; Dubner, S.; Scapin, O.; Perrone, S.V. Nonsustained ventricular tachycardia in severe heart failure. Independent marker of increased mortality due to sudden death. GESICA-GEMA Investigators. *Circulation* **1996**, *94*, 3198–3203. [CrossRef] [PubMed]
27. Engelstein, E.D.; Zipes, D.P. Sudden cardiac death. In *Hurst's The Heart*; Alexander, R.W., Schlant, R.C., Fuster, V., Eds.; McGraw Hill: New York, NY, USA, 1998; Volume 9, pp. 1081–1112.

28. Huikuri, H.V.; Castellanos, A.; Myerburg, R.J. Sudden death due to cardiac arrhythmias. *N. Engl. J. Med.* **2001**, *345*, 1473–1482. [CrossRef]
29. Jost, A.; Rauch, B.; Hochadel, M.; Winkler, R.; Schneider, S.; Jacobs, M.; Kilkowski, C.; Kilkowski, A.; Lorenz, H.; Muth, K.; et al. Beta-blocker treatment of chronic systolic heart failure improves prognosis even in patients meeting one or more exclusion criteria of the MERIT-HF study. *Eur. Heart J.* **2005**, *26*, 2689–2697. [CrossRef]
30. Maskin, C.S.; Siskind, S.J.; LeJemtel, T.H. High prevalence of nonsustained ventricular tachycardia in severe congestive heart failure. *Am. Heart J.* **1984**, *107*, 896–901. [CrossRef]
31. Podrid, P.J.; Fogel, R.I.; Fuchs, T.T. Ventricular arrhythmia in congestive heart failure. *Am. J. Cardiol.* **1992**, *69*, 82G–95G. [CrossRef]
32. Sami, M.H. Sudden death in congestive heart failure. *J. Clin. Pharmacol.* **1991**, *31*, 1081–1084. [CrossRef]
33. Singh, B.N. Significance and control of cardiac arrhythmias in patients with congestive cardiac failure. *Heart Fail. Rev.* **2002**, *7*, 285–300. [CrossRef]
34. Singh, S.N.; Carson, P.E.; Fisher, S.G. Nonsustained ventricular tachycardia in severe heart failure. *Circulation* **1997**, *96*, 3794–3795. [PubMed]
35. Teerlink, J.R.; Jalaluddin, M.; Anderson, S.; Kukin, M.L.; Eichhorn, E.J.; Francis, G.; Packer, M.; Massie, B.M. Ambulatory ventricular arrhythmias in patients with heart failure do not specifically predict an increased risk of sudden death. PROMISE (Prospective Randomized Milrinone Survival Evaluation) Investigators. *Circulation* **2000**, *101*, 40–46. [CrossRef] [PubMed]
36. Al-Gobari, M.; El, K.C.; Pillon, F.; Gueyffier, F. Beta-blockers for the prevention of sudden cardiac death in heart failure patients: A meta-analysis of randomized controlled trials. *BMC Cardiovasc. Disord.* **2013**, *13*, 52. [CrossRef] [PubMed]
37. Babick, A.; Elimban, V.; Zieroth, S.; Dhalla, N.S. Reversal of cardiac dysfunction and subcellular alterations by metoprolol in heart failure due to myocardial infarction. *J. Cell. Physiol.* **2013**, *228*, 2063–2070. [CrossRef]
38. De Ferrari, G.M.; Schwartz, P.J. Left cardiac sympathetic denervation in patients with heart failure: A new indication for an old intervention? *J. Cardiovasc. Transl. Res.* **2014**, *7*, 338–346. [CrossRef]
39. Fiuzat, M.; Wojdyla, D.; Kitzman, D.; Fleg, J.; Keteyian, S.J.; Kraus, W.E.; Pina, I.L.; Whellan, D.; O'Connor, C.M. Relationship of beta-blocker dose with outcomes in ambulatory heart failure patients with systolic dysfunction: Results from the HF-ACTION (Heart Failure: A Controlled Trial Investigating Outcomes of Exercise Training) trial. *J. Am. Coll. Cardiol.* **2012**, *60*, 208–215. [CrossRef]
40. Gheorghiade, M.; Colucci, W.S.; Swedberg, K. Beta-blockers in chronic heart failure. *Circulation* **2003**, *107*, 1570–1575. [CrossRef]
41. Nevzorov, R.; Porath, A.; Henkin, Y.; Kobal, S.L.; Jotkowitz, A.; Novack, V. Effect of beta blocker therapy on survival of patients with heart failure and preserved systolic function following hospitalization with acute decompensated heart failure. *Eur. J. Intern. Med.* **2012**, *23*, 374–378. [CrossRef]
42. Shah, R.; Assis, F.; Alugubelli, N.; Okada, D.R.; Cardoso, R.; Shivkumar, K.; Tandri, H. Cardiac sympathetic denervation for refractory ventricular arrhythmias in patients with structural heart disease: A systematic review. *Heart Rhythm* **2019**, *16*, 1499–1505. [CrossRef]
43. Vaseghi, M.; Barwad, P.; Malavassi Corrales, F.J.; Tandri, H.; Mathuria, N.; Shah, R.; Sorg, J.M.; Gima, J.; Mandal, K.; Saenz Morales, L.C.; et al. Cardiac sympathetic denervation for refractory ventricular arrhythmias. *J. Am. Coll. Cardiol.* **2017**, *69*, 3070–3080. [CrossRef] [PubMed]
44. Bhatt, A.S.; DeVore, A.D.; DeWald, T.A.; Swedberg, K.; Mentz, R.J. Achieving a maximally tolerated beta-blocker dose in heart failure patients: Is there room for improvement? *J. Am. Coll. Cardiol.* **2017**, *69*, 2542–2550. [CrossRef] [PubMed]
45. Bos, J.M.; Bos, K.M.; Johnson, J.N.; Moir, C.; Ackerman, M.J. Left cardiac sympathetic denervation in long QT syndrome: Analysis of therapeutic nonresponders. *Circ. Arrhythm. Electrophysiol.* **2013**, *6*, 705–711. [CrossRef]
46. Coleman, M.A.; Bos, J.M.; Johnson, J.N.; Owen, H.J.; Deschamps, C.; Moir, C.; Ackerman, M.J. Videoscopic left cardiac sympathetic denervation for patients with recurrent ventricular fibrillation/malignant ventricular arrhythmia syndromes besides congenital long-QT syndrome. *Circ. Arrhythm. Electrophysiol.* **2012**, *5*, 782–788. [CrossRef]
47. Moss, A.J.; Zareba, W.; Hall, W.J.; Schwartz, P.J.; Crampton, R.S.; Benhorin, J.; Vincent, G.M.; Locati, E.H.; Priori, S.G.; Napolitano, C.; et al. Effectiveness and limitations of beta-blocker therapy in congenital long-QT syndrome. *Circulation* **2000**, *101*, 616–623. [CrossRef] [PubMed]
48. Napolitano, C.; Priori, S.G. Diagnosis and treatment of catecholaminergic polymorphic ventricular tachycardia. *Heart Rhythm* **2007**, *4*, 675–678. [CrossRef] [PubMed]
49. Smith, K.V.; Dunning, J.R.; Fischer, C.M.; MacLean, T.E.; Bosque-Hamilton, J.W.; Fera, L.E.; Grant, J.Y.; Zelle, D.J.; Matta, L.; Gaziano, T.A.; et al. Evaluation of the usage and dosing of guideline-directed medical therapy for heart failure with reduced ejection fraction patients in clinical practice. *J. Pharm. Pract.* **2022**, *35*, 747–751. [CrossRef]
50. Veenis, J.F.; Rocca, H.B.; Linssen, G.C.M.; Erol-Yilmaz, A.; Pronk, A.C.B.; Engelen, D.J.M.; van Tooren, R.M.; Koornstra-Wortel, H.J.J.; de Boer, R.A.; van der Meer, P.; et al. Impact of sex-specific target dose in chronic heart failure patients with reduced ejection fraction. *Eur. J. Prev. Cardiol.* **2021**, *28*, 957–965. [CrossRef]
51. Hofferberth, S.C.; Cecchin, F.; Loberman, D.; Fynn-Thompson, F. Left thoracoscopic sympathectomy for cardiac denervation in patients with life-threatening ventricular arrhythmias. *J. Thorac. Cardiovasc. Surg.* **2014**, *147*, 404–409. [CrossRef]
52. Schneider, H.E.; Steinmetz, M.; Krause, U.; Kriebel, T.; Ruschewski, W.; Paul, T. Left cardiac sympathetic denervation for the management of life-threatening ventricular tachyarrhythmias in young patients with catecholaminergic polymorphic ventricular tachycardia and long QT syndrome. *Clin. Res. Cardiol.* **2013**, *102*, 33–42. [CrossRef]

53. Rathinam, S.; Nanjaiah, P.; Sivalingam, S.; Rajesh, P.B. Excision of sympathetic ganglia and the rami communicantes with histological confirmation offers better early and late outcomes in Video assisted thoracoscopic sympathectomy. *J. Cardiothorac. Surg.* **2008**, *3*, 50–53. [CrossRef] [PubMed]
54. Webster, G.; Monge, M.C. Left cardiac sympathetic denervation: Should we sweat the side effects? *Circ. Arrhythm. Electrophysiol.* **2015**, *8*, 1007–1009. [CrossRef] [PubMed]
55. Lathro, D.A.; Spooner, P.M. On the neural connection. *J. Cardiovasc. Electrophysiol.* **2001**, *12*, 841–844. [CrossRef] [PubMed]
56. Verrier, R.L.; Antzelevitch, C. Autonomic aspects of arrhythmogenesis: The enduring and the new. *Curr. Opin. Cardiol.* **2004**, *19*, 2–11. [CrossRef]
57. Zucker, I.H.; Xiao, L.; Haack, K.K. The central renin-angiotensin system and sympathetic nerve activity in chronic heart failure. *Clin. Sci.* **2014**, *126*, 695–706. [CrossRef]
58. Zucker, I.H.; Schultz, H.D.; Patel, K.P.; Wang, W.; Gao, L. Regulation of central angiotensin type 1 receptors and sympathetic outflow in heart failure. *Am. J. Physiol. Heart Circ. Physiol.* **2009**, *297*, H1557–H1566. [CrossRef]
59. Zhang, D.; Liu, J.; Tu, H.; Melleman, R.L.; Cornish, K.G.; Li, Y.L. In-vivo transfection of manganese superoxide dismutase gene or NFkB shRNA in nodose ganglia improves aortic baroreceptor function in heart failure rats. *Hypertension* **2014**, *63*, 88–95. [CrossRef]
60. Wang, W.; Chen, J.S.; Zucker, I.H. Carotid sinus baroreceptor sensitivity in experimental heart failure. *Circulation* **1990**, *81*, 1959–1966. [CrossRef]
61. Schultz, H.D.; Li, Y.L.; Ding, Y. Arterial chemoreceptors and sympathetic nerve activity: Implications for hypertension and heart failure. *Hypertension* **2007**, *50*, 6–13. [CrossRef]
62. Patel, K.P.; Zheng, H. Central neural control of sympathetic nerve activity in heart failure following exercise training. *Am. J. Physiol. Heart Circ. Physiol.* **2012**, *302*, H527–H537. [CrossRef]
63. May, C.N.; Yao, S.T.; Booth, L.C.; Ramchandra, R. Cardiac sympathoexcitation in heart failure. *Auton. Neurosci.* **2013**, *175*, 76–84. [CrossRef] [PubMed]
64. Lymperopoulos, A.; Rengo, G.; Koch, W.J. Adrenergic nervous system in heart failure: Pathophysiology and therapy. *Circ. Res.* **2013**, *113*, 739–753. [CrossRef] [PubMed]
65. Felder, R.B. Mineralocorticoid receptors, inflammation and sympathetic drive in a rat model of systolic heart failure. *Exp. Physiol.* **2010**, *95*, 19–25. [CrossRef] [PubMed]
66. Xu, B.; Li, H. Brain mechanisms of sympathetic activation in heart failure: Roles of the renin-angiotensin system, nitric oxide and pro-inflammatory cytokines (Review). *Mol. Med. Rep.* **2015**, *12*, 7823–7829. [CrossRef]
67. Doehner, W.; Ural, D.; Haeusler, K.G.; Čelutkienė, J.; Bestetti, R.; Cavusoglu, Y.; Peña-Duque, M.A.; Glavas, D.; Iacoviello, M.; Laufs, U.; et al. Heart and brain interaction in patients with heart failure: Overview and proposal for a taxonomy. A position paper from the Study Group on Heart and Brain Interaction of the Heart Failure Association. *Eur. J. Heart Fail.* **2018**, *20*, 199–215. [CrossRef]
68. Cuevas, J. Molecular mechanisms of dysautonomia during heart failure. Focus on "Heart failure-induced changes of voltage-gated Ca2+ channels and cell excitability in rat cardiac postganglionic neurons". *Am. J. Physiol. Cell Physiol.* **2014**, *306*, C121–C122. [CrossRef]
69. Wallis, D.; Watson, A.H.; Mo, N. Cardiac neurons of autonomic ganglia. *Microsc. Res. Tech.* **1996**, *35*, 69–79. [CrossRef]
70. Wehrwein, E.A.; Orer, H.S.; Barman, S.M. Overview of the anatomy, physiology, and pharmacology of the autonomic nervous system. *Compr. Physiol.* **2016**, *6*, 1239–1278.
71. Wink, J.; van Delft, R.; Notenboom, R.G.E.; Wouters, P.F.; DeRuiter, M.C.; Plevier, J.W.M.; Jongbloed, M.R.M. Human adult cardiac autonomic innervation: Controversies in anatomical knowledge and relevance for cardiac neuromodulation. *Auton. Neurosci.* **2020**, *227*, 102674. [CrossRef]
72. Hasan, W. Autonomic cardiac innervation: Development and adult plasticity. *Organogenesis* **2013**, *9*, 176–193. [CrossRef]
73. Pardini, B.J.; Lund, D.D.; Schmid, P.G. Organization of the sympathetic postganglionic innervation of the rat heart. *J. Auton. Nerv. Syst.* **1989**, *28*, 193–201. [CrossRef]
74. Hoang, J.D.; Salavatian, S.; Yamaguchi, N.; Swid, M.A.; David, H.; Vaseghi, M. Cardiac sympathetic activation circumvents high-dose beta blocker therapy in part through release of neuropeptide Y. *JCI Insight* **2020**, *5*, e135519. [CrossRef] [PubMed]
75. Borovac, J.A.; D'Amario, D.; Bozic, J.; Glavas, D. Sympathetic nervous system activation and heart failure: Current state of evidence and the pathophysiology in the light of novel biomarkers. *World J. Cardiol.* **2020**, *12*, 373–408. [CrossRef] [PubMed]
76. de Lucia, C.; Piedepalumbo, M.; Paolisso, G.; Koch, W.J. Sympathetic nervous system in age-related cardiovascular dysfunction: Pathophysiology and therapeutic perspective. *Int. J. Biochem. Cell Biol.* **2019**, *108*, 29–33. [CrossRef] [PubMed]
77. Coote, J.H.; Chauhan, R.A. The sympathetic innervation of the heart: Important new insights. *Auton. Neurosci.* **2016**, *199*, 17–23. [CrossRef] [PubMed]
78. Meng, L.; Shivkumar, K.; Ajijola, O. Autonomic regulation and ventricular arrhythmias. *Curr. Treat Options Cardiovasc. Med.* **2018**, *20*, 38. [CrossRef]
79. Chen, P.S.; Chen, L.S.; Cao, J.M.; Sharifi, B.; Karagueuzian, H.S.; Fishbein, M.C. Sympathetic nerve sprouting, electrical remodeling and the mechanisms of sudden cardiac death. *Cardiovasc. Res.* **2001**, *50*, 409–416. [CrossRef]
80. Huang, W.A.; Boyle, N.G.; Vaseghi, M. Cardiac innervation and the autonomic nervous system in sudden cardiac death. *Card. Electrophysiol. Clin.* **2017**, *9*, 665–679. [CrossRef]

81. Kimura, K.; Ieda, M.; Fukuda, K. Development, maturation, and transdifferentiation of cardiac sympathetic nerves. *Circ. Res.* **2012**, *110*, 325–336. [CrossRef]
82. Backs, J.; Haunstetter, A.; Gerber, S.H.; Metz, J.; Borst, M.M.; Strasser, R.H.; Kübler, W.; Haass, M. The neuronal norepinephrine transporter in experimental heart failure: Evidence for a posttranscriptional downregulation. *J. Mol. Cell. Cardiol.* **2001**, *33*, 461–472. [CrossRef]
83. Eisenhofer, G.; Friberg, P.; Rundqvist, B.; Quyyumi, A.A.; Lambert, G.; Kaye, D.M.; Kopin, I.J.; Goldstein, D.S.; Esler, M.D. Cardiac sympathetic nerve function in congestive heart failure. *Circulation* **1996**, *93*, 1667–1676. [CrossRef] [PubMed]
84. Liang, C.S. Cardiac sympathetic nerve terminal function in congestive heart failure. *Acta Pharmacol. Sin.* **2007**, *28*, 921–927. [CrossRef] [PubMed]
85. Hanani, M.; Spray, D.C. Emerging importance of satellite glia in nervous system function and dysfunction. *Nat. Rev. Neurosci.* **2020**, *21*, 485–498. [CrossRef] [PubMed]
86. Hanani, M. Satellite glial cells in sympathetic and parasympathetic ganglia: In search of function. *Brain Res. Rev.* **2010**, *64*, 304–327. [CrossRef]
87. Hanani, M.; Verkhratsky, A. Satellite glial cells and astrocytes, a comparative review. *Neurochem. Res.* **2021**, *46*, 2525–2537. [CrossRef]
88. Durham, P.L.; Garrett, F.G. Development of functional units within trigeminal ganglia correlates with increased expression of proteins involved in neuron-glia interactions. *Neuron Glia Biol.* **2010**, *6*, 171–181. [CrossRef]
89. Grubišić, V.; McClain, J.L.; Fried, D.E.; Grants, I.; Rajasekhar, P.; Csizmadia, E.; Ajijola, O.A.; Watson, R.E.; Poole, D.P.; Robson, S.C.; et al. Enteric glia modulate macrophage phenotype and visceral sensitivity following inflammation. *Cell Rep.* **2020**, *32*, 108100. [CrossRef]
90. Kurata, S.; Goto, T.; Gunjigake, K.K.; Kataoka, S.; Kuroishi, K.N.; Ono, K.; Toyono, T.; Kobayashi, S.; Yamaguchi, K. Nerve growth factor involves mutual interaction between neurons and satellite glial cells in the rat trigeminal ganglion. *Acta Histochem. Cytochem.* **2013**, *46*, 65–73. [CrossRef]
91. van Weperen, V.Y.H.; Littman, R.J.; Arneson, D.V.; Contreras, J.; Yang, X.; Ajijola, O.A. Single-cell transcriptomic profiling of satellite glial cells in stellate ganglia reveals developmental and functional axial dynamics. *Glia* **2021**, *69*, 1281–1291. [CrossRef]
92. Nguyen, B.L.; Li, H.; Fishbein, M.C.; Lin, S.F.; Gaudio, C.; Chen, P.S.; Chen, L.S. Acute myocardial infarction induces bilateral stellate ganglia neural remodeling in rabbits. *Cardiovasc. Pathol.* **2012**, *21*, 143–148. [CrossRef]
93. Nakamura, K.; Ajijola, O.A.; Aliotta, E.; Armour, J.A.; Ardell, J.L.; Shivkumar, K. Pathological effects of chronic myocardial infarction on peripheral neurons mediating cardiac neurotransmission. *Auton. Neurosci.* **2016**, *197*, 34–40. [CrossRef] [PubMed]
94. Han, S.; Kobayashi, K.; Joung, B.; Piccirillo, G.; Maruyama, M.; Vinters, H.V.; March, K.; Lin, S.F.; Shen, C.; Fishbein, M.C.; et al. Electroanatomic remodeling of the left stellate ganglion after myocardial infarction. *J. Am. Coll. Cardiol.* **2012**, *59*, 954–961. [CrossRef] [PubMed]
95. Tan, A.Y.; Elharrif, K.; Cardona-Guarache, R.; Mankad, P.; Ayers, O.; Joslyn, M.; Das, A.; Kaszala, K.; Lin, S.F.; Ellenbogen, K.A.; et al. Persistent proarrhythmic neural remodeling despite recovery from premature ventricular contraction-induced cardiomyopathy. *J. Am. Coll. Cardiol.* **2020**, *75*, 1–13. [CrossRef] [PubMed]
96. Ajijola, O.A.; Yagishita, D.; Reddy, N.K.; Yamakawa, K.; Vaseghi, M.; Downs, A.M.; Hoover, D.B.; Ardell, J.L.; Shivkumar, K. Remodeling of stellate ganglion neurons after spatially targeted myocardial infarction: Neuropeptide and morphologic changes. *Heart Rhythm* **2015**, *12*, 1027–1035. [CrossRef] [PubMed]
97. Ajijola, O.A.; Wisco, J.J.; Lambert, H.W.; Mahajan, A.; Stark, E.; Fishbein, M.C.; Shivkumar, K. Extracardiac neural remodeling in humans with cardiomyopathy. *Circ. Arrhythm. Electrophysiol.* **2012**, *5*, 1010–1116. [CrossRef] [PubMed]
98. Ogawa, M.; Zhou, S.; Tan, A.Y.; Song, J.; Gholmieh, G.; Fishbein, M.C.; Luo, H.; Siegel, R.J.; Karagueuzian, H.S.; Chen, L.S.; et al. Left stellate ganglion and vagal nerve activity and cardiac arrhythmias in ambulatory dogs with pacing-induced congestive heart failure. *J. Am. Coll. Cardiol.* **2007**, *50*, 335–343. [CrossRef]
99. Tu, H.; Liu, J.; Zhang, D.; Zheng, H.; Patel, K.P.; Cornish, K.G.; Wang, W.Z.; Muelleman, R.L.; Li, Y.L. Heart failure-induced changes of voltage-gated Ca2+ channels and cell excitability in rat cardiac postganglionic neurons. *Am. J. Physiol. Cell Physiol.* **2014**, *306*, C132–C142. [CrossRef]
100. Augustine, G.J. How does calcium trigger neurotransmitter release? *Curr. Opin. Neurobiol.* **2001**, *11*, 320–326. [CrossRef]
101. Borst, J.G.; Sakmann, B. Calcium influx and transmitter release in a fast CNS synapse. *Nature* **1996**, *383*, 431–434. [CrossRef]
102. Zucker, R.S. Calcium and transmitter release. *J. Physiol. Paris* **1993**, *87*, 25–36. [CrossRef]
103. Tsien, R.W.; Lipscombe, D.; Madison, D.; Bley, K.; Fox, A. Reflections on Ca(2+)-channel diversity, 1988–1994. *Trends Neurosci.* **1995**, *18*, 52–54. [PubMed]
104. Tsien, R.W.; Lipscombe, D.; Madison, D.V.; Bley, K.R.; Fox, A.P. Multiple types of neuronal calcium channels and their selective modulation. *Trends Neurosci.* **1988**, *11*, 431–438. [CrossRef]
105. Benarroch, E.E. Neuronal voltage-gated calcium channels: Brief overview of their function and clinical implications in neurology. *Neurology* **2010**, *74*, 1310–1315. [CrossRef] [PubMed]
106. Catterall, W.A. Structure and regulation of voltage-gated Ca2+ channels. *Annu. Rev. Cell Dev. Biol.* **2000**, *16*, 521–555. [CrossRef] [PubMed]

107. Ino, M.; Yoshinaga, T.; Wakamori, M.; Miyamoto, N.; Takahashi, E.; Sonoda, J.; Kagaya, T.; Oki, T.; Nagasu, T.; Nishizawa, Y.; et al. Functional disorders of the sympathetic nervous system in mice lacking the alpha 1B subunit (Cav 2.2) of N-type calcium channels. *Proc. Natl. Acad. Sci. USA* **2001**, *98*, 5323–5328. [CrossRef] [PubMed]
108. Molderings, G.J.; Likungu, J.; Gothert, M. N-Type calcium channels control sympathetic neurotransmission in human heart atrium. *Circulation* **2000**, *101*, 403–407. [CrossRef]
109. Yamada, Y.; Kinoshita, H.; Kuwahara, K.; Nakagawa, Y.; Kuwabara, Y.; Minami, T.; Yamada, C.; Shibata, J.; Nakao, K.; Cho, K.; et al. Inhibition of N-type Ca2+ channels ameliorates an imbalance in cardiac autonomic nerve activity and prevents lethal arrhythmias in mice with heart failure. *Cardiovasc. Res.* **2014**, *104*, 183–193. [CrossRef]
110. Zhang, D.; Tu, H.; Wang, C.; Cao, L.; Hu, W.; Hackfort, B.T.; Muelleman, R.L.; Wadman, M.C.; Li, Y.L. Inhibition of N-type calcium channels in cardiac sympathetic neurons attenuates ventricular arrhythmogenesis in heart failure. *Cardiovasc. Res.* **2021**, *117*, 137–148. [CrossRef]
111. Barber, M.J.; Mueller, T.M.; Henry, D.P.; Felten, S.Y.; Zipes, D.P. Transmural myocardial infarction in the dog produces sympathectomy in noninfarcted myocardium. *Circulation* **1983**, *67*, 787–796. [CrossRef]
112. Li, W.; Knowlton, D.; Van Winkle, D.M.; Habecker, B.A. Infarction alters both the distribution and noradrenergic properties of cardiac sympathetic neurons. *Am. J. Physiol. Heart Circ. Physiol.* **2004**, *286*, H2229–H2236. [CrossRef]
113. Zipes, D.P. Influence of myocardial ischemia and infarction on autonomic innervation of heart. *Circulation* **1990**, *82*, 1095–1105. [CrossRef] [PubMed]
114. Cao, J.M.; Fishbein, M.C.; Han, J.B.; Lai, W.W.; Lai, A.C.; Wu, T.J.; Czer, L.; Wolf, P.L.; Denton, T.A.; Shintaku, I.P.; et al. Relationship between regional cardiac hyperinnervation and ventricular arrhythmia. *Circulation* **2000**, *101*, 1960–1969. [CrossRef] [PubMed]
115. Oh, Y.S.; Jong, A.Y.; Kim, D.T.; Li, H.; Wang, C.; Zemljic-Harpf, A.; Ross, R.S.; Fishbein, M.C.; Chen, P.S.; Chen, L.S. Spatial distribution of nerve sprouting after myocardial infarction in mice. *Heart Rhythm* **2006**, *3*, 728–736. [CrossRef] [PubMed]
116. Zhou, S.; Chen, L.S.; Miyauchi, Y.; Miyauchi, M.; Kar, S.; Kangavari, S.; Fishbein, M.C.; Sharifi, B.; Chen, P.S. Mechanisms of cardiac nerve sprouting after myocardial infarction in dogs. *Circ. Res.* **2004**, *95*, 76–83. [CrossRef]
117. Kimura, K.; Kanazawa, H.; Ieda, M.; Kawaguchi-Manabe, H.; Miyake, Y.; Yagi, T.; Arai, T.; Sano, M.; Fukuda, K. Norepinephrine-induced nerve growth factor depletion causes cardiac sympathetic denervation in severe heart failure. *Auton. Neurosci.* **2010**, *156*, 27–35. [CrossRef]
118. Lorentz, C.U.; Parrish, D.C.; Alston, E.N.; Pellegrino, M.J.; Woodward, W.R.; Hempstead, B.L.; Habecker, B.A. Sympathetic denervation of peri-infarct myocardium requires the p75 neurotrophin receptor. *Exp. Neurol.* **2013**, *249*, 111–119. [CrossRef]
119. Clyburn, C.; Sepe, J.J.; Habecker, B.A. What gets on the nerves of cardiac patients? Pathophysiological changes in cardiac innervation. *J. Physiol.* **2022**, *600*, 451–461. [CrossRef]
120. Herring, N.; Kalla, M.; Paterson, D.J. The autonomic nervous system and cardiac arrhythmias: Current concepts and emerging therapies. *Nat. Rev. Cardiol.* **2019**, *16*, 707–726. [CrossRef]
121. Grkovski, M.; Zanzonico, P.B.; Modak, S.; Humm, J.L.; Narula, J.; Pandit-Taskar, N. F-18 meta-fluorobenzylguanidine PET imaging of myocardial sympathetic innervation. *J. Nucl. Cardiol.* 2022, *online ahead of print*.
122. Li, J.; Zheng, L. The mechanism of cardiac sympathetic activity assessment methods: Current knowledge. *Front. Cardiovasc. Med.* **2022**, *9*, 931219. [CrossRef]
123. Orimo, S.; Yogo, M.; Nakamura, T.; Suzuki, M.; Watanabe, H. (123)I-meta-iodobenzylguanidine (MIBG) cardiac scintigraphy in α-synucleinopathies. *Ageing Res. Rev.* **2016**, *30*, 122–133. [CrossRef]
124. Rubart, M.; Zipes, D.P. Mechanisms of sudden cardiac death. *J. Clin. Investig.* **2005**, *115*, 2305–2315. [CrossRef] [PubMed]
125. Yokoyama, T.; Lee, J.K.; Miwa, K.; Opthof, T.; Tomoyama, S.; Nakanishi, H.; Yoshida, A.; Yasui, H.; Iida, T.; Miyagawa, S.; et al. Quantification of sympathetic hyperinnervation and denervation after myocardial infarction by three-dimensional assessment of the cardiac sympathetic network in cleared transparent murine hearts. *PLoS ONE* **2017**, *12*, e0182072. [CrossRef] [PubMed]
126. Aggarwal, A.; Elser, M.D.; Socratous, F.; Kaye, D.M. Evidence for functional presynaptic alpha-2 adrenoceptors and their down-regulation in human heart failure. *J. Am. Coll. Cardiol.* **2001**, *37*, 1246–1251. [CrossRef]
127. Kiuchi, M.G.; Nolde, J.M.; Villacorta, H.; Carnagarin, R.; Chan, J.J.S.; Lugo-Gavidia, L.M.; Ho, J.K.; Matthews, V.B.; Dwivedi, G.; Schlaich, M.P. New approaches in the management of sudden cardiac death in patients with heart failure-targeting the sympathetic nervous system. *Int. J. Mol. Sci.* **2019**, *20*, 2430. [CrossRef]
128. Latini, R.; Masson, S.; Jeremic, G.; LuvarAÿ, G.; Fiordaliso, F.; Calvillo, L.; Bernasconi, R.; Torri, M.; Rondelli, I.; Razzetti, R.; et al. Comparative efficacy of a DA2/alpha2 agonist and a beta-blocker in reducing adrenergic drive and cardiac fibrosis in an experimental model of left ventricular dysfunction after coronary artery occlusion. *J. Cardiovasc. Pharmacol.* **1998**, *31*, 601–608. [CrossRef]
129. Minatoguchi, S. Heart failure and its treatment from the perspective of sympathetic nerve activity. *J. Cardiol.* **2022**, *79*, 691–697. [CrossRef] [PubMed]
130. Ramchandra, R.; Hood, S.G.; Xing, D.; Lambert, G.W.; May, C.N. Mechanisms underlying the increased cardiac norepinephrine spillover in heart failure. *Am. J. Physiol. Heart Circ. Physiol.* **2018**, *315*, H340–H347. [CrossRef] [PubMed]
131. Tygesen, H.; Rundqvist, B.; Waagstein, F.; Wennerblom, B. Heart rate variability measurement correlates with cardiac norepin ephrine spillover in congestive heart failure. *Am. J. Cardiol.* **2001**, *87*, 1308–1311. [CrossRef]

132. Meredith, I.T.; Eisenhofer, G.; Lambert, G.W.; Dewar, E.M.; Jennings, G.L.; Esler, M.D. Cardiac sympathetic nervous activity in congestive heart failure. Evidence for increased neuronal norepinephrine release and preserved neuronal uptake. *Circulation* **1993**, *88*, 136–145. [CrossRef]
133. Bateman, T.M.; Ananthasubramaniam, K.; Berman, D.S.; Gerson, M.; Gropler, R.; Henzlova, M.; Mendoza, F.; Miyamoto, M.; Shah, M.; Weiland, F. Reliability of the (123)I-mIBG heart/mediastinum ratio: Results of a multicenter test-retest reproducibility study. *J. Nucl. Cardiol.* **2019**, *26*, 1555–1565. [CrossRef]
134. Jacobson, A.F.; Senior, R.; Cerqueira, M.D.; Wong, N.D.; Thomas, G.S.; Lopez, V.A.; Agostini, D.; Weiland, F.; Chandna, H.; Narula, J. Myocardial iodine-123 meta-iodobenzylguanidine imaging and cardiac events in heart failure. Results of the prospective ADMIRE-HF (AdreView Myocardial Imaging for Risk Evaluation in Heart Failure) study. *J. Am. Coll. Cardiol.* **2010**, *55*, 2212–2221. [CrossRef] [PubMed]
135. Merlet, P.; Valette, H.; Dubois-Randé, J.L.; Moyse, D.; Duboc, D.; Dove, P.; Bourguignon, M.H.; Benvenuti, C.; Duval, A.M.; Agostini, D.; et al. Prognostic value of cardiac metaiodobenzylguanidine imaging in patients with heart failure. *J. Nucl. Med.* **1992**, *33*, 471–477. [PubMed]
136. Rajapreyar, I.; Pamboukian, S.V. Cardiac sympathetic imaging in heart failure: Is revival possible? *J. Nucl. Cardiol.* **2021**, *28*, 86–89. [CrossRef]
137. Silverio, A.; Polito, M.V.; Pace, L.; D'Auria, F.; Vitulano, G.; Scarano, M.; Citro, R.; Galasso, G.; Piscione, F. Predictors of outcome in patients with de novo diagnosis of heart failure with reduced ejection fraction: Role of combined myocardial and lung Iodine-123 Meta–Iodobenzylguanidine imaging. *J. Nucl. Cardiol.* **2021**, *28*, 72–85. [CrossRef] [PubMed]
138. Ardell, J.L.; Foreman, R.D.; Armour, J.A.; Shivkumar, K. Cardiac sympathectomy and spinal cord stimulation attenuate reflex-mediated norepinephrine release during ischemia preventing ventricular fibrillation. *JCI Insight* **2019**, *4*, e131648. [CrossRef] [PubMed]
139. Chan, S.A.; Vaseghi, M.; Kluge, N.; Shivkumar, K.; Ardell, J.L.; Smith, C. Fast in vivo detection of myocardial norepinephrine levels in the beating porcine heart. *Am. J. Physiol. Heart Circ. Physiol.* **2020**, *318*, H1091–H1099. [CrossRef]
140. Gilinsky, M.A.; Faibushevish, A.A.; Lunte, C.E. Determination of myocardial norepinephrine in freely moving rats using in vivo microdialysis sampling and liquid chromatography with dual-electrode amperometric detection. *J. Pharm. Biomed. Anal.* **2001**, *24*, 929–935. [CrossRef]
141. Zhang, Q.; Liu, B.; Wu, Q.; Liu, B.; Li, Y.; Sun, S.; Wang, Y.; Wu, X.; Chai, Z.; Jiang, X.; et al. Differential co-release of two neurotransmitters from a vesicle fusion pore in mammalian adrenal chromaffin cells. *Neuron* **2019**, *102*, 173–183. [CrossRef]
142. Yagishita, D.; Chui, R.W.; Yamakawa, K.; Rajendran, P.S.; Ajijola, O.A.; Nakamura, K.; So, E.L.; Mahajan, A.; Shivkumar, K.; Vaseghi, M. Sympathetic nerve stimulation, not circulating norepinephrine, modulates T-peak to T-end interval by increasing global dispersion of repolarization. *Circ. Arrhythm. Electrophysiol.* **2015**, *8*, 174–185. [CrossRef]
143. Zekios, K.C.; Mouchtouri, E.T.; Lekkas, P.; Nikas, D.N.; Kolettis, T.M. Sympathetic activation and arrhythmogenesis after myocardial infarction: Where do we stand? *J. Cardiovasc. Dev. Dis.* **2021**, *8*, 57. [CrossRef]
144. Ajijola, O.A.; Chatterjee, N.A.; Gonzales, M.J.; Gornbein, J.; Liu, K.; Li, D.; Paterson, D.J.; Shivkumar, K.; Singh, J.P.; Herring, N. Coronary sinus neuropeptide Y levels and adverse outcomes in patients with stable chronic heart failure. *JAMA Cardiol.* **2020**, *5*, 318–325. [CrossRef] [PubMed]
145. Gibbs, T.; Tapoulal, N.; Shanmuganathan, M.; Burrage, M.K.; Borlotti, A.; Banning, A.P.; Choudhury, R.P.; Neubauer, S.; Kharbanda, R.K.; Ferreira, V.M.; et al. Neuropeptide-Y levels in ST-segment-elevation myocardial infarction: Relationship with coronary microvascular function, heart Failure, and mortality. *J. Am. Heart Assoc.* **2022**, *11*, e024850. [CrossRef] [PubMed]
146. Kalla, M.; Hao, G.; Tapoulal, N.; Tomek, J.; Liu, K.; Woodward, L.; Dall'Armellina, E.; Banning, A.P.; Choudhury, R.P.; Neubauer, S.; et al. The cardiac sympathetic co-transmitter neuropeptide Y is pro-arrhythmic following ST-elevation myocardial infarction despite beta-blockade. *Eur. Heart J.* **2020**, *41*, 2168–2179. [CrossRef] [PubMed]
147. Özkaramanli-Gür, D.; Sağbaş, M.; Akyüz, A.; Güzel, S.; Alpsoy, Ş.; Güler, N. Role of sympathetic cotransmitter galanin on autonomic balance in heart failure: An active player or a bystander? *Anatol. J. Cardiol.* **2017**, *18*, 281–288. [CrossRef]
148. Tan, C.M.J.; Green, P.; Tapoulal, N.; Lewandowski, A.J.; Leeson, P.; Herring, N. The role of neuropeptide Y in cardiovascular health and disease. *Front. Physiol.* **2018**, *9*, 1281. [CrossRef]
149. Gardner, R.T.; Ripplinger, C.M.; Myles, R.C.; Habecker, B.A. Molecular mechanisms of sympathetic remodeling and arrhythmias. *Circ. Arrhythm. Electrophysiol.* **2016**, *9*, e001359. [CrossRef]
150. Pannese, E.; Procacci, P. Ultrastructural localization of NGF receptors in satellite cells of the rat spinal ganglia. *J. Neurocytol.* **2002**, *31*, 755–763. [CrossRef]
151. Korsching, S.; Thoenen, H. Nerve growth factor in sympathetic ganglia and corresponding target organs of the rat: Correlation with density of sympathetic innervation. *Proc. Natl. Acad. Sci. USA* **1983**, *80*, 3513–3516. [CrossRef]
152. Scott-Solomon, E.; Boehm, E.; Kuruvilla, R. The sympathetic nervous system in development and disease. *Nat. Rev. Neurosci.* **2021**, *22*, 685–702. [CrossRef]
153. Hagan, N.; Kane, J.L.; Grover, D.; Woodworth, L.; Madore, C.; Saleh, J.; Sancho, J.; Liu, J.; Li, Y.; Proto, J.; et al. CSF1R signaling is a regulator of pathogenesis in progressive MS. *Cell Death Dis.* **2020**, *11*, 904. [CrossRef]
154. Pixley, F.J.; Stanley, E.R. CSF-1 regulation of the wandering macrophage: Complexity in action. *Trends Cell Biol.* **2004**, *14*, 628–638. [CrossRef] [PubMed]

155. Bacmeister, L.; Schwarzl, M.; Warnke, S.; Stoffers, B.; Blankenberg, S.; Westermann, D.; Lindner, D. Inflammation and fibrosis in murine models of heart failure. *Basic Res. Cardiol.* **2019**, *114*, 19. [PubMed]
156. Dick, S.A.; Epelman, S. Chronic heart failure and inflammation: What do we really know? *Circ. Res.* **2016**, *119*, 159–176. [CrossRef] [PubMed]
157. Thackeray, J.T.; Hupe, H.C.; Wang, Y.; Bankstahl, J.P.; Berding, G.; Ross, T.L.; Bauersachs, J.; Wollert, K.C.; Bengel, F.M. Myocardial inflammation predicts remodeling and neuroinflammation after myocardial infarction. *J. Am. Coll. Cardiol.* **2018**, *71*, 263–275. [CrossRef]
158. Hulsmans, M.; Clauss, S.; Xiao, L.; Aguirre, A.D.; King, K.R.; Hanley, A.; Hucker, W.J.; Walfers, E.M.; Seemann, G.; Courties, G.; et al. Macrophages facilitate electrical conduction in the heart. *Cell* **2017**, *169*, 510–522. [CrossRef] [PubMed]
159. Lambert, J.M.; Lopez, E.F.; Lindsey, M.L. Macrophage roles following myocardial infarction. *Int. J. Cardiol.* **2008**, *130*, 147–158. [CrossRef] [PubMed]
160. Zhang, D.; Hu, W.; Tu, H.; Hackfort, B.T.; Duan, B.; Xiong, W.; Wadman, M.C.; Li, Y.L. Macrophage depletion in stellate ganglia alleviates cardiac sympathetic overactivation and ventricular arrhythmogenesis by attenuating neuroinflammation in heart failure. *Basic Res. Cardiol.* **2021**, *116*, 28. [CrossRef]
161. Nohara, A.; Okada, S.; Ohshima, K.; Pessin, J.E.; Mori, M. Cyclin-dependent kinase-5 is a key molecule in tumor necrosis factor-alpha-induced insulin resistance. *J. Biol. Chem.* **2011**, *286*, 33457–33465. [CrossRef]
162. Rozas, P.; Lazcano, P.; Pina, R.; Cho, A.; Terse, A.; Pertusa, M.; Madrid, R.; Gonzalez-Billault, C.; Kulkarni, A.B.; Utreras, E. Targeted overexpression of tumor necrosis factor-alpha increases cyclin-dependent kinase 5 activity and TRPV1-dependent Ca2+ influx in trigeminal neurons. *Pain* **2016**, *157*, 1346–1362. [CrossRef]
163. Utreras, E.; Futatsugi, A.; Rudrabhatla, P.; Keller, J.; Iadarola, M.J.; Pant, H.C.; Kulkarni, A.B. Tumor necrosis factor-alpha regulates cyclin-dependent kinase 5 activity during pain signaling through transcriptional activation of p35. *J. Biol. Chem.* **2009**, *284*, 2275–2284. [CrossRef]
164. Su, S.C.; Seo, J.; Pan, J.Q.; Samuels, B.A.; Rudenko, A.; Ericsson, M.; Neve, R.L.; Yue, D.T.; Tsai, L.H. Regulation of N-type voltage-gated calcium channels and presynaptic function by cyclin-dependent kinase 5. *Neuron* **2012**, *75*, 675–687. [CrossRef] [PubMed]
165. Zhang, D.; Tu, H.; Hu, W.; Wadman, M.C.; Li, Y.L. CDK5 promotes ventricular arrhythmogenesis through phosphorylation of N-type calcium channels in cardiac sympathetic postganglionic neurons. *FASEB J.* **2020**, *34*, 03088. [CrossRef]
166. Ajijola, O.A.; Hoover, D.B.; Simerly, T.M.; Brown, T.C.; Yanagawa, J.; Biniwale, R.M.; Lee, J.M.; Sadeghi, A.; Khanlou, N.; Ardell, J.L.; et al. Inflammation, oxidative stress, and glial cell activation characterize stellate ganglia from humans with electrical storm. *JCI Insight* **2017**, *2*, e94715. [CrossRef] [PubMed]
167. Shanks, J.; Gao, L.; Zucker, I.H. Sympathomodulation in heart failure: A role for stellate ganglia Nrf2. *FASEB J.* **2019**, *33*, 564–565. [CrossRef]
168. Singh, S.; Sayers, S.; Walter, J.S.; Thomas, D.; Dieter, R.S.; Nee, L.M.; Wurster, R.D. Hypertrophy of neurons within cardiac ganglia in human, canine, and rat heart failure: The potential role of nerve growth factor. *J. Am. Heart. Assoc.* **2013**, *2*, e000210. [CrossRef]
169. Meloni, M.; Caporali, A.; Graiani, G.; Lagrasta, C.; Katare, R.; Van Linthout, S.; Spillmann, F.; Campesi, I.; Madeddu, P.; Quaini, F.; et al. Nerve growth factor promotes cardiac repair following myocardial infarction. *Circ. Res.* **2010**, *106*, 1275–1284. [CrossRef]
170. Atwal, J.K.; Massie, B.; Miller, F.D.; Kaplan, D.R. The TrkB-Shc site signals neuronal survival and local axon growth via MEK and P13-kinase. *Neuron* **2000**, *27*, 265–277. [CrossRef]
171. Bodmer, D.; Ascaño, M.; Kuruvilla, R. Isoform-specific dephosphorylation of dynamin1 by calcineurin couples neurotrophin receptor endocytosis to axonal growth. *Neuron* **2011**, *70*, 1085–1099. [CrossRef]
172. Spillane, M.; Ketschek, A.; Donnelly, C.J.; Pacheco, A.; Twiss, J.L.; Gallo, G. Nerve growth factor-induced formation of axonal filopodia and collateral branches involves the intra-axonal synthesis of regulators of the actin-nucleating Arp2/3 complex. *J. Neurosci.* **2012**, *32*, 17671–17689. [CrossRef]
173. Kisiswa, L.; Osório, C.; Erice, C.; Vizard, T.; Wyatt, S.; Davies, A.M. TNFα reverse signaling promotes sympathetic axon growth and target innervation. *Nat. Neurosci.* **2013**, *16*, 865–873. [CrossRef]
174. O'Keeffe, G.W.; Gutierrez, H.; Howard, L.; Laurie, C.W.; Osorio, C.; Gavaldà, N.; Wyatt, S.L.; Davies, A.M. Region-specific role of growth differentiation factor-5 in the establishment of sympathetic innervation. *Neural Dev.* **2016**, *11*, 4. [CrossRef] [PubMed]
175. Pellegrino, M.J.; Habecker, B.A. STAT3 integrates cytokine and neurotrophin signals to promote sympathetic axon regeneration. *Mol. Cell. Neurosci.* **2013**, *56*, 272–282. [CrossRef] [PubMed]
176. Pellegrino, M.J.; McCully, B.H.; Habecker, B.A. Leptin stimulates sympathetic axon outgrowth. *Neurosci. Lett.* **2014**, *566*, 1–5. [CrossRef]
177. Gamage, K.K.; Cheng, I.; Park, R.E.; Karim, M.S.; Edamura, K.; Hughes, C.; Spano, A.J.; Erisir, A.; Deppmann, C.D. Death receptor 6 promotes wallerian degeneration in peripheral axons. *Curr. Biol.* **2017**, *27*, 890–896. [CrossRef] [PubMed]
178. Kohn, J.; Aloyz, R.S.; Toma, J.G.; Haak-Frendscho, M.; Miller, F.D. Functionally antagonistic interactions between the TrkA and p75 neurotrophin receptors regulate sympathetic neuron growth and target innervation. *J. Neurosci.* **1999**, *19*, 5393–5408. [CrossRef] [PubMed]
179. Majdan, M.; Walsh, G.S.; Aloyz, R.; Miller, F.D. TrkA mediates developmental sympathetic neuron survival in vivo by silencing an ongoing p75NTR-mediated death signal. *J. Cell Biol.* **2001**, *155*, 1275–1285. [CrossRef] [PubMed]

180. Nikolaev, A.; McLaughlin, T.; O'Leary, D.D.; Tessier-Lavigne, M. APP binds DR6 to trigger axon pruning and neuron death via distinct caspases. *Nature* **2009**, *457*, 981–989. [CrossRef]
181. Singh, K.K.; Park, K.J.; Hong, E.J.; Kramer, B.M.; Greenberg, M.E.; Kaplan, D.R.; Miller, F.D. Developmental axon pruning mediated by BDNF-p75NTR-dependent axon degeneration. *Nat. Neurosci.* **2008**, *11*, 649–658. [CrossRef]
182. Teng, K.K.; Felice, S.; Kim, T.; Hempstead, B.L. Understanding proneurotrophin actions: Recent advances and challenges. *Dev. Neurobiol.* **2010**, *70*, 350–359. [CrossRef]
183. Yong, Y.; Gamage, K.; Cheng, I.; Barford, K.; Spano, A.; Winckler, B.; Deppmann, C. p75NTR and DR6 regulate distinct phases of axon degeneration demarcated by spheroid rupture. *J. Neurosci.* **2019**, *39*, 9503–9520. [CrossRef]
184. Sepe, J.J.; Gardner, R.T.; Blake, M.R.; Brooks, D.M.; Staffenson, M.A.; Betts, C.B.; Sivagnanam, S.; Larson, W.; Kumar, S.; Bayles, R.G.; et al. Therapeutics that promote sympathetic reinnervation modulate the inflammatory response after myocardial infarction. *JACC Basic Transl. Sci.* **2022**, *7*, 915–930. [CrossRef]
185. Huang, R.; Wang, Y.; Li, J.; Jiang, X.; Li, Y.; Liu, B.; Wu, X.; Du, X.; Hang, Y.; Jin, M.; et al. Ca(2+)-independent but voltage-dependent quantal catecholamine secretion (CiVDS) in the mammalian sympathetic nervous system. *Proc. Natl. Acad. Sci. USA* **2019**, *116*, 20201–20209. [CrossRef] [PubMed]
186. Beau, S.L.; Saffitz, J.E. Transmural heterogeneity of norepinephrine uptake in failing human hearts. *J. Am. Coll. Cardiol.* **1994**, *23*, 579–585. [CrossRef]
187. Böhm, M.; La Rosée, K.; Schwinger, R.H.; Erdmann, E. Evidence for reduction of norepinephrine uptake sites in the failing human heart. *J. Am. Coll. Cardiol.* **1995**, *25*, 146–153. [CrossRef]
188. Liang, C.S.; Fan, T.H.; Sullebarger, J.T.; Sakamoto, S. Decreased adrenergic neuronal uptake activity in experimental right heart failure. A chamber-specific contributor to beta-adrenoceptor downregulation. *J. Clin. Investig.* **1989**, *84*, 1267–1275. [CrossRef]
189. Mao, W.; Iwai, C.; Qin, F.; Liang, C.S. Norepinephrine induces endoplasmic reticulum stress and downregulation of norepinephrine transporter density in PC12 cells via oxidative stress. *Am. J. Physiol. Heart Circ. Physiol.* **2005**, *288*, H2381–H2389. [CrossRef]
190. Mao, W.; Qin, F.; Iwai, C.; Vulapalli, R.; Keng, P.C.; Liang, C.S. Extracellular norepinephrine reduces neuronal uptake of norepinephrine by oxidative stress in PC12 cells. *Am. J. Physiol. Heart Circ. Physiol.* **2004**, *287*, H29–H39. [CrossRef]
191. Vatta, M.S.; Bianciotti, L.G.; Guil, M.J.; Hope, S.I. Regulation of the norepinephrine transporter by endothelins: A potential therapeutic target. *Vitam. Horm.* **2015**, *98*, 371–405.
192. Backs, J.; Bresch, E.; Lutz, M.; Kristen, A.V.; Haass, M. Endothelin-1 inhibits the neuronal norepinephrine transporter in hearts of male rats. *Cardiovasc. Res.* **2005**, *67*, 283–290. [CrossRef]
193. Mapps, A.A.; Thomsen, M.B.; Boehm, E.; Zhao, H.; Hattar, S.; Kuruvilla, R. Diversity of satellite glia in sympathetic and sensory ganglia. *Cell Rep.* **2022**, *38*, 110328. [CrossRef]

International Journal of *Molecular Sciences*

MDPI

Review

Sympathetic System in Wound Healing: Multistage Control in Normal and Diabetic Skin

Evgenii Ivanov *, Marina Akhmetshina, Aleksei Erdiakov and Svetlana Gavrilova

Faculty of Medicine, Lomonosov Moscow State University, 119991 Moscow, Russia
* Correspondence: ivanovev102@yandex.ru

Abstract: In this review, we discuss sympathetic regulation in normal and diabetic wound healing. Experimental denervation studies have confirmed that sympathetic nerve endings in skin have an important and complex role in wound healing. Vasoconstrictor neurons secrete norepinephrine (NE) and neuropeptide Y (NPY). Both mediators decrease blood flow and interact with inflammatory cells and keratinocytes. NE acts in an ambiguous way depending on receptor type. Beta2-adrenoceptors could be activated near sympathetic endings; they suppress inflammation and re-epithelialization. Alpha1- and alpha2-adrenoceptors induce inflammation and activate keratinocytes. Sudomotor neurons secrete acetylcholine (ACh) and vasoactive intestinal peptide (VIP). Both induce vasodilatation, angiogenesis, inflammation, keratinocytes proliferation and migration. In healthy skin, all effects are important for successful healing. In treatment of diabetic ulcers, mediator balance could be shifted in different ways. Beta2-adrenoceptors blockade and nicotinic ACh receptors activation are the most promising directions in treatment of diabetic ulcers with neuropathy, but they require further research.

Keywords: diabetes mellitus; wound healing; sympathetic system; norepinephrine; acetylcholine; NPY; VIP; NGF; keratinocytes; angiogenesis

Citation: Ivanov, E.; Akhmetshina, M.; Erdiakov, A.; Gavrilova, S. Sympathetic System in Wound Healing: Multistage Control in Normal and Diabetic Skin. *Int. J. Mol. Sci.* **2023**, *24*, 2045. https://doi.org/10.3390/ijms24032045

Academic Editors: Yutang Wang, Kate Denton and David A. Hart

Received: 30 November 2022
Revised: 14 January 2023
Accepted: 18 January 2023
Published: 20 January 2023

1. Introduction

Diabetes mellitus is associated with several common comorbidities. Diabetic neuropathy is very common among patients with severe disease. Up to 50% of patients experience diabetic neuropathy at some point in their life. In addition to direct clinical manifestations, neuropathy contributes to the development of other complications of diabetes. Among other complications, diabetic neuropathy increases foot ulcer risk to 25% [1].

Diabetic ulcers are severe chronic wounds that often occur on the distal parts of lower extremities. They are resistant to treatment and contribute to the majority of nontraumatic leg amputations [2–4]. Some surgical and orthopedic treatment tactics improved diabetic ulcer healing rates [5,6]. The new generation of glycaemia-controlling medications has also alleviated the burden of diabetic ulcers. Despite these advances, these treatments are often not enough to improve patients' conditions [7,8]. There is an urgent need for new pathogenetic ways to treat diabetic ulcers. Diabetic wound pathogenesis involves almost all major diabetic pathways and complications. Largely, wound progression is correlated with skin ischemia. Clinically, wounds with reduced skin blood flow are called angiopathic [9,10]. Some severe ulcers have a weak correlation with diabetic angiopathy. This subset of diabetic ulcers is thought to be completely neuropathic [11–13]. One of the most prominent clinical forms of neuropathic wound is Charcot neuroarthropathy [12,14]. But neuropathic factors can reveal themselves in different clinical forms. In any diabetic wound, both local ischemia and neuropathy are important, but in different proportions. Major surgical interventions and some medications were designed to restore microcirculation in wound area, yet they are less effective in neuropathic wounds [11,15]. Also, there are several important intermediate factors, like excessive inflammation, reactive oxygen species hyperproduction, metabolic disturbances and epigenetic changes [16–19].

Different components of diabetic ulcer pathogenesis have received unequal attention. There are more than 5000 reviews about diabetic ulcers in Medline database. Many of those are related to therapy options, wound infections, pathological inflammation, and microcirculation. Among them we have found nine reviews dedicated to neuropathy's involvement in diabetic wound healing. The oldest one lists some general ways by which neuropathy could affect diabetic ulcers without pathogenetic mechanisms [20]. One review summarizes information about neurovascular control neuropathic impairments [21]. Five works give a detailed picture of neuropeptide and neurotrophic factors' involvement in diabetic wound healing [22–26]. S. Sun et al. in a recent review focused β-adrenergic involvement and β-blockers treatment, though they did not touch a-adrenoceptors and cholinergic fibers [27]. In the most recent work, N.C. Nowak et al. gave a comprehensive review for sensory fibers' involvement and neuropeptides. We unequivocally recommend this paper as the most interesting and complete picture of neuropathy in diabetic wound healing [28]. There are also a few reviews about nervous system involvement in normal wound healing or in different pathologies. L. Pan et al. described sympathetic interaction with angiogenesis in detail. Subsequently, we will give less credit to this aspect of autonomic nervous system role, but for more information, we recommend this article [29]. M. Ashrafi et al. and D. Gupta et al. mostly describe neuropeptide functions [30,31]. Important information was collected in reviews of intraepidermal noradrenaline and acetylcholine synthesis. They are complementary to our work, but either do not have information about wound healing and diabetes or do not make clear distinctions between nerve-derived and keratinocytes-derived neurotransmitters [32–34]. There is much information in the literature about the general sympathetic nervous system's role as inflammation modulator. While those articles do not cover wound healing, they provide valuable concepts of adrenergic inflammation control. Additional to our work, information is reviewed by G. Pongratz and R. Straub [35].

Overall, the autonomic nervous system's role in wound healing in normal and diabetic wounds has not been completely explained in all the current reviews. It is not clear how adrenergic, cholinergic and neuropeptide regulation mechanisms are integrated during the wound healing process. At the same time, there are many fragmentary studies and much broad evidence that these mechanisms are important. Autonomic diabetic neuropathy affects more than 20% of patients with diabetes [36]. Cardiovascular diabetic neuropathy is diagnosed most frequently, but studies provide clues that diabetes affects all parts of the autonomic nervous system [37]. While cardiac autonomic neuropathy is present in 43–66% of patients with diabetic wounds, other forms of autonomic neuropathy are probably prevalent as well [38]. Beta-blockers underwent some trials in diabetic ulcers, but we could better target the autonomic nervous system with a clearer view of underlying mechanisms. Neuropeptides' involvement and angiogenesis control were thoroughly reviewed, and we will mention them briefly. We focused on articles about norepinephrine, acetylcholine, related nerve fibers and their interactions with effector cells on different steps of the healing process. Our aims are to compare known mechanisms with denervation studies results and clinical data, and to find important empty spaces in the current state of knowledge.

2. Methods

We conducted a literature (narrative) review; therefore, strict methodological criteria were not applicable. The search was performed in several steps for different parts of review in Medline, Scopus, and Web of Science databases. Below, we give search queries in Medline format, total results without duplicates through Medline database and final addition from all databases minus results from previous queries. Preclinical and clinical articles with any relevant disease model or disease were picked for analysis. Any article type with full text available was analyzed, and non-English works were auto-translated to extract their main positions. The search was conducted between 10 November 2022 and 10 December 2022.

(1) ("wound healing" [Mesh] OR skin wound* OR diabet* wound* OR diabet* ulcer*) AND ("Autonomic Nervous System" [Mesh] OR autonom* nerv*)—458 results, 78 to analysis
(2) ("wound healing" [Mesh] OR skin wound* OR diabet* wound* OR diabet* ulcer*) AND ("Receptors, Adrenergic" [Mesh])—59 results, 32 to analysis
(3) ("wound healing" [Mesh] OR skin wound* OR diabet* wound* OR diabet* ulcer*) AND ("Epinephrine" [Mesh] OR "Norepinephrine" [Mesh] OR epinephrine OR norepinephrine)—365 results, 61 to analysis
(4) ("wound healing" [Mesh] OR skin wound* OR diabet* wound* OR diabet* ulcer*) AND ("Receptors, Cholinergic" [Mesh])—41 results, 18 to analysis
(5) ("wound healing" [Mesh] OR skin wound* OR diabet* wound* OR diabet* ulcer*) AND ("Acetylcholine" [Mesh])—50 results, 2 to analysis
(6) ("wound healing" [Mesh] OR skin wound* OR diabet* wound* OR diabet* ulcer*) AND ("Neuropeptide Y" [Mesh] OR NPY)—33 results, 5 to analysis
(7) ("wound healing" [Mesh] OR skin wound* OR diabet* wound* OR diabet* ulcer*) AND ("Vasoactive Intestinal Peptide" [Mesh] OR VIP)—257 results, 7 to analysis

3. Sympathetic Neurotransmitters: Neuronal and Paracrine Sources

3.1. Sympathetic Regulation in Normal Skin

Autonomic nerves lie in dermis near blood and lymphatic vessels and dermal appendages and except for facial skin are only sympathetic. These fibers belong to cutaneous vasoconstrictor neurons, piloerector neurons and sudomotor neurons (Figure 1) [39]. Both types have C-type unmyelinated axons and are located in paravertebral ganglia [40]. Microneurographic studies have shown that one sympathetic unit innervates 24–350 mm^2 of skin [41]. Immunolabeling histological studies allowed the differentiation of noradrenergic sympathetic fibers near arterioles or erector pili muscles and apocrine sweat glands and cholinergic sympathetic fibers near eccrine sweat glands [42]. These fibers lay in dermis at different depths. Contrary to sensory fibers, sympathetic fibers do not directly penetrate epithelium basal membrane [43].

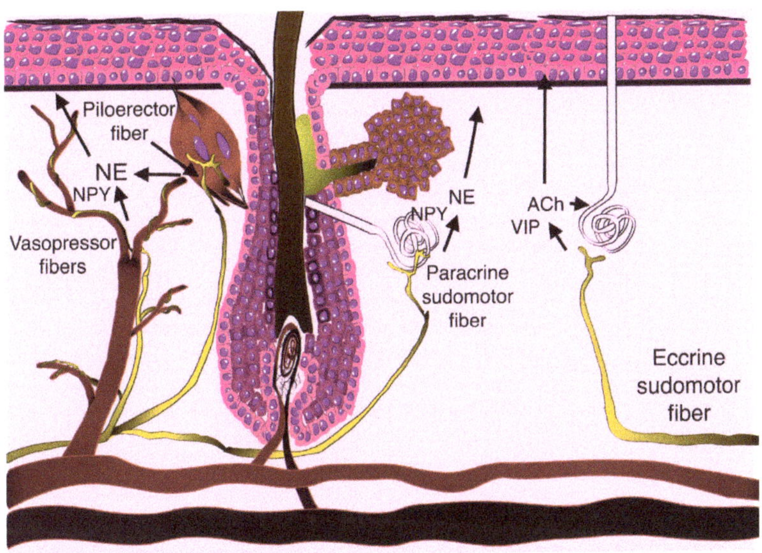

Figure 1. Normal autonomic skin innervation.

Vasoconstrictor terminals innervate every arteriole in the skin; therefore, they are rather equally distributed. Some studies have also found their divisions alongside capillaries, where they could change endothelial integrity and conduct retrograde signals. In normal skin, vasomotor sympathetic fibers reduce blood flow. Other adrenergic fibers activate piloerection and regulate hair follicle cells' activity. Also, some fibers innervate veins and collect lymphatic ducts, though their innervation is scarce. All adrenergic nerve endings primarily release epinephrine and neuropeptide Y (NPY), also different POMK—derived peptides. Sudomotor fibers increase sweating in stress reactions and heat dissipation. Cholinergic sudomotor neurons among acetylcholine release vasoactive intestinal peptide (VIP) [44]. Further we will see that sympathetic fibers have additional roles in wound resolving.

Deep soft tissue wounds can involve subcutaneous adipose tissue and superficial muscles. Adipose tissue has dense sympathetic innervation that directly regulates energetic balance and heat production [45,46]. Skeletal muscles contain only sympathetic fibers that innervate blood vessels, though recent evidence indicates their role in modulation of muscle contraction force [47,48]. We lack data on possible involvement of adipose and skeletal muscle sympathetic fibers in soft tissue wound healing. Therefore, we will further discuss dermal sympathetic fiber role and general sympathetic activation or denervation.

3.2. Neurotransmitter Synthesis in Skin Cells

Resident skin cells regulate normal tissue properties by paracrine release of signaling molecules. Among other, they produce catecholamines and neuropeptides and present different cell surface receptors. Keratinocytes express enzymes for norepinephrine (NE) and acetylcholine (ACh) synthesis and various receptors for them through all epidermis layers [33,49]. Keratinocytes vesicles contain tyrosine hydroxylase and phenylethanolamine-*N*-methyl transferase (PNMT), norepinephrine synthesis enzymes. Norepinephrine was also found in an isolated keratinocytes culture media [50]. In normal skin, the most immature basal keratinocytes produce more norepinephrine. NE increases calcium influx in the nearest cells through β_2-AR, thus promoting keratinocyte maturation. Beta-adrenoreceptors are the best described in skin. They were found on keratinocytes from different skin localizations and on fibroblasts. Epidermis is densely labeled for β_2-adrenoceptors in all layers with no β_1-AR presentation. Sweat glands also have high β_2-AR presentation; sebaceous glands are not labeled [51].

Therefore, it is proposed that intrinsic skin NE release is crucial to the native epidermis architecture. Several diseases, like atopic dermatitis and psoriasis, both impair keratinocytes differentiation and decrease norepinephrine levels [52].

Almost all human cells produce acetylcholine to a certain degree; therefore, active ACh depends on synthesis to acetylcholinesterase (AChE) hydrolysis ratio. All viable keratinocytes have equal ACh synthetic end exocytosis capacity, but AChE activity is localized in basal layer. Therefore, in intact epidermis, ACh concentration increases from basal to upper layers. Keratinocytes express a wide variety of ACh receptor genes—different non-muscular nAChR subunits and M_1–M_5 mAChR subtypes. The acetylcholine receptors are unequally distributed through the epidermis, and different acetylcholine actions in the different maturation stages have been speculated [35]. In the literature, a_3, a_5, b_2, b_4 nAChR subunits were detected in the epidermal basal layer and lower stratum granulosum, a_9 nAChR subunits—in the basal layer and in the lowest suprabasal epidermis, a_7, a_{10} and b_1 nAChR subunits—in the upper stratum spinosum and stratum granulosum. In addition, M_1 and M_4 muscarinic acetylcholine receptors were described in the suprabasal layers, M_2, M_3 and M_5—in the lower layers [53]. While different receptor subtypes have opposing effects, keratinocytes could respond to ACh differently dependent on maturation stage. Overall, normal ACh stimulation in intact epidermis promotes keratinocyte maturation through different phenotypes. Also, ACh is important for steady sebaceous and sweat gland secretion.

Catecholamines and acetylcholine regulation of skin homeostasis can be explained by both in-place paracrine regulation and neurogenic regulation [54]. We need more evidence to evaluate effects from two sources separately. As some authors propose, the high rate of neurotransmitter metabolism in basal keratinocytes is important to separate the two regulatory compartments. NE, ACh from sympathetic fibers could freely interact with skin appendages, basal epidermis layer and dermal cells. Because of uptake by basal keratinocytes, neuronal transmitters have a little effect on upper epidermis layers and vice versa. But in soft tissue wounds, natural barriers are compromised. All cells that migrate into the wound area are equally exposed to neurotransmitters and their regulation is important.

4. Sympathetic Regulation in Normal Wound Healing

Soft tissue healing, like any acute inflammation, has several stages: vessel dilatation and permeabilization, proliferation and reparation [55]. The autonomic nervous system regulates all stages of wound healing in the wound area, in intact skin near the wound and in the whole body. Sympathetic mediators could constrict arteries in the skin to prevent excessive blood loss. In intact skin around the wound, sympathetic fibers are crucial in maintaining physical properties of the skin. Without normal regulation, sweating skin can become dry and vulnerable to infections. In the wound area catecholamines, acetylcholine and neuropeptides modulate leukocyte activity, reepithelization, wound contraction, and other important processes.

4.1. Inflammation and Immune Cells

In the first days after wounding, acute inflammation decides the fate of the healing process. In minutes after initial damage injured cells, resident immune cells and tissue debris affect local blood vessels and attract granulocytes from blood stream. Vasodilatation and increased vessel permeability enhance leukocyte transcytosis. Sympathetic fibers modulate acute inflammation in several ways with different neurotransmitters and receptors.

As any serious trauma usually acts as a stressor, the catecholamine level in blood quickly increases, simultaneous with local sympathetic fibers' discharge. Catecholamines increase skin arterioles' resistance, shunting more blood to vital organs. This effect could delay leukocyte migration, but it also prevents excessive blood loss. Some older works have stated that catecholamines also increase capillary permeability [56,57]. Later research contradicted this finding. Lack of sympathetic activation, for example in diabetic skin, rather increased capillary leak through lower vessel tone [58,59]. While most cytokines and inflammation mediators in the wound area increased capillary permeability, catecholamines act as antagonists.

Catecholamines effectively modulate leukocytes' activity [60]. Adrenoceptors have been found on all types of leukocytes with different density, most abundant are β_2-adrenergic receptors [61]. Information about other types of adrenergic receptors is controversial in different species and cell types. Next to β_2 by quantity are a_1, then a_2 and β_1-receptors, respectively [62]. a-adrenoceptors and β-adrenoceptors enhance opposite processes in leukocytes, but they have different affinity and abundance. β_2 adrenergic regulation of immune response is the most studied and probably the most prevalent and important. β_2-agonists decrease TNFa and other pro-inflammatory cytokines release, leukocyte migration and chemotaxis [63]. IL-10 concentration is elevated rapidly after β_2-AR activation leading to immunosuppression or localized inflammation [64]. Possible mechanisms include protein kinase A (PKA)—mediated NF-kB suppression and β-arrestin 2 protein synthesis [64,65]. A. Gosain et al. studied norepinephrine effects on activated neutrophils and macrophages, isolated from rat wounds. Both types of cells have shown significantly reduced phagocytic activity by PKA activation. Macrophages were inhibited both by physiological and pharmacological NE doses, while neutrophils were inhibited only by pharmacological NE dose [66,67].

Less data is available about a-adrenoceptor's role in soft tissue wound healing. In most cases, they have shown pro-inflammatory properties. a-adrenergic agonists increase pro-inflammatory cytokines production, immune progenitor proliferation and reactive oxygen species production [68,69]. In other inflammatory situations a-adrenoceptors stimulation increases many pathological reactions. In systemic inflammation, a_1-adrenergic stimulation increases the release of TLR-driven cytokines from macrophages [70]. a_{1b} receptors also have shown ability to form heterodimeric complexes with chemokine receptors and regulate their activity [71,72].

Overall, acute catecholamines release reduce inflammation through β_2-adrenergic receptors. This is interesting, because inflammation is vital in normal wound healing [73,74]. Despite thousands of works it is still impossible to predict when inflammation becomes deleterious or when anti-inflammatory agents become harmful. For example, glucocorticoids have been shown to prevent wound healing or even to induce wounding in different situations [75,76]. On the other hand, in some wound models, glucocorticoids improve healing [77]. Similarly, β-AR affecting drugs are very inconsistent in different inflammation models and diseases [78–86]. In simple wounding model cell proliferation rate, neutrophiles and mast cells migration, myofibroblast density and the blood vessels volume density were increased by a beta-blocker, while healing was delayed overall [87]. To the contrary, after propranolol administration in streptozotocin-induced rat diabetes model, the wound area was smaller 7 and 14 days after wounding in propranolol group, and inflammatory cells number and MMP-9 level were reduced [88].

Therefore, we should study pro- and anti-inflammatory agents in more detail to find finer switching mechanisms. Because β_2-AR related effects are dominant, Pongratz et al. hypothesize that sympathetic system nerve endings prevent deleterious inflammation spreading and tissue damage. β-adrenergic receptors bind NE with lower affinity, than a-adrenergic receptors. Sympathetic nerve terminals release NE, and in proximity it binds with both β- and a-adrenergic receptors on leukocytes with subsequent β_2-AR induced IL-10 release. Farther from NE source, a high-affinity a-adrenergic receptor binds more mediator, than β-adrenergic receptors and TNFa release prevails over IL-10 production. Therefore, intact sympathetic fibers reduce inflammatory response in intact wound margins and increase closer to the wound's center. Possible repulsion of sympathetic fibers from the inflammation area would also positively modulate inflammation [35]. This concept only includes catecholamines, though other sympathetic transmitters in the skin also could be important for net inflammation balance.

As sudomotor sympathetic nerves release acetylcholine, they could trigger acetylcholine responses aside from vasodilation. As ubiquitous paracrine agent, acetylcholine is an important immune system mediator. Acetylcholine has been revealed as a pro-inflammatory mediator in many studies [89,90]. Leukocytes produce different types of cholinergic receptors: M_1–M_5 muscarinic receptors, a_3, a_5, a_7, a_9, a_{10} nAChR subunits [91,92]. Both cholinergic receptor types inhibit cytokine secretion from leukocytes, though mAChR are better-researched [93–95]. Considering that, acetylcholine release also could decrease inflammation in wound area. Contrary to catecholamines, ACh promotes vasodilation that stimulates leukocyte extravasation [53,96].

Vasoactive intestinal peptide (VIP) is characteristic for cholinergic neurons, including sympathetic sudomotor neurons. In the acute phase of inflammation, VIP induces histamine and bradykinin release from mast cells. Therefore, VIP itself and vasoactive inflammatory mediators induced by its actions promote vasodilatation in wound margins. In later stages, VIP could realize its anti-inflammatory properties. In different circumstances VIP could increase Treg cells level and protect them from apoptosis, inhibit TNFa and IL-6 secretion [97–99].

Neuropeptide Y (NPY) is produced in skin primarily by vasomotor sympathetic nerve endings. Its role in soft tissue wound healing is not completely understood. In other pathologies NPY acts as pro-inflammatory agent [100–103]. NPY induces cytokine production in leukocytes through Y1 and Y5 receptors, but other types of NPY receptors with other roles also were scarcely described [102,104].

Among typical skin sympathetic neurotransmitters, only NPY can possibly act as a pro-inflammatory without immunosuppressive functions in normal wound healing. To focus inflammation near the wound's margins immune cells need to repel nearby sympathetic fibers. Indeed, inflammatory sympathetic repulsion is a well-known phenomenon. A special class of semaphorin molecules called nerve repellent factors regulates neurite outgrowth. They include semaphorin 3F(SEMA3F), plexin-A2, neuropilin-2 and other factors [105]. In inflamed tissues, including diabetic Charcot foot, semaphorin 3C is highly expressed with lower sympathetic fibers' density [106]. S. Clatt et al. tested whether cytokines and hormonal factors released in inflamed tissue also have repellent properties. TNF-α repelled nerve fibers with moderate to strong effects (0–100%). High concentrations of dopamine and norepinephrine (10^{-6} M) induced weak but significant nerve fiber repulsion (up to 20%). Stimulation with low concentrations of 17β-estradiol (10^{-10} M, but not 10^{-8} M) repelled SNFs [107]. Systemic inflammatory responses in sepsis also induced nerve repulsion in primary immune organs. D. Hoover et al. have found that in patients with sepsis there are about 16% of normal sympathetic fibers. While spleen innervation provides ambiguous immune regulation, it's dysfunction could both decrease inflammation specificity and increase detrimental effects [108]. Major aspects of sympathetic neurotransmitters' interaction with immune cells were placed in Table 1.

Table 1. Overview of sympathetic neurotransmitter interactions with immune cells.

Neurotransmitters	**Norepinephrine/ Epinephrine**	**Acetylcholine**	**VIP**	**NPY**
Primary effect	Immunosuppressive	Immunosuppressive	Immunosuppressive	Pro-inflammatory
Source	Vasomotor fibers/keratinocytes	Sudomotor fibers/keratinocytes	Sudomotor fibers/keratinocytes	Vasomotor fibers/keratinocytes
Wound healing role—inflammation	Low concentration in wound area, high in healthy tissue around	Not clear	Not clear	Not clear
Primary receptor blockade	Switch to pro-inflammatory, better healing or hyperinflammation	Inflammation increase, better healing?	Inflammation increase, better healing?	Studied in some hyperinflammatory conditions only

4.2. Keratinocytes

Keratinocytes (KC) are activated after wounding among other cells through the first minutes and hours. Keratinocytes are characterized by keratin expression profile, which defines the KC phenotype. In healthy skin there are basal KC, that proliferate and replenish lost corneocytes through different maturation stages. After wounding basal KC and some mature KC switch their phenotype to activated or contractible. Activated KC can migrate, proliferate and are crucial for re-epithelialization. Contractible KC pull the extracellular matrix to make wound area smaller. Without any kind of epidermis, the wound has a great risk of becoming infected or chronic [109,110].

Because intact epidermis cells synthetize NE, they could affect each other and prevent inflammatory activation. In vitro β_2-adrenergic receptors activation inhibits keratinocyte migration [111,112]. Epinephrine is a more potent keratinocyte migration inhibitor than norepinephrine [113]. Cellular events after keratinocytes β_2-AR activation include serine/threonine phosphatase PP2A activation, extracellular signal-related kinase (ERK) dephosphorylation and promigratory signaling cascade blockade [114]. Keratinocyte proliferation is also inhibited by β_2-AR agonist isoproterenol [115].

Unlike in immune cells, keratinocytes β_2-AR contribute to pro-inflammatory cellular response as well. Epinephrine increases interleukin production, and β_2-AR—TLR crosstalk significantly augments inflammatory response [116]. Because epidermis is always in contact with the outer world and its germs, cytokines could be important in maintaining skin defenses. While β_2-AR blocks KC activation, an increased cytokines level possibly could overcome this effect [117]. For now, this question remains open.

Wound modeling in keratinocytes culture leads to rapid norepinephrine release and persistent downregulation of β_2-AR protein and catecholamine synthesis enzymes gene expression [118]. Keratinocyte culture scratch wounding downregulated expression of β-adrenoceptors genes, tyrosine hydroxylase and PNMT genes. While β_2-adrenoceptors functional activity remained depressed, their gene expression returned to the baseline. With decreased β_2-AR stimulation, keratinocytes produced more norepinephrine, which impaired their migration activity in wound edges. Effects were diminished by β_2-AR selective antagonist ICI-118,551, β_1-AR selective antagonist bisoprolol did not change them [118]. R. Sivamani et al. in vitro and in mouse burn model have found detrimental role of β_2-AR after epinephrine exposure. In vitro keratinocytes blockade was achieved by physiological stress-induced epinephrine concentrations [119].

As with immune cells, alpha-adrenoceptors have opposite effects after KC activation. In studies, a_{2A}/a_{2C}-adrenoceptors knockout transgenic mice have shown accelerated wound contraction and re-epithelialization. On the other hand, a_2-adrenoceptors on the presynaptic membrane reduce catecholamine release, therefore external a_2-AR sympathetic activation could improve wound healing through inhibited NE release and lower β_2-AR activation [120]. In vitro a_2-ARs increase keratinocyte migration. Under low norepinephrine concentration a_2-ARs overcome β_2-adrenoceptors and their stimulation induces rapid migration [121]. Probably, a-adrenoceptors activation prevents KC from switching back to a stable basal or mature phenotype.

Acetylcholine also is abundant in epidermis and can act on the keratinocytes directly via cell receptors. Muscarinic receptors of five molecular subtypes and nicotinic receptors were found in keratinocytes and melanocytes [33]. Combined acetylcholine receptors blockade in vitro leads to the complete organotypic skin culture growth and proliferation inhibition. The nAChR receptors blockage led to less prominent changes than did the mAChR blockage in terms of culture thickness and maturation marker genes' expression [122,123]. Important acetylcholine function is a keratinocytes cohesion stimulation [53,124]. Like NE, ACh also increases cytokine synthesis in keratinocytes. We propose that, by the same logic, cytokine stimulation can compensate direct inhibition [125]. In some reports M_3 mAChR activation inhibits KC migration, while M_4 mAChR activates migration [54]. In vitro experiments establish that a_9 nAChR is important for migration start; without it KC remained attached to the surface. A huge number of receptors with opposite effects reduced the chances to successfully target cholinergic system in wound healing [34,126]. Neuropeptides' interactions with KC are less researched. VIP induces keratinocytes migration and proliferation, and probably it is one of the most promising targets for study [127–129]. NPY receptors Y1 and Y4 were detected in all epidermal layers of the human skin [130]. Interestingly, while CGRP and VIP activate cAMP in keratinocytes culture, leading to increased cell proliferation, NPY downregulates cAMP with the opposite effects. It is probable that NPY blockers also could have some useful implications [131]. Major aspects of sympathetic neurotransmitters' interaction with keratinocytes were placed in Table 2.

Table 2. Overview of sympathetic neurotransmitter interactions with keratinocytes.

Neurotransmitters	Norepinephrine/ Epinephrine	Acetylcholine	VIP	NPY
Primary effect	Inhibits activation, stimulates cytokines	Mixed through different receptors, stimulates cytokines	Activates KC	Inhibits activation
Source	Vasomotor fibers/keratinocytes	Sudomotor fibers/keratinocytes	Sudomotor fibers/keratinocytes	Vasomotor fibers/keratinocytes
Wound healing role—re-epithelialization	Stimulates KC to produce cytokines, probably to stop migration	Stimulates KC to produce cytokines, to start migration	Stimulates KC	Inhibits KC
Primary receptor blockade	Ambiguous data	Ambiguous data	Worse healing?	Better healing?

4.3. Fibroblasts

In later stages of the wound healing process, fibroblasts became key cellular elements. Their growth factors terminate inflammation, and their work defines the degree of scarring and functional restoration [132,133].

Fibroblasts produce almost all types of adrenergic and cholinergic receptors. In proliferation phase, β_2-receptors become more beneficent, as they activate fibroblasts. In zebrafishes and porcine skin wound model β_2-AR agonists inhibit contraction and fibrosis, reduce scar area, and improve scar quality [134]. β_2-AR activates fibroblasts migration and attenuates cAMP-dependent matrix contraction [135,136]. After the beta-adrenoceptors blockade wound contraction and epidermal healing were delayed, decreased hydroxyproline levels, collagen density and neo-epidermal thickness were in evidence [87]. After propranolol administration in streptozotocin-induced rat diabetes model, the wound area was smaller 7 and 14 days after wounding in propranolol group; MMP-9 level was reduced and cell proliferation, mast cells number, collagen deposition, blood vessels density and nitric oxide levels were increased [88]. In Pullar et al. work, β_2-AR antagonism increased angiogenesis, fibroblast functions, re-epithelialization [137]. Therefore, real data is inconsistent and more high-quality research is needed.

Less data is available about alpha-adrenoceptors, but in most cases, they have shown pro-inflammatory properties. Alpha-adrenergic agonists increase pro-inflammatory cytokines production, immune progenitor proliferation and reactive oxygen species production, as well as TGFb synthesis [68,69]. Dermal fibroblasts also express several acetylcholine receptors: a_3b_2, a_5, a_7, a_9 nAChRs, M_2, M_4, and M_5 mAChRs [53,124]. Cellular effects were not properly studied, but mostly, acetylcholine receptors activation promotes matrix formation or remodeling [138].

4.4. Blood Vessel Cells

Finally, fast, and functional restoration requires increased blood supply. All sympathetic neurotransmitters and neuropeptides affect angiogenesis. As this topic is described in many reviews, we briefly discuss several points [9,29,139–142].

Both in murine wound models and in human skin wounds, β_2-AR activation prevents phospho-ERK cytoskeleton remodeling and delays wound re-epithelialization and healing [143]. Interestingly, the same effects could be seen in vascular smooth muscle cell culture [144,145]. β_2-AR activation also decreases angiogenesis, and endothelial cells' migration via cAMP-dependent mechanisms [146]. It is possible that a_1-AR gene overexpression in vascular cells could lead to altered circulatory dynamics. Indirectly, these alterations could contribute to dysfunctional keloid scars that maintain high a_1-AR production [147].

High SNS activity leads to stable NPY increase, hence vascular tone is permanently elevated, like in arterial hypertension. Diabetes mellitus is often accompanied by lower NPY production in skin [148]. NPY released by sympathetic nerve fibers stimulates endothelial cells proliferation and migration [149,150]. NPY Y2 receptors' deletion in mice delays the wound healing by an angiogenesis blockade [148].

5. Wound Healing in Denervation Models

Experiments with denervated animals revealed the significant role of nerve endings in wound healing, though results are not always consistent. Partial surgical denervation delays wound healing in some studies and does not affect healing process in others. Complete denervation worsens skin regeneration in almost all studies [23]. As an exception, Ranne et al. did not find histological changes in rat groin skin after surgical denervation, though there could be functional impairments [150]. Further we will compare denervation experiments with information about different parts of sympathetic interactions with wounds.

Peripheral sympathetic denervation is executed with high dose 6-hydroxydopamine (6-OHDA). K. Saburo et al. described morphological changes in rat burns healing after 6-OHDA sympathectomy. There were fewer capillaries in granulation tissue, vessels were dilated, and the collagen was fine fibrous, rather than the thick bundles of the control group [151]. Probably, these results go in line with NE deficiency. L. Kim et al. reported decreased linear skin incisions healing in rats after 6-OHDA sympathectomy (two-times more rats in the control group were healed by day 14). Functional sympathetic blockade in rats with propranolol (beta-blocker) and phentolamine (alpha-blocker) reduced wound contraction and re-epithelialization but increased total cell proliferation, possibly by inflammatory cells [152]. It is interesting because catecholamine receptor blockade should prevent KC inhibition. Probably, as we mentioned before, NE, importantly, increases cytokine production. In previous work authors have found that general sympathetic denervation with 6-OHDA accelerates wound contraction in rats but delays epidermal restoration [153]. Z. Zheng et al. in similar experiment confirmed increased wound contraction and delayed reepithelization and have shown decreased levels of norepinephrine, epidermal growth factor, IL1-beta, NG2 proteoglycan and desmin [154]. Interestingly, A. Jurjus et al. reported other results in rat burn model with guanethidine denervation, that induce postganglionic neurons depletion. Wound surface area was reduced faster after sympathectomy, contrary to decreased healing rate in 6-OHDA models [155]. One experiment has shown that exogenous local SNS activation with low doses of 6-OHDA increased epidermal wound healing by 35% and dermal strength was increased by 43% [156].

Autonomic response could be predicted by comparing results of total and sympathetic denervation. Sensory denervation with capsaicin has a wide range of results [23]. In immature rat capsaicin treatment delayed cutaneous wound healing with increased proliferation and decreased apoptosis [157]. In some works, selective capsaicin sensory denervation had no impact on wound healing [158]. J. Wallengren et al. studied both capsaicin and surgical denervation wound healing models with negative results. With 70% nerve fibers depletion, the researchers found normal wound closure. Additionally, by day 10 they reported nerve fiber density recovery [159].

6. Sympathetic System in Diabetic Wounds Regeneration—Beneficial or Deleterious?

Wound healing is severely compromised in diabetes. There are multiple factors that decrease skin reparation capability in the diabetic condition. Diabetes mellitus is often accompanied by a diabetic neuropathy (DN). In the skin, DN leads to reduced nerve fiber density, neuropeptides deficiency and failed nerve regeneration. C. Chan et al. have shown that delayed wound healing in diabetic mice is related to 30–50% fewer axons at the wound margins, compared to 10–15% fewer axons in the skin before the wounding. Moreover, wound healing was accompanied by 70% adjacent hair follicles innervation reduction and worse axon plasticity [160]. D. Levy et al. described reduced CGRP+, SP+, VIP+ and NPY+ nerve fibers density in the skin of diabetic patients [161].

One of the major diabetic neuropathy forms is autonomic neuropathy. This could affect any part of the autonomic nervous system, though the most obvious clinical signs include dysmotility, urine incontinence and arrhythmias. The majority of thematic research has been focused on the sensory component of neuropathy in diabetic ulcer pathogenesis. But sensory neuropathy becomes prominent only after severe neuronal loss [162]. Despite normal basal line activity, stress tests could reveal the sympathetic system decrease much earlier. In type I diabetic subjects, exercises led to smaller and shorter norepinephrine and NPY elevation in blood [163]. Sensory neuropathy also could indirectly trigger altered sympathetic activity. Painful neuropathy is always accompanied by somatic changes, and mediated by sympathetic hyperactivation [164]. Also, while much information is transferred by C-type nerve fibers, both sympathetic and sensory nerve systems often suffer simultaneously. Perception and sympathetic skin response impairment are correlated in patients with diabetic polyneuropathies [165].

There could be different skin changes in diabetic patients, though in most cases all regulatory and trophic functions are reduced. Skin sympathetic reaction to intra-epidermal electrical stimulation correlates with diabetes severity, independent of clinical signs of neuropathy [166]. Sudomotor dysfunction is described as a high-sensitivity screening test to diagnose diabetes complications [167]. Diabetic neuropathy is accompanied by a_1-AR gene overexpression in cutaneous vessels. Possibly, denervated structures overexpress a_2-adrenoceptors and become more sensitive to other catecholamine sources [168]. Diabetes severely compromises microcirculation. Sympathetic fibers deficiency independently increases capillary leakage and reduces vasomotion. Indeed, the microcirculation state is dependent on both vessel wall thickness and local blood flow regulation [58].

Z. Zheng et al. studied the effects of chemical denervation with 6-OHDA in diabetic mice wounds. Sympathectomy was accompanied by smaller wound areas. To the contrary, histological regeneration scores were reduced in 6-OHDA group on day 21. Sympathectomy decreased the number of mast cells and norepinephrine, epidermal growth factor (EGF), interleukin-1 beta, NG2 proteoglycan, and desmin gene expression on the days 17 and 21t. Also, pericyte proliferation was reduced, which could explain vascular dysfunction. The authors concluded that 6-OHDA in diabetic mice delays re-epithelialization during wound healing by decreasing EGF but increases wound contraction by reducing IL-1β levels and the number of mast cells [154].

Diabetes is accompanied by a hyperinflammatory state. This is especially important in obesity-related conditions. White adipose tissue is a major source of inflammatory cytokines: IL-1, TNF-a, IL-6, chemokines etc. Dyslipidemia and insulin resistance have direct impacts on leukocytes and promote their activation. High glucose levels in blood also directly stimulate inflammation. Acute hyperglycemia causes acute IL-1a, IL-4 and IL-6 elevation. Even in healthy individuals, transient hyperglycemia induces major inflammatory transcription factor NF-kb gene expression. Reactive oxygen species are overproduced by leukocytes, and therefore many processes are compromised by oxidation [169,170].

Like other hyperinflammatory states, diabetes diminishes antimicrobial defenses. In the steady-state the immune system is always hyperactive, but its real defense power is low. Many studies report signs of pathological inflammation in diabetic ulcers [171–173]. Neutrophils have lower phagocytic activity index and intracellular killing activity. In early stages of wounding leukocyte infiltration could be decreased, but in chronic state neutrophiles and macrophages persist in wound margins. Finally, granulation tissue is produced slower and ulcer re-epithelialization is worse [174].

7. Targeting Sympathetic Nervous System in Diabetic Ulcers

SNS-targeting drugs were studied in several works to improve wound healing in different pathological conditions, but few of them were clinical trials. Timolol is one of the most extensively studied β-blockers for wound healing. H. Albrecht et al. studied combined thermal and radiation wounds in isolated human skin flaps. Beta-blocker timolol increased wound epithelialization by 5–20% [175].

The first clinical trials of systemic timolol administration refer to burn healing, though systemic effects in burns are distinct from regular wounds [176]. Topical timolol was studied in a couple of case studies with different nosology and in a few small clinical trials [177–179]. B. Thomas et al. conducted case-control study of topical timolol in chronic leg ulcers of diabetic or venous origin. They have found significant changes in mean wound area at 4, 8 and 12 weeks of complex treatment. Wound closure was almost two times as fast in the timolol group in all three checkpoints. Diabetic wounds and venous ulcer have shown similar response to treatment [180]. A. Ghanbarzamani et al. report positive results treating 64 patients' graft sites after burn treatment [181]. T. Baltazard et al. treated 40 patients with venous leg ulcers with timolol gel or control. Twice as many patients in the timolol group demonstrated high ulcer area reduction [182]. For now, there are no completed clinical studies examining, with a clear design, timolol in diabetic wound healing. R. Caur et al. registered a phase two randomized clinical trial to test timolol in diabetic ulcers treatment. As of 2022, there are 30 patients enrolled, and the study could be completed in 2024 [183].

Topical muscarinic antagonists have shown preclinical effectiveness in reducing diabetic neuropathy symptoms, possibly through interaction with sympathetic nerve membranes. C.G. Jolival et al. reported results of 2% pirenzepine treatment in mice model of streptozotocin-induced diabetes. Topical delivery dose-dependently prevented tactile allodynia, thermal hypoalgesia and loss of epidermal nerve fibers. Drug withdrawal or frequency reduction reversed the effects, so muscarinic antagonists provide local symptomatic relief to diabetic neuropathy [184]. As far as we know, muscarinic antagonists were not tested in wound healing. There are few reports about nAChR agonists and other cholinergic agents in diabetic wounds healing. R.S. Dillon reported slower wound healing in six patients with diabetic neuropathy after atropine application and faster healing after methacholine (muscarinic agonist) application [185].

M. Kishibe et al. have studied topical nAChR agonists in diabetic mouse wounds. They have found diminished bacterial survival and systemic dissemination, as well as reduced wound TLR2 production. Reporting a_7 nAChR impaired synthesis in human diabetic skin, authors proposed that a reduced cholinergic response accounted for the excessive inflammation [186]. M. Dong et al. have found that selective a_7 nAChR agonist PNU282987 inhibits AGE-induced NF-kB activation and RAGE gene expression in diabetic wound macrophages. A reduced TNF-a level in diabetic mice was accompanied by accelerated healing rate, elevated fibroblast number and collagen deposition [187]. J-Y. Li et al. further discovered the same results in uncovered (non-diabetic) wounds. In wounds, covered with dressing, results were the opposite. nAChR7 activator inhibited re-epithelialization, angiogenesis and epithelial proliferation. Possibly, the effects are opposite because inflammatory processes are very different in covered and uncovered wounds [188]. Nicotine itself in low doses topically has shown promising results in accelerating healing of different wound types by angiogenesis promotion [140,189–191].

8. Conclusions

The sympathetic nervous system regulates skin homeostasis in different ways and at multiple levels. Therefore, we tried to classify the existing knowledge in this topic. We propose that there are three main ways SNS could interact with skin and wounds.

Firstly, vasomotor nerves suppress keratinocyte migration and proliferation, reducing inflammation by catecholamines through β_2-adrenergic receptors. They are activated only by high norepinephrine concentrations; therefore, effects are prominent near vasoconstrictor nerve endings. NPY released near the vessels has similar effects and prevents inflammation and proliferation. Anti-inflammatory effects could be deleterious in the early stages of normal wound healing and skin cells in wound area could repel nerve endings. Intact skin vasoconstrictor nerve endings have a protective function and NPY also stimulates angiogenesis, which is important for distant wound margin. In the reparation phase, semaphorin concentration should fall, and regrowing nerve endings stimulate fibroblasts activity and matrix deposition through β_2-AR.

Secondly, the same norepinephrine from distant nerve endings could stimulate cells in active wound centers through a_1- and a_2-adrenergic receptors. They display prominent pro-inflammatory properties and induce an activated keratinocytes phenotype. This pathway could be dangerous for patients with neuropathies, because a-adrenoceptors diminish blood flow and inhibit angiogenesis. Therefore, phase transition could be compromised and hyperinflammation could support chronic wounds. a-adrenoceptors also inhibit fibroblast activity and delay functional restoration.

The third way is sudomotor activation by acetylcholine and VIP. Simultaneously, they increase blood flow, induce angiogenesis, reduce inflammation, and stimulate keratinocytes migration and proliferation. Some of these effects are mixed by minor types of cholinergic receptors with opposite functions. Unfortunately, it is not clear when and where sudomotor neurons take part in wound healing. We can speculate that they could be beneficial in the proliferation and reparation phases. Probably, they are also repelled in acute inflammation phase and do not have distant mode of action.

With this concept we could predict severe healing dysregulation after sympathetic denervation. Lack of alpha-adrenergic stimulation will impair keratinocytes' migration and decrease the healing rate. Early selective beta-adrenergic blockade could improve fast re-epithelialization but could be detrimental later. Experiments with sympathetic denervation generally lay in line with this hypothesis. Re-epithelialization is delayed, and the wound healing rate is lower, while inflammation is increased after 6-OHDA denervation. Functional beta-blockade impairs normal wound healing in some works, improves it in others, while alpha-blockade always improves it. Because results do not go in the same direction with 6-OHDA experiments, we propose that the sudomotor nerve endings effects are missing as a major positive contribution. In diabetic wounds beta blockade and cholinergic stimulation provide the best results in improving wound healing. We propose that diabetic keratinocytes lack stimulation to the point that inhibitory β_2-adrenergic signals totally prevent re-epithelialization. While there are a few serious trials, we must be cautious in our interpretations and keep in mind that β_2-effects are controversial in later stages of wound reparation. It is probable that experiments with VIP or NPY could give us more hope in the future.

Author Contributions: Conceptualization: E.I. and S.G.; methodology, E.I.; software, A.E.; validation: A.E., M.A. and S.G.; formal analysis: E.I.; investigation: E.I.; resources; E.I., A.E., M.A. and S.G.; data curation, E.I. and A.E.; writing, original draft preparation, E.I., writing, review and editing: E.I., A.E., M.A. and S.G.; visualization, E.I. and S.G.; supervision, S.G.; project administration, E.I. All authors have read and agreed to the published version of the manuscript.

Funding: This research received no external funding.

Institutional Review Board Statement: Not applicable.

Informed Consent Statement: Not applicable.

Data Availability Statement: Not applicable.

Conflicts of Interest: The authors declare no conflict of interest.

References

1. Hicks, C.W.; Selvin, E. Epidemiology of Peripheral Neuropathy and Lower Extremity Disease in Diabetes. *Curr. Diabetes Rep.* **2019**, *19*, 86. [CrossRef] [PubMed]
2. Rekha, P.-D.; Rao, S.S.; Sahana, T.G.; Prabhu, A. Diabetic wound management. *Br. J. Community Nurs.* **2018**, *23*, S16–S22. [CrossRef] [PubMed]
3. Boulton, A.J.; Armstrong, D.G.; Kirsner, R.S.; Attinger, C.E.; Lavery, L.A.; Lipsky, B.A.; Mills, J.L., Sr.; Steinberg, J.S. Diagnosis and management of diabetic foot complications. *Diabetes* **2018**, *2018*, 1–20. [CrossRef] [PubMed]
4. Akkus, G.; Sert, M. Diabetic foot ulcers: A devastating complication of diabetes mellitus continues non-stop in spite of new medical treatment modalities. *World J. Diabetes* **2022**, *13*, 1106–1121. [CrossRef]
5. Ghotaslou, R.; Memar, M.Y.; Alizadeh, N. Classification, microbiology and treatment of diabetic foot infections. *J. Wound Care* **2018**, *27*, 434–441. [CrossRef] [PubMed]
6. Matoori, S.; Veves, A.; Mooney, D.J. Advanced bandages for diabetic wound healing. *Sci. Transl. Med.* **2021**, *13*, eabe4839. [CrossRef]
7. Jalilian, M.; Sarbarzeh, P.A.; Oubari, S. Factors Related to Severity of Diabetic Foot Ulcer: A Systematic Review. *Diabetes Metab. Syndr. Obes. Targets Ther.* **2020**, *13*, 1835–1842. [CrossRef]
8. Wang, X.; Yuan, C.-X.; Xu, B.; Yu, Z. Diabetic foot ulcers: Classification, risk factors and management. *World J. Diabetes* **2022**, *13*, 1049–1065. [CrossRef]
9. Okonkwo, U.A.; DiPietro, L.A. Diabetes and Wound Angiogenesis. *Int. J. Mol. Sci.* **2017**, *18*, 1419. [CrossRef] [PubMed]
10. Mieczkowski, M.; Mrozikiewicz-Rakowska, B.; Kowara, M.; Kleibert, M.; Czupryniak, L. The Problem of Wound Healing in Diabetes—From Molecular Pathways to the Design of an Animal Model. *Int. J. Mol. Sci.* **2022**, *23*, 7930. [CrossRef]
11. Urso, B.; Ghias, M.; John, A.; Khachemoune, A. Neuropathic ulcers: A focused review. *Int. J. Dermatol.* **2020**, *60*, e383–e389. [CrossRef]
12. Rathur, H.M.; Boulton, A.J.M. The neuropathic diabetic foot. *Nat. Clin. Pract. Endocrinol. Metab.* **2007**, *3*, 14–25. [CrossRef] [PubMed]
13. Aldana, P.C.; Cartron, A.M.; Khachemoune, A. Reappraising Diabetic Foot Ulcers: A Focus on Mechanisms of Ulceration and Clinical Evaluation. *Int. J. Low. Extremity Wounds* **2020**, *21*, 294–302. [CrossRef] [PubMed]
14. Boulton, A.J.M. Diabetic Neuropathy and Foot Complications. In *Handbook of Clinical Neurology*; Elsevier B.V.: Amsterdam, The Netherlands, 2014; Volume 126, pp. 97–107.

15. Guo, Q.; Ying, G.; Jing, O.; Zhang, Y.; Liu, Y.; Deng, M.; Long, S. Influencing factors for the recurrence of diabetic foot ulcers: A meta-analysis. *Int. Wound J.* **2022**. [CrossRef]
16. Stachura, A.; Khanna, I.; Krysiak, P.; Paskal, W.; Włodarski, P. Wound Healing Impairment in Type 2 Diabetes Model of Leptin-Deficient Mice—A Mechanistic Systematic Review. *Int. J. Mol. Sci.* **2022**, *23*, 8621. [CrossRef] [PubMed]
17. Stino, A.M.; Smith, A.G. Peripheral neuropathy in prediabetes and the metabolic syndrome. *J. Diabetes Investig.* **2017**, *8*, 646–655. [CrossRef]
18. Noor, S.; Zubair, M.; Ahmad, J. Diabetic foot ulcer—A review on pathophysiology, classification and microbial etiology. *Diabetes Metab. Syndr. Clin. Res. Rev.* **2015**, *9*, 192–199. [CrossRef]
19. Deng, L.; Du, C.; Song, P.; Chen, T.; Rui, S.; Armstrong, D.G.; Deng, W. The Role of Oxidative Stress and Antioxidants in Diabetic Wound Healing. *Oxidative Med. Cell. Longev.* **2021**, *2021*, 8852759. [CrossRef]
20. Steeper, R. A critical review of the aetiology of diabetic neuropathic ulcers. *J. Wound Care* **2005**, *14*, 101–103. [CrossRef]
21. Schaper, N.C.; Huijberts, M.; Pickwell, K. Neurovascular control and neurogenic inflammation in diabetes. *Diabetes/Metab. Res. Rev.* **2008**, *24*, S40–S44. [CrossRef]
22. Pradhan, L.; Nabzdyk, C.; Andersen, N.D.; LoGerfo, F.W.; Veves, A. Inflammation and neuropeptides: The connection in diabetic wound healing. *Expert Rev. Mol. Med.* **2009**, *11*, e2. [CrossRef] [PubMed]
23. Barker, A.R.; Rosson, G.; Dellon, A.L. Wound Healing in Denervated Tissue. *Ann. Plast. Surg.* **2006**, *57*, 339–342. [CrossRef] [PubMed]
24. Theocharidis, G.; Veves, A. Autonomic Nerve Dysfunction and Impaired Diabetic Wound Healing: The Role of Neuropep-tides. *Auton. Neurosci.* **2020**, *223*, 102610. [CrossRef] [PubMed]
25. Anand, P. Neurotrophic factors and their receptors in human sensory neuropathies. *Prog. Brain Res.* **2004**, *146*, 477–492. [CrossRef] [PubMed]
26. Da Silva, L.; Carvalho, E.; Cruz, M.T. Role of neuropeptides in skin inflammation and its involvement in diabetic wound healing. *Expert Opin. Biol. Ther.* **2010**, *10*, 1427–1439. [CrossRef]
27. Sun, S.; Ma, J.; Ran, X. Mechanisms of Adrenergic β-Antagonist for Wounds and Its Application Prospect in Diabetic Foot Ulcers. *Zhongguo Xiu Fu Chong Jian Wai Ke Za Zhi* **2020**, *34*, 1630–1634. [CrossRef]
28. Nowak, N.C.; Menichella, D.M.; Miller, R.; Paller, A.S. Cutaneous innervation in impaired diabetic wound healing. *Transl. Res.* **2021**, *236*, 87–108. [CrossRef]
29. Pan, L.; Tang, J.; Liu, H.; Cheng, B. Sympathetic nerves: How do they affect angiogenesis, particularly during wound healing of soft tissues? *Clin. Hemorheol. Microcirc.* **2016**, *62*, 181–191. [CrossRef]
30. Ashrafi, M.; Baguneid, M.; Bayat, A. The Role of Neuromediators and Innervation in Cutaneous Wound Healing. *Acta Dermato-Venereol.* **2016**, *96*, 587–597. [CrossRef]
31. Gupta, D.; Kaushik, D.; Mohan, V. Role of neurotransmitters in the regulation of cutaneous wound healing. *Exp. Brain Res.* **2022**, *240*, 1649–1659. [CrossRef]
32. Slominski, A.T.; Slominski, R.M.; Raman, C.; Chen, J.Y.; Athar, M.; Elmets, C. Neuroendocrine signaling in the skin with a special focus on the epidermal neuropeptides. *Am. J. Physiol. Physiol.* **2022**, *323*, C1757–C1776. [CrossRef] [PubMed]
33. Grando, S.A.; Pittelkow, M.R.; Schallreuter, K.U. Adrenergic and Cholinergic Control in the Biology of Epidermis: Physiological and Clinical Significance. *J. Investig. Dermatol.* **2006**, *126*, 1948–1965. [CrossRef] [PubMed]
34. Kurzen, H.; Schallreuter, K.U. Novel aspects in cutaneous biology of acetylcholine synthesis and acetylcholine receptors. *Exp. Dermatol.* **2004**, *13*, 27–30. [CrossRef]
35. Pongratz, G.; Straub, R.H. The sympathetic nervous response in inflammation. *Arthritis Res. Ther.* **2014**, *16*, 504. [CrossRef]
36. Vinik, A.I.; Maser, R.E.; Mitchell, B.D.; Freeman, R. Diabetic Autonomic Neuropathy. *Diabetes Care* **2003**, *26*, 1553–1579. [CrossRef] [PubMed]
37. Sharma, J.K.; Rohatgi, A.; Sharma, D. Diabetic Autonomic Neuropathy: A Clinical Update. *J. R. Coll. Physicians Edinb.* **2020**, *50*, 269–273. [CrossRef] [PubMed]
38. Basra, R.; Papanas, N.; Farrow, F.; Karalliedde, J.; Vas, P. Diabetic Foot Ulcers and Cardiac Autonomic Neuropathy. *Clin. Ther.* **2022**, *44*, 323–330. [CrossRef]
39. Anderson, C.; Bergner, A.; Murphy, S. How many types of cholinergic sympathetic neuron are there in the rat stellate ganglion? *Neuroscience* **2006**, *140*, 567–576. [CrossRef]
40. Macefield, V.G. Sympathetic Microneurography. In *Handbook of Clinical Neurology*; Elsevier B.V.: Amsterdam, The Netherlands, 2013; Volume 117, pp. 353–364.
41. Schmelz, M.; Schmidt, R.; Bickel, A.; Torebjörk, H.E.; Handwerker, H.O. Innervation territories of single sympathetic C fibers in human skin. *J. Neurophysiol.* **1998**, *79*, 1653–1660. [CrossRef]
42. Donadio, V.; Incensi, A.; Vacchiano, V.; Infante, R.; Magnani, M.; Liguori, R. The autonomic innervation of hairy skin in humans: An in vivo confocal study. *Sci. Rep.* **2019**, *9*, 16982. [CrossRef]
43. von Duuring, M.; Fricke, B. Organization of Peripheral Nerves in Skin, Musculoskeletal System and Viscera. In *Neuronal Activity in Tumor Tissue*; KARGER: Basel, Switzerland, 2007; Volume 39, pp. 30–44.
44. Benarroch, E.E. Neuropeptides in the sympathetic system: Presence, plasticity, modulation, and implications. *Ann. Neurol.* **1994**, *36*, 6–13. [CrossRef] [PubMed]

45. Li, X.; Mao, Z.; Yang, L.; Sun, K. Co-staining Blood Vessels and Nerve Fibers in Adipose Tissue. *J. Vis. Exp.* **2019**, *144*, e59266. [CrossRef] [PubMed]
46. Bartness, T.J.; Ryu, V. Neural control of white, beige and brown adipocytes. *Int. J. Obes. Suppl.* **2015**, *5*, S35–S39. [CrossRef] [PubMed]
47. Katayama, K.; Saito, M. Muscle sympathetic nerve activity during exercise. *J. Physiol. Sci.* **2019**, *69*, 589–598. [CrossRef] [PubMed]
48. Di Bona, A.; Vita, V.; Costantini, I.; Zaglia, T. Towards a clearer view of sympathetic innervation of cardiac and skeletal muscles. *Prog. Biophys. Mol. Biol.* **2020**, *154*, 80–93. [CrossRef]
49. Grando, S.A. Cholinergic control of epidermal cohesion. *Exp. Dermatol.* **2006**, *15*, 265–282. [CrossRef]
50. Pullar, C.E.; Rizzo, A.; Isseroff, R.R. β-Adrenergic receptor antagonists accelerate skin wound healing: Evidence for a catecholamine synthesis network in the epidermis. *J. Biol. Chem.* **2006**, *281*, 21225–21235. [CrossRef]
51. Steinkraus, V.; Mak, J.C.W.; Pichlmeier, U.; Mensing, H.; Ring, J.; Barnes, P.J. Autoradiographic mapping of beta-adrenoceptors in human skin. *Arch. Dermatol. Res.* **1996**, *288*, 549–553. [CrossRef]
52. Sivamani, R.K.; Lam, S.T.; Isseroff, R.R. Beta Adrenergic Receptors in Keratinocytes. *Dermatol. Clin.* **2007**, *25*, 643–653. [CrossRef]
53. Kurzen, H.; Wessler, I.; Kirkpatrick, C.J.; Kawashima, K.; Grando, S.A. The Non-neuronal Cholinergic System of Human Skin. *Horm. Metab. Res.* **2007**, *39*, 125–135. [CrossRef]
54. Sivamani, R.K.; Garcia, M.S.; Rivkah Isseroff, R. Wound Re-Epithelialization: Modulating Keratinocyte Migration in Wound Healing. *Front. Biosci.* **2007**, *12*, 2849–2868.
55. Suh, D.Y.; Hunt, T.K. Time line of wound healing. *Clin. Podiatr. Med. Surg.* **1998**, *15*, 1–9. [PubMed]
56. Engel, D. The influence of the sympathetic nervous system on capillary permeability. *Res. Exp. Med.* **1978**, *173*, 1–8. [CrossRef] [PubMed]
57. Coderre, T.J.; Basbaum, A.I.; Levine, J.D. Neural control of vascular permeability: Interactions between primary afferents, mast cells, and sympathetic efferents. *J. Neurophysiol.* **1989**, *62*, 48–58. [CrossRef]
58. Lefrandt, J.D.; Bosma, E.; Oomen, P.H.N.; van der Hoeven, J.H.; van Roon, A.M.; Smit, A.J.; Hoogenberg, K. Sympathetic mediated vasomotion and skin capillary permeability in diabetic patients with peripheral neuropathy. *Diabetologia* **2003**, *46*, 40–47. [CrossRef]
59. Sulakvelidze, I.; Baluk, P.; McDonald, D.M. Plasma extravasation induced in rat trachea by 6-OHDA is mediated by sensory nerves, not by sympathetic nerves. *J. Appl. Physiol.* **1994**, *76*, 701–707. [CrossRef]
60. Rough, J.; Engdahl, R.; Opperman, K.; Yerrum, S.; Monroy, M.A.; Daly, J.M. β2 Adrenoreceptor blockade attenuates the hyperinflammatory response induced by traumatic injury. *Surgery* **2009**, *145*, 235–242. [CrossRef]
61. Beta-Adrenergic Receptors in Human Leukocyte Subpopulations–PubMed. Available online: https://pubmed.ncbi.nlm.nih.gov/1333965/ (accessed on 12 June 2022).
62. Scanzano, A.; Cosentino, M. Adrenergic regulation of innate immunity: A review. *Front. Pharmacol.* **2015**, *6*, 171. [CrossRef]
63. Galvan, D.L.; Danesh, F.R. β2-adrenergic receptors in inflammation and vascular complications of diabetes. *Kidney Int.* **2017**, *92*, 14–16. [CrossRef]
64. Ağaç, D.; Estrada, L.D.; Maples, R.; Hooper, L.V.; Farrar, J.D. The β2-adrenergic receptor controls inflammation by driving rapid IL-10 secretion. *Brain Behav. Immun.* **2018**, *74*, 176–185. [CrossRef]
65. Kolmus, K.; Tavernier, J.; Gerlo, S. β2-Adrenergic receptors in immunity and inflammation: Stressing NF-κB. *Brain Behav. Immun.* **2015**, *45*, 297–310. [CrossRef]
66. Gosain, A.; Gamelli, R.L.; DiPietro, L.A. Norepinephrine-Mediated Suppression of Phagocytosis by Wound Neutrophils. *J. Surg. Res.* **2009**, *152*, 311–318. [CrossRef] [PubMed]
67. Gosain, A.; Muthu, K.; Gamelli, R.L.; DiPietro, L.A. Norepinephrine suppresses wound macrophage phagocytic efficiency through alpha- and beta-adrenoreceptor dependent pathways. *Surgery* **2007**, *142*, 170–179. [CrossRef] [PubMed]
68. Miksa, M.; Das, P.; Zhou, M.; Wu, R.; Dong, W.; Ji, Y.; Goyert, S.; Ravikumar, T.S.; Wang, P. Pivotal Role of the α2A-Adrenoceptor in Producing Inflammation and Organ Injury in a Rat Model of Sepsis. *PLoS ONE* **2009**, *4*, e5504. [CrossRef]
69. Sharma, D.; Farrar, J.D. Adrenergic regulation of immune cell function and inflammation. *Semin. Immunopathol.* **2020**, *42*, 709–717. [CrossRef]
70. Grisanti, L.A.; Woster, A.P.; Dahlman, J.; Sauter, E.R.; Combs, C.K.; Porter, J.E. α_1-Adrenergic Receptors Positively Regulate Toll-Like Receptor Cytokine Production from Human Monocytes and Macrophages. *J. Pharmacol. Exp. Ther.* **2011**, *338*, 648–657. [CrossRef] [PubMed]
71. Enten, G.A.; Gao, X.; Strzelinski, H.R.; Weche, M.; Liggett, S.B.; Majetschak, M. $\alpha_{1B/D}$ -adrenoceptors regulate chemokine receptor–mediated leukocyte migration via formation of heteromeric receptor complexes. *Proc. Natl. Acad. Sci. USA* **2022**, *119*, e2123511119. [CrossRef]
72. Gao, X.; Albee, L.J.; Volkman, B.F.; Gaponenko, V.; Majetschak, M. Asymmetrical ligand-induced cross-regulation of chemokine (C-X-C motif) receptor 4 by α1-adrenergic receptors at the heteromeric receptor complex. *Sci. Rep.* **2018**, *8*, 2730. [CrossRef]
73. Hübnera, G.; Brauchlea, M.; Smolab, H.; Madlenera, M.; Fässlerc, R.; Wernera, S. Differential regulation of pro-inflammatory cytokines during wound healing in normal and glucocorticoid-treated mice. *Cytokine* **1996**, *8*, 548–556. [CrossRef]
74. Valls, M.D.; Cronstein, B.N.; Montesinos, M.C. Adenosine receptor agonists for promotion of dermal wound healing. *Biochem. Pharmacol.* **2009**, *77*, 1117–1124. [CrossRef]

75. Almeida, T.; Pires, T.D.C.; Monte-Alto-Costa, A. Blockade of glucocorticoid receptors improves cutaneous wound healing in stressed mice. *Exp. Biol. Med.* **2015**, *241*, 353–358. [CrossRef]
76. Nguyen, V.T.; Ngo, Q.T.; Ramirez, R.P.; Nakamura, T.; Farman, N.; Aractingi, S.; Jaisser, F. The myeloid mineralocorticoid receptor regulates dermal angiogenesis and inflammation in glucocorticoid-induced impaired wound healing. *Br. J. Pharmacol.* **2022**, *179*, 5222–5232. [CrossRef]
77. Tu, H.; Zhang, D.; Barksdale, A.N.; Wadman, M.C.; Muelleman, R.L.; Li, Y.-L. Dexamethasone Improves Wound Healing by Decreased Inflammation and Increased Vasculogenesis in Mouse Skin Frostbite Model. *Wilderness Environ. Med.* **2020**, *31*, 407–417. [CrossRef]
78. Durand, M.; Hagimont, E.P.; Louis, H.; Asfar, P.M.; Frippiat, J.-P.; Singer, M.M.; Gauchotte, G.M.; Labat, C.B.; Lacolley, P.M.; Levy, B.M.; et al. The β1-Adrenergic Receptor Contributes to Sepsis-Induced Immunosuppression Through Modulation of Regulatory T-Cell Inhibitory Function. *Crit. Care Med.* **2022**, *50*, e707–e718. [CrossRef]
79. Sitkauskiene, B. The Role of β2-Adrenergic Receptors in Inflammation and Allergy. *Curr. Drug Targets Inflamm. Allergy* **2005**, *4*, 157–162. [CrossRef]
80. Van Der Jagt, M.; Miranda, D.R. Beta-blockers in Intensive Care Medicine: Potential Benefit in Acute Brain Injury and Acute Respiratory Distress Syndrome. *Recent Patents Cardiovasc. Drug Discov.* **2012**, *7*, 141–151. [CrossRef]
81. Lira, A.; Pinsky, M.R. Should β-blockers be used in septic shock? *Crit. Care* **2014**, *18*, 304. [CrossRef]
82. Nguyen, L.P.; Omoluabi, O.; Parra, S.; Frieske, J.M.; Clement, C.; Ammar-Aouchiche, Z.; Ho, S.B.; Ehre, C.; Kesimer, M.; Knoll, B.J.; et al. Chronic Exposure to Beta-Blockers Attenuates Inflammation and Mucin Content in a Murine Asthma Model. *Am. J. Respir. Cell Mol. Biol.* **2008**, *38*, 256–262. [CrossRef]
83. Novotny, N.M.; Lahm, T.; Markel, T.A.; Crisostomo, P.R.; Wang, M.; Wang, Y.; Ray, R.; Tan, J.; Al-Azzawi, D.; Meldrum, D.R. β-Blockers in Sepsis: Reexamining the evidence. *Shock* **2009**, *31*, 113–119. [CrossRef]
84. Al-Kuraishy, H.M.; Al-Gareeb, A.I.; Mostafa-Hedeab, G.; Kasozi, K.I.; Zirintunda, G.; Aslam, A.; Allahyani, M.; Welburn, S.C.; Batiha, G.E.-S. Effects of β-Blockers on the Sympathetic and Cytokines Storms in Covid-19. *Front. Immunol.* **2021**, *12*. [CrossRef]
85. Hasegawa, D.; Sato, R.; Prasitlumkum, N.; Nishida, K.; Takahashi, K.; Yatabe, T.; Nishida, O. Effect of Ultrashort-Acting β-Blockers on Mortality in Patients With Sepsis With Persistent Tachycardia Despite Initial Resuscitation: A Systematic Review and Meta-Analysis of Randomized Controlled Trials. *Chest* **2021**, *159*, 2289–2300. [CrossRef]
86. Henriquez, A.R.; Snow, S.J.; Schladweiler, M.C.; Miller, C.N.; Dye, J.A.; Ledbetter, A.D.; Richards, J.E.; Mauge-Lewis, K.; McGee, M.A.; Kodavanti, U.P. Adrenergic and glucocorticoid receptor antagonists reduce ozone-induced lung injury and inflammation. *Toxicol. Appl. Pharmacol.* **2017**, *339*, 161–171. [CrossRef]
87. Souza, B.R.; Santos, J.S.; Costa, A.M. Blockade of beta1- and beta2-adrenoceptors delays wound contraction and re-epithelialization in rats. *Clin. Exp. Pharmacol. Physiol.* **2006**, *33*, 421–430. [CrossRef]
88. Romana-Souza, B.; Nascimento, A.P.; Monte-Alto-Costa, A. Propranolol improves cutaneous wound healing in streptozotocin-induced diabetic rats. *Eur. J. Pharmacol.* **2009**, *611*, 77–84. [CrossRef]
89. Pavlov, V.A.; Wang, H.; Czura, C.J.; Friedman, S.G.; Tracey, K.J. The Cholinergic Anti-inflammatory Pathway: A Missing Link in Neuroimmunomodulation. *Mol. Med.* **2003**, *9*, 125–134. [CrossRef]
90. Sato, E.; Koyama, S.; Okubo, Y.; Kubo, K.; Sekiguchi, M. Acetylcholine stimulates alveolar macrophages to release inflammatory cell chemotactic activity. *Am. J. Physiol. Cell. Mol. Physiol.* **1998**, *274*, L970–L979. [CrossRef]
91. Fujii, T.; Mashimo, M.; Moriwaki, Y.; Misawa, H.; Ono, S.; Horiguchi, K.; Kawashima, K. Expression and Function of the Cholinergic System in Immune Cells. *Front. Immunol.* **2017**, *8*, 1085. [CrossRef]
92. Koarai, A.; Traves, S.L.; Fenwick, P.S.; Brown, S.M.; Chana, K.K.; Russell, R.; Nicholson, A.G.; Barnes, P.J.; Donnelly, L.E. Expression of muscarinic receptors by human macrophages. *Eur. Respir. J.* **2011**, *39*, 698–704. [CrossRef]
93. Kawashima, K.; Fujii, T.; Moriwaki, Y.; Misawa, H. Critical roles of acetylcholine and the muscarinic and nicotinic acetylcholine receptors in the regulation of immune function. *Life Sci.* **2012**, *91*, 1027–1032. [CrossRef]
94. St-Pierre, S.; Jiang, W.; Roy, P.; Champigny, C.; Leblanc, E.; Morley, B.J.; Hao, J.; Simard, A.R. Nicotinic Acetylcholine Receptors Modulate Bone Marrow-Derived Pro-Inflammatory Monocyte Production and Survival. *PLoS ONE* **2016**, *11*, e0150230. [CrossRef]
95. Lu, J.; Wu, W. Cholinergic modulation of the immune system–A novel therapeutic target for myocardial inflammation. *Int. Immunopharmacol.* **2021**, *93*, 107391. [CrossRef] [PubMed]
96. Razani-Boroujerdi, S.; Singh, S.P.; Knall, C.; Hahn, F.F.; Peña-Philippides, J.C.; Kalra, R.; Langley, R.J.; Sopori, M.L. Chronic nicotine inhibits inflammation and promotes influenza infection. *Cell. Immunol.* **2004**, *230*, 1–9. [CrossRef]
97. Delgado, M.; Pozo, D.; Ganea, D. The Significance of Vasoactive Intestinal Peptide in Immunomodulation. *Pharmacol. Rev.* **2004**, *56*, 249–290. [CrossRef]
98. Gonzalez-Rey, E.; Chorny, A.; Fernandez-Martin, A.; Ganea, D.; Delgado, M. Vasoactive intestinal peptide generates human tolerogenic dendritic cells that induce CD4 and CD8 regulatory T cells. *Blood* **2006**, *107*, 3632–3638. [CrossRef]
99. Delgado, M.; Chorny, A.; Gonzalez-Rey, E.; Ganea, D. Vasoactive intestinal peptide generates CD4+CD25+ regulatory T cells in vivo. *J. Leukoc. Biol.* **2005**, *78*, 1327–1338. [CrossRef] [PubMed]
100. Dimitrijević, M.; Stanojević, S. The intriguing mission of neuropeptide Y in the immune system. *Amino Acids* **2011**, *45*, 41–53. [CrossRef] [PubMed]
101. Wheway, J.; Herzog, H.; Mackay, F. NPY and Receptors in Immune and Inflammatory Diseases. *Curr. Top. Med. Chem.* **2007**, *7*, 1743–1752. [CrossRef] [PubMed]

102. Chandrasekharan, B.; Nezami, B.G.; Srinivasan, S. Emerging neuropeptide targets in inflammation: NPY and VIP. *Am. J. Physiol. Gastrointest. Liver Physiol.* **2013**, *304*, G949–G957. [CrossRef]
103. El-Salhy, M.; Hausken, T. The role of the neuropeptide Y (NPY) family in the pathophysiology of inflammatory bowel disease (IBD). *Neuropeptides* **2016**, *55*, 137–144. [CrossRef]
104. Makinde, T.O.; Steininger, R.; Agrawal, D.K. NPY and NPY receptors in airway structural and inflammatory cells in allergic asthma. *Exp. Mol. Pathol.* **2013**, *94*, 45–50. [CrossRef]
105. Kunath, J.; Delaroque, N.; Szardenings, M.; Neundorf, I.; Straub, R.H. Sympathetic nerve repulsion inhibited by designer molecules in vitro and role in experimental arthritis. *Life Sci.* **2017**, *168*, 47–53. [CrossRef] [PubMed]
106. Koeck, F.-X.; Bobrik, V.; Fassold, A.; Grifka, J.; Kessler, S.; Straub, R.H. Marked loss of sympathetic nerve fibers in chronic Charcot foot of diabetic origin compared to ankle joint osteoarthritis. *J. Orthop. Res.* **2008**, *27*, 736–741. [CrossRef] [PubMed]
107. Klatt, S.; Fassold, A.; Straub, R.H. Sympathetic nerve fiber repulsion: Testing norepinephrine, dopamine, and 17β-estradiol in a primary murine sympathetic neurite outgrowth assay. *Ann. N. Y. Acad. Sci.* **2012**, *1261*, 26–33. [CrossRef] [PubMed]
108. Hoover, D.B.; Brown, T.C.; Miller, M.K.; Schweitzer, J.B.; Williams, D.L. Loss of Sympathetic Nerves in Spleens from Pa-tients with End Stage Sepsis. *Front. Immunol.* **2017**, *8*, 1712. [CrossRef]
109. Usui, M.L.; Underwood, R.A.; Mansbridge, J.N.; Muffley, L.A.; Carter, W.G.; Olerud, J.E. Morphological evidence for the role of suprabasal keratinocytes in wound reepithelialization. *Wound Repair Regen.* **2005**, *13*, 468–479. [CrossRef]
110. Galkowska, H.; Wojewodzka, U.; Olszewski, W.L. Chemokines, cytokines, and growth factors in keratinocytes and dermal endothelial cells in the margin of chronic diabetic foot ulcers. *Wound Repair Regen.* **2006**, *14*, 558–565. [CrossRef]
111. Chen, J.; Hoffman, B.B.; Isseroff, R.R. β-Adrenergic Receptor Activation Inhibits Keratinocyte Migration via a Cyclic Adenosine Monophosphate-independent Mechanism. *J. Investig. Dermatol.* **2002**, *119*, 1261–1268. [CrossRef]
112. Pullar, C.E.; Manabat-Hidalgo, C.G.; Bolaji, R.S.; Isseroff, R.R. β-Adrenergic receptor modulation of wound repair. *Pharmacol. Res.* **2008**, *58*, 158–164. [CrossRef]
113. Donaldson, D.J.; Mahan, J.T. Influence of catecholamines on epidermal cell migration during wound closure in adult newts. *Comp. Biochem. Physiol. Part C Comp. Pharmacol.* **1984**, *78*, 267–270. [CrossRef]
114. Pullar, C.E.; Chen, J.; Isseroff, R.R. PP2A Activation by β2-Adrenergic Receptor Agonists: Novel Regulatory Mecha-nism of Keratinocyte Migration. *J. Biol. Chem.* **2003**, *278*, 22555–22562. [CrossRef]
115. Wu, C.-S.; Tsao, D.-A.; Chang, H.-R. Beta2-adrenergic receptor agonist inhibits keratinocyte proliferation by mechanisms involving nitric oxide. *Adv. Dermatol. Allergol.* **2021**, *38*, 396–403. [CrossRef] [PubMed]
116. Dasu, M.R.; Ramirez, S.R.; La, T.D.; Gorouhi, F.; Nguyen, C.; Lin, B.R.; Mashburn, C.; Stewart, H.; Peavy, T.R.; Nolta, J.A.; et al. Crosstalk Between Adrenergic and Toll-Like Receptors in Human Mesenchymal Stem Cells and Keratinocytes: A Recipe for Impaired Wound Healing. *STEM CELLS Transl. Med.* **2014**, *3*, 745–759. [CrossRef] [PubMed]
117. Parrado, A.C.; Canellada, A.; Gentile, T.; Rey-Roldán, E.B. Dopamine Agonists Upregulate IL-6 and IL-8 Production in Human Keratinocytes. *Neuroimmunomodulation* **2012**, *19*, 359–366. [CrossRef] [PubMed]
118. Sivamani, R.K.; Shi, B.; Griffiths, E.; Vu, S.M.; Lev-Tov, H.A.; Dahle, S.; Chigbrow, M.; La, T.D.; Mashburn, C.; Peavy, T.R.; et al. Acute Wounding Alters the Beta2-Adrenergic Signaling and Catecholamine Synthetic Pathways in Keratinocytes. *J. Investig. Dermatol.* **2014**, *134*, 2258–2266. [CrossRef]
119. Sivamani, R.K.; Pullar, C.E.; Manabat-Hidalgo, C.G.; Rocke, D.M.; Carlsen, R.C.; Greenhalgh, D.G.; Isseroff, R.R. Stress-Mediated Increases in Systemic and Local Epinephrine Impair Skin Wound Healing: Potential New Indication for Beta Blockers. *PLoS Med.* **2009**, *6*, e1000012. [CrossRef]
120. Romana-Souza, B.; Nascimento, A.P.; Brum, P.C.; Monte-Alto-Costa, A. Deletion of the α2A/α2C-adrenoceptors accelerates cutaneous wound healing in mice. *Int. J. Exp. Pathol.* **2014**, *95*, 330–341. [CrossRef]
121. Yang, H.-Y.; Steenhuis, P.; Glucksman, A.M.; Gurenko, Z.; La, T.D.; Isseroff, R.R. Alpha and beta adrenergic receptors modulate keratinocyte migration. *PLoS ONE* **2021**, *16*, e0253139. [CrossRef]
122. Kurzen, H.; Berger, H.; Jäger, C.; Hartschuh, W.; Maas-Szabowski, N. Alpha 9 acetylcholine receptors are essential for epidermal differentiation. *Exp. Dermatol.* **2005**, *14*, 155. [CrossRef]
123. Nguyen, V.T.; Chernyavsky, A.I.; Arredondo, J.; Bercovich, D.; Orr-Urtreger, A.; Vetter, D.; Wess, J.; Beaudet, A.L.; Kitajima, Y.; Grando, S.A. Synergistic control of keratinocyte adhesion through muscarinic and nicotinic acetylcholine receptor subtypes. *Exp. Cell Res.* **2004**, *294*, 534–549. [CrossRef]
124. Buchli, R.; Ndoye, A.; Rodriguez, J.G.; Zia, S.; Webber, R.J.; Grando, S.A. Human Skin Fibroblasts Express M2, M4, and M5 Subtypes of Muscarinic Acetylcholine Receptors. *J. Cell. Biochem.* **1999**, *74*, 264–277. [CrossRef]
125. Kishibe, M.; Griffin, T.M.; Radek, K.A. Keratinocyte nicotinic acetylcholine receptor activation modulates early TLR2-mediated wound healing responses. *Int. Immunopharmacol.* **2015**, *29*, 63–70. [CrossRef] [PubMed]
126. Ndoye, A.; Buchli, R.; Greenberg, B.; Nguyen, V.T.; Zia, S.; Rodriguez, J.G.; Webber, R.J.; Lawry, M.A.; Grando, S.A. Identification and Mapping of Keratinocyte Muscarinic Acetylcholine Receptor Subtypes in Human Epidermis. *J. Investig. Dermatol.* **1998**, *111*, 410–416. [CrossRef]
127. Wollina, U.; Huschenbeck, J.; Knöll, B.; Sternberg, B.; Hipler, U.-C. Vasoactive intestinal peptide supports induced migration of human keratinocytes and their colonization of an artificial polyurethane matrix. *Regul. Pept.* **1997**, *70*, 29–36. [CrossRef] [PubMed]
128. Haegerstrand, A.; Jonzon, B.; Dalsgaard, C.J.; Nilsson, J. Vasoactive intestinal polypeptide stimulates cell proliferation and adenylate cyclase activity of cultured human keratinocytes. *Proc. Natl. Acad. Sci. USA* **1989**, *86*, 5993–5996. [CrossRef]

129. Bennett, L.A.T.; Johnson, J.M.; Stephens, D.P.; Saad, A.R.; Kellogg, D.L. Evidence for a Role for Vasoactive Intestinal Peptide in Active Vasodilatation in the Cutaneous Vasculature of Humans. *J. Physiol.* **2003**, *552*, 223–232. [CrossRef] [PubMed]
130. Dumont, Y.; Bastianetto, S.; Duranton, A.; Breton, L.; Quirion, R. Immunohistochemical distribution of neuropeptide Y, peptide YY, pancreatic polypeptide-like immunoreactivity and their receptors in the epidermal skin of healthy women. *Peptides* **2015**, *70*, 7–16. [CrossRef] [PubMed]
131. Takahashi, K.; Nakanishi, S.; Imamura, S. Direct Effects of Cutaneous Neuropeptides on Adenylyl Cyclase Activity and Proliferation in a Keratinocyte Cell Line: Stimulation of Cyclic AMP Formation by CGRP and VIP/PHM, and Inhibition by NPY Through G Protein-Coupled Receptors. *J. Investig. Dermatol.* **1993**, *101*, 646–651. [CrossRef]
132. Jiang, C.K.; Tomić-Canić, M.; Lucas, D.J.; Simon, M.; Blumenberg, M. TGF beta promotes the basal phenotype of epidermal keratinocytes: Transcriptional induction of K#5 and K#14 keratin genes. *Growth Factors* **1995**, *12*, 87–97.
133. Hameedaldeen, A.; Liu, J.; Batres, A.; Graves, G.S.; Graves, D.T. FOXO1, TGF-β Regulation and Wound Healing. *Int. J. Mol. Sci.* **2014**, *15*, 16257–16269. [CrossRef]
134. Le Provost, G.S.; Pullar, C.E. β2-Adrenoceptor Activation Modulates Skin Wound Healing Processes to Reduce Scarring. *J. Investig. Dermatol.* **2015**, *135*, 279–288. [CrossRef]
135. Pullar, C.E.; Isseroff, R.R. ß2-adrenergic receptor activation delays dermal fibroblast-mediated contraction of collagen gels via a cAMP-dependent mechanism. *Wound Repair Regen.* **2005**, *13*, 405–411. [CrossRef] [PubMed]
136. Pullar, C.E.; Isseroff, R.R. The β2-adrenergic receptor activates pro-migratory and pro-proliferative pathways in dermal fibroblasts via divergent mechanisms. *J. Cell Sci.* **2006**, *119*, 592–602. [CrossRef] [PubMed]
137. Pullar, C.E.; Le Provost, G.S.; O'Leary, A.P.; Evans, S.E.; Baier, B.S.; Isseroff, R.R. β2AR Antagonists and β2AR Gene Deletion Both Promote Skin Wound Repair Processes. *J. Investig. Dermatol.* **2012**, *132*, 2076–2084. [CrossRef]
138. Arredondo, J.; Nguyen, V.T.; Chernyavsky, A.I.; Bercovich, D.; Orr-Urtreger, A.; Vetter, D.E.; Grando, S.A. Functional role of α7 nicotinic receptor in physiological control of cutaneous homeostasis. *Life Sci.* **2003**, *72*, 2063–2067. [CrossRef] [PubMed]
139. Shome, S.; Rana, T.; Ganguly, S.; Basu, B.; Choudhury, S.C.; Sarkar, C.; Chakroborty, D.; Dasgupta, P.S.; Basu, S. Dopamine Regulates Angiogenesis in Normal Dermal Wound Tissues. *PLoS ONE* **2011**, *6*, e25215. [CrossRef] [PubMed]
140. Jacobi, J.; Jang, J.J.; Sundram, U.; Dayoub, H.; Fajardo, L.F.; Cooke, J.P. Nicotine Accelerates Angiogenesis and Wound Healing in Genetically Diabetic Mice. *Am. J. Pathol.* **2002**, *161*, 97–104. [CrossRef] [PubMed]
141. Chakroborty, D.; Goswami, S.; Basu, S.; Sarkar, C. Catecholamines in the regulation of angiogenesis in cutaneous wound healing. *FASEB J.* **2020**, *34*, 14093–14102. [CrossRef]
142. Li, J.; Zhang, Y.-P.; Kirsner, R.S. Angiogenesis in wound repair: Angiogenic growth factors and the extracellular matrix. *Microsc. Res. Tech.* **2003**, *60*, 107–114. [CrossRef]
143. Pullar, C.E.; Grahn, J.C.; Liu, W.; Isseroff, R.R. β2-Adrenergic receptor activation delays wound healing. *FASEB J.* **2006**, *20*, 76–86. [CrossRef]
144. Johnson, R.; Webb, J.G.; Newman, W.H.; Wang, Z. Regulation of Human Vascular Smooth Muscle Cell Migration by Beta-Adrenergic Receptors. *Am. Surg.* **2006**, *72*, 51–54. [CrossRef]
145. O'Leary, A.P.; Fox, J.M.; Pullar, C.E. Beta-Adrenoceptor Activation Reduces Both Dermal Microvascular Endothelial Cell Migration via a cAMP-Dependent Mechanism and Wound Angiogenesis. *J. Cell. Physiol.* **2014**, *230*, 356–365. [CrossRef] [PubMed]
146. Drummond, P.D.; Dawson, L.F.; Wood, F.M.; Fear, M.W. Up-regulation of α 1 -adrenoceptors in burn and keloid scars. *Burns* **2018**, *44*, 582–588. [CrossRef] [PubMed]
147. Pradhan, L.; Cai, X.; Wu, S.; Andersen, N.D.; Martin, M.; Malek, J.; Guthrie, P.; Veves, A.; LoGerfo, F.W. Gene Expression of Pro-Inflammatory Cytokines and Neuropeptides in Diabetic Wound Healing. *J. Surg. Res.* **2009**, *167*, 336–342. [CrossRef] [PubMed]
148. Ekstrand, A.J.; Cao, R.; Björndahl, M.; Nyström, S.; Jönsson-Rylander, A.-C.; Hassani, H.; Hallberg, B.; Nordlander, M.; Cao, Y. Deletion of neuropeptide Y (NPY) 2 receptor in mice results in blockage of NPY-induced angiogenesis and delayed wound healing. *Proc. Natl. Acad. Sci. USA* **2003**, *100*, 6033–6038. [CrossRef]
149. Polak, J.; Bloom, S. Regulatory peptides—The distribution of two newly discovered peptides: PHI and NPY. *Peptides* **1984**, *5*, 79–89. [CrossRef]
150. Ranne, J.; Kalimo, H.; Pyykkö, K.; Scheinin, M.; Aaltonen, V.; Niinikoski, J.; Laato, M. Wound Healing in Denervated Rat Groin Skin Flap. *Eur. Surg. Res.* **2000**, *32*, 197–202. [CrossRef]
151. Saburo, K.; Hiroshi, K.; Toshihiro, M. The role of sympathetic catecholaminergic nerves in wound healing. *Burns* **1982**, *9*, 135–141. [CrossRef]
152. Romana-Souza, B.; Monte-Alto-Costa, A. Simultaneous blockade of alpha and beta adrenoceptors impairs cutaneous wound healing in rats. *J. Eur. Acad. Dermatol. Venereol.* **2010**, *24*, 349–352. [CrossRef]
153. Souza, B.R.; Cardoso, J.F.; Amadeu, T.P.; Desmoulière, A.; Costa, A.M.A. Sympathetic denervation accelerates wound contraction but delays reepithelialization in rats. *Wound Repair Regen.* **2005**, *13*, 498–505. [CrossRef]
154. Zheng, Z.; Wan, Y.; Liu, Y.; Yang, Y.; Tang, J.; Huang, W.; Cheng, B. Sympathetic Denervation Accelerates Wound Contraction but Inhibits Reepithelialization and Pericyte Proliferation in Diabetic Mice. *J. Diabetes Res.* **2017**, *2017*, 7614685. [CrossRef]
155. Jurjus, A.; Hourani, R.; Daouk, H.; Youssef, L.; Bou-Khalil, P.; Haidar, H.; Atiyeh, B.; Saade, N. Effect of denervation on burn wound healing. *Ann. Burn. Fire Disasters* **2018**, *31*, 278–291.

156. Kim, L.R.; Pomeranz, B. The sympathomimetic agent, 6-hydroxydopamine, accelerates cutaneous wound healing. *Eur. J. Pharmacol.* **1999**, *376*, 257–264. [CrossRef] [PubMed]
157. Smith, P.G.; Liu, M. Impaired cutaneous wound healing after sensory denervation in developing rats: Effects on cell proliferation and apoptosis. *Cell Tissue Res.* **2002**, *307*, 281–291. [CrossRef]
158. Kim, L.R.; Whelpdale, K.; Zurowski, M.; Pomeranz, B. Sympathetic denervation impairs epidermal healing in cutaneous wounds. *Wound Repair Regen.* **1998**, *6*, 194–201. [CrossRef] [PubMed]
159. Wallengren, J.; Chen, D.; Sundler, F. Neuropeptide-containing C-fibres and wound healing in rat skin. Neither capsaicin nor peripheral neurotomy affect the rate of healing. *Br. J. Dermatol.* **1999**, *140*, 400–408. [CrossRef]
160. Cheng, C.; Singh, V.; Krishnan, A.; Kan, M.; Martinez, J.A.; Zochodne, D.W. Loss of Innervation and Axon Plasticity Accompanies Impaired Diabetic Wound Healing. *PLoS ONE* **2013**, *8*, e75877. [CrossRef]
161. Levy, D.M.; Karanth, S.S.; Springall, D.R.; Polak, J.M. Depletion of cutaneous nerves and neuropeptides in diabetes mellitus: An immunocytochemical study. *Diabetologia* **1989**, *32*, 427–433. [CrossRef]
162. Reiber, G.E.; Vileikyte, L.; Boyko, E.J.; del Aguila, M.; Smith, D.G.; Lavery, L.A.; Boulton, A.J. Causal pathways for incident lower-extremity ulcers in patients with diabetes from two settings. *Diabetes Care* **1999**, *22*, 157–162. [CrossRef]
163. Ahlborg, G.; Lundbergt, J.M. Exercise-induced changes in neuropeptide Y, noradrenaline and endothelin-1 levels in young people with type I diabetes. *Clin. Physiol. Funct. Imaging* **1996**, *16*, 645–655. [CrossRef]
164. Burton, A.R.; Fazalbhoy, A.; Macefield, V.G. Sympathetic Responses to Noxious Stimulation of Muscle and Skin. *Front. Neurol.* **2016**, *7*, 109. [CrossRef]
165. Oishi, M.; Mochizuki, Y.; Suzuki, Y.; Ogawa, K.; Naganuma, T.; Nishijo, Y.; Mizutani, T. Current Perception Threshold and Sympathetic Skin Response in Diabetic and Alcoholic Polyneuropathies. *Intern. Med.* **2002**, *41*, 819–822. [CrossRef] [PubMed]
166. Ono, S.; Nishijo, Y.; Oishi, M.; Mizutani, T. Comparison of the Utility of Sympathetic Skin Response and Current Perception Threshold Examinations with Conventional Examinations for the Early Electrophysiological Diagnosis of Diabetic Poly-neuropathy. *Electromyogr. Clin. Neurophysiol.* **2006**, *46*, 401–407. [PubMed]
167. Ziegler, D.; Papanas, N.; Roden, M.; GDC Study Group. Neuropad: Evaluation of three cut-off points of sudomotor dysfunction for early detection of polyneuropathy in recently diagnosed diabetes. *Diabet. Med.* **2011**, *28*, 1412–1415. [CrossRef] [PubMed]
168. Schlereth, T.; Morellini, N.; Lismont, N.C.; Lemper, C.; Birklein, F.; Drummond, P.D. Alpha 1 adrenoceptor expression in skin, nerves and blood vessels of patients with painful diabetic neuropathy. *Auton. Neurosci.* **2021**, *234*, 102814. [CrossRef] [PubMed]
169. Premkumar, L.S.; Pabbidi, R.M. Diabetic Peripheral Neuropathy: Role of Reactive Oxygen and Nitrogen Species. *Cell Biochem. Biophys.* **2013**, *67*, 373–383. [CrossRef] [PubMed]
170. Di Marco, E.; Jha, J.C.; Sharma, A.; Wilkinson-Berka, J.L.; Jandeleit-Dahm, K.A.; de Haan, J.B. Are reactive oxygen species still the basis for diabetic complications? *Clin. Sci.* **2015**, *129*, 199–216. [CrossRef] [PubMed]
171. Zhao, R.; Liang, H.; Clarke, E.; Jackson, C.; Xue, M. Inflammation in Chronic Wounds. *Int. J. Mol. Sci.* **2016**, *17*, 2085. [CrossRef]
172. Zhao, G.; Usui, M.L.; Lippman, S.I.; James, G.A.; Stewart, P.S.; Fleckman, P.; Olerud, J.E. Biofilms and Inflammation in Chronic Wounds. *Adv. Wound Care* **2013**, *2*, 389–399. [CrossRef]
173. Tellechea, A.; Kafanas, A.; Leal, E.; Tecilazich, F.; Kuchibhotla, S.; Auster, M.E.; Kontoes, I.; Paolino, J.; Carvalho, E.; Nabzdyk, L.P.; et al. Increased Skin Inflammation and Blood Vessel Density in Human and Experimental Diabetes. *Int. J. Low. Extrem. Wounds* **2013**, *12*, 4–11. [CrossRef]
174. Wetzler, C.; Kämpfer, H.; Stallmeyer, B.; Pfeilschifter, J.; Frank, S. Large and Sustained Induction of Chemokines during Impaired Wound Healing in the Genetically Diabetic Mouse: Prolonged Persistence of Neutrophils and Macrophages during the Late Phase of Repair. *J. Investig. Dermatol.* **2000**, *115*, 245–253. [CrossRef]
175. Albrecht, H.; Yang, H.-Y.; Kiuru, M.; Maksaereekul, S.; Durbin-Johnson, B.; Wong, M.S.; Stevenson, T.R.; Rocke, D.M.; Isseroff, R.R. The Beta 2 Adrenergic Receptor Antagonist Timolol Improves Healing of Combined Burn and Radiation Wounds. *Radiat. Res.* **2018**, *189*, 441–445. [CrossRef] [PubMed]
176. Arbabi, S.; Ahrns, K.S.; Wahl, W.L.; Hemmila, M.R.; Wang, S.C.; Brandt, M.-M.; Taheri, P.A. Beta-Blocker Use Is Associated with Improved Outcomes in Adult Burn Patients. *J. Trauma Inj. Infect. Crit. Care* **2004**, *56*, 265–271. [CrossRef] [PubMed]
177. Beroukhim, K.; Rotunda, A.M. Topical 0.5% timolol heals a recalcitrant irradiated surgical scalp wound. *Dermatol. Surg.* **2014**, *40*, 924–926. [CrossRef] [PubMed]
178. Tang, J.C.; Dosal, J.; Kirsner, R.S. Topical Timolol for a Refractory Wound. *Dermatol. Surg.* **2012**, *38*, 135–138. [CrossRef]
179. Braun, L.R.; Lamel, S.A.; Richmond, N.A.; Kirsner, R.S. Topical Timolol for Recalcitrant Wounds. *JAMA Dermatol.* **2013**, *149*, 1400–1402. [CrossRef] [PubMed]
180. Thomas, B.; Kurien, J.S.; Jose, T.; Ulahannan, S.E.; Varghese, S.A. Topical timolol promotes healing of chronic leg ulcer. *J. Vasc. Surg. Venous Lymphat. Disord.* **2017**, *5*, 844–850. [CrossRef]
181. Ghanbarzamani, A.; Salehifar, E.; Jafarirad, A.; Hesamirostami, M.H.; Bagherzadehsaba, A.; Saeedi, M.; Ghazaeian, M.; Khorasani, G.; Moosazadeh, M. Efficacy and Safety of 0.25% Timolol Gel in HealinSplit-Thickness Skin Graft Site. *Iran. J. Pharm. Res.* **2021**, *20*, 178–186. [CrossRef]
182. Baltazard, T.; Senet, P.; Momar, D.; Picard, C.; Joachim, C.; Adas, A.; Lok, C.; Chaby, G. Evaluation of timolol maleate gel for management of hard-to-heal chronic venous leg ulcers. Phase II randomised-controlled study. In *Annales de Dermatologie et de Vénéréologie*; Elsevier Masson: Paris, France, 2021; Volume 148, pp. 228–232. [CrossRef]

183. Kaur, R.; Tchanque-Fossuo, C.; West, K.; Hadian, Y.; Gallegos, A.; Yoon, D.; Ismailyan, L.; Schaefer, S.; Dahle, S.E.; Isseroff, R.R. Beta-adrenergic antagonist for the healing of chronic diabetic foot ulcers: Study protocol for a prospective, randomized, double-blinded, controlled and parallel-group study. *Trials* **2020**, *21*, 496. [CrossRef]
184. Jolivalt, C.G.; Frizzi, K.E.; Han, M.M.; Mota, A.J.; Guernsey, L.S.; Kotra, L.P.; Fernyhough, P.; Calcutt, N.A. Topical Delivery of Muscarinic Receptor Antagonists Prevents and Reverses Peripheral Neuropathy in Female Diabetic Mice. *J. Pharmacol. Exp. Ther.* **2020**, *374*, 44–51. [CrossRef]
185. Dillon, R.S. Role of Cholinergic Nervous System in Healing Neuropathic Lesions: Preliminary Studies and Prospective, Double-Blinded, Placebo-Controlled Studies. *Angiology* **1991**, *42*, 767–778. [CrossRef]
186. Kishibe, M.; Griffin, T.M.; Goslawski, M.; Sinacore, J.; Kristian, S.A.; Radek, K.A. Topical nicotinic receptor activation improves wound bacterial infection outcomes and TLR2-mediated inflammation in diabetic mouse wounds. *Wound Repair Regen.* **2018**, *26*, 403–412. [CrossRef] [PubMed]
187. Dong, M.-W.; Li, M.; Chen, J.; Fu, T.-T.; Lin, K.-Z.; Ye, G.-H.; Han, J.-G.; Feng, X.-P.; Li, X.-B.; Yu, L.-S.; et al. Activation of α7nAChR Promotes Diabetic Wound Healing by Suppressing AGE-Induced TNF-α Production. *Inflammation* **2015**, *39*, 687–699. [CrossRef] [PubMed]
188. Li, J.-Y.; Jiang, S.-K.; Wang, L.-L.; Zhang, M.-Z.; Wang, S.; Jiang, Z.-F.; Liu, Y.-L.; Cheng, H.; Zhang, M.; Zhao, R.; et al. α7-nAChR Activation Has an Opposite Effect on Healing of Covered and Uncovered Wounds. *Inflammation* **2017**, *41*, 474–484. [CrossRef] [PubMed]
189. Masuoka, H.; Morimoto, N.; Sakamoto, M.; Ogino, S.; Suzuki, S. Exploration of the wound healing effect of topical administration of nicotine in combination with collagen scaffold in a rabbit model. *J. Artif. Organs* **2015**, *19*, 167–174. [CrossRef]
190. Morimoto, N.; Takemoto, S.; Kawazoe, T.; Suzuki, S. Nicotine at a Low Concentration Promotes Wound Healing. *J. Surg. Res.* **2008**, *145*, 199–204. [CrossRef] [PubMed]
191. Liem, P.H.; Morimoto, N.; Ito, R.; Kawai, K.; Suzuki, S. Treating a collagen scaffold with a low concentration of nicotine promoted angiogenesis and wound healing. *J. Surg. Res.* **2013**, *182*, 353–361. [CrossRef] [PubMed]

International Journal of
Molecular Sciences

MDPI

Review

Leptin Increases: Physiological Roles in the Control of Sympathetic Nerve Activity, Energy Balance, and the Hypothalamic–Pituitary–Thyroid Axis

Davide Martelli [1] and Virginia L. Brooks [2,*]

[1] Department of Biomedical and NeuroMotor Sciences, 40126 Bologna, Italy
[2] Department of Chemical Physiology and Biochemistry, Oregon Health & Science University, Portland, OR 97239, USA
* Correspondence: brooksv@ohsu.edu

Abstract: It is well established that decreases in plasma leptin levels, as with fasting, signal starvation and elicit appropriate physiological responses, such as increasing the drive to eat and decreasing energy expenditure. These responses are mediated largely by suppression of the actions of leptin in the hypothalamus, most notably on arcuate nucleus (ArcN) orexigenic neuropeptide Y neurons and anorexic pro-opiomelanocortin neurons. However, the question addressed in this review is whether the effects of increased leptin levels are also significant on the long-term control of energy balance, despite conventional wisdom to the contrary. We focus on leptin's actions (in both lean and obese individuals) to decrease food intake, increase sympathetic nerve activity, and support the hypothalamic–pituitary–thyroid axis, with particular attention to sex differences. We also elaborate on obesity-induced inflammation and its role in the altered actions of leptin during obesity.

Keywords: energy expenditure; brown adipose tissue; obesity; diet-induced thermogenesis; sex differences; arcuate nucleus; paraventricular nucleus; obesity-induced inflammation; selective leptin resistance; weight regain

Citation: Martelli, D.; Brooks, V.L. Leptin Increases: Physiological Roles in the Control of Sympathetic Nerve Activity, Energy Balance, and the Hypothalamic–Pituitary–Thyroid Axis. *Int. J. Mol. Sci.* **2023**, *24*, 2684. https://doi.org/10.3390/ijms24032684

Academic Editors: Yutang Wang and Kate Denton

Received: 29 December 2022
Revised: 17 January 2023
Accepted: 21 January 2023
Published: 31 January 2023

1. Introduction

Energy balance is defined as food or energy intake minus energy expenditure, which in turn is determined by internal work (energy expended by normal organ function, which is dependent on ATP formation and ultimately results in heat) plus external work (skeletal muscle activity). If food intake exceeds energy expenditure, because energy cannot be created or destroyed (first law of thermodynamics), the excess energy is stored largely as fat in adipose tissue. However, in young to middle-aged healthy individuals, body weight remains remarkably constant over the long-term (months to years), despite widely varying food intakes and energy expenditures [1–3]. Clearly, homeostatic mechanisms must exist to ensure a constant body weight, which, ignoring changes in total body water content, is proportional to fat content. This constancy makes teleological sense from an evolutionary perspective, as a low fat content limits energy reserves during starvation and impedes reproduction, whereas a high fat content impairs predatory survival.

While the mechanisms that underlie the control of food intake and energy expenditure in the short-term (meal-to-meal) are fairly well understood, the mechanisms contributing to the precision of long-term body weight control remain largely unresolved, with one key exception: decreased levels of the hormone leptin, produced in and released from adipose tissue in direct proportion to adipose size, signal starvation to the body, leading appropriately to hunger and reduced energy expenditure [4,5]. Nevertheless, the normal long-term stability of fat mass and body weight suggests that mechanisms must also exist to suppress food intake and increase energy expenditure when the energy balance becomes positive (albeit for relatively short but sustained periods). However, whether increases in leptin levels physiologically signal an adipose surfeit to suppress food intake and increase energy expenditure is often disputed [5]. One argument against such a role is that acutely large, non-physiological doses of exogenous leptin are required [5,6].

The purpose of this review was to consider evidence for and against a role for increases in endogenous leptin in the long-term control of energy balance. Although leptin circulates in plasma, the vast majority of its actions are mediated via binding to receptors in the brain [4]. After central neuronal binding, leptins can influence the function of many organ systems in addition to those involved in the regulation of energy balance, including the hypothalamic–pituitary–thyroid (HPT) axis and the cardiovascular system [4]. The effects of leptin on energy balance as well as on the cardiovascular system are mediated at least in part by changes in the activity of autonomic nervous innervation of organs important in the control of energy balance [4]. Therefore, we particularly focus on its actions to increase sympathetic nerve activity (SNA).

2. Dissimilar Impact of Decreases versus Increases in Leptin

The discovery of leptin and the leptin receptor (LepR) pivoted on the generation of two mouse models exhibiting extreme obesity: the *ob/ob* mouse (carrying mutations in the gene responsible for the production of leptin) and the *db/db* mouse (carrying mutations in the gene responsible for the production of LepR). With its discovery, it quickly became clear that decreased leptin levels signal a negative energy balance (predominantly via the JAK-Stat pathway [4]), largely mediated by reduced binding of leptin to its receptors in the hypothalamus [4,5]. Hence, leptin is often called the "starvation hormone." On the other hand, normal plasma leptin concentrations are near a "threshold" level in most individuals, such that the relationships between leptin and its actions are steep when leptin levels fall, as during fasting, but further elevations in leptin seem to have minimal effects [5–8]. Explanations for these reduced effects include maximal receptor occupancy at physiological leptin concentrations and, following leptin binding to its receptor, activation of short-loop negative feedback systems via increased levels of the negative cellular regulator, suppressor of cytokine signaling 3 (SOCS3) [9,10]. Nevertheless, it is important to note that most studies documenting the relative resistance to increases in leptin levels utilized large, supraphysiological, and often acute doses of leptin. Do these data, therefore, preclude a role for increases in leptin within the physiological range?

Early research suggests not. Halaas et al. [11] convincingly demonstrated that long-term (2–4 week) subcutaneous leptin infusions in rats, which produced an increase in plasma leptin concentrations by as little as 40%, significantly decreased body weight by both suppressing food intake and increasing energy expenditure. How is this possible if maximal receptor occupancy already exists at normal plasma leptin concentrations? Recent data highlights one explanation. With time (hours to days), leptin induces the expression of its own receptor in two key hypothalamic sites, the arcuate nucleus (ArcN) and the paraventricular nucleus (PVN) [12,13]. Leptin also induces the expression of its receptor in cultured microglia [14]. Thus, sustained increases in leptin could effectively increase its threshold level. While increases in most hormones typically trigger a negative feedback decrease in receptor expression, the ability of leptin to increase the expression of its receptor makes teleological sense in the light of leptin's role in the long-term control of energy balance. Given, as indicated above, that leptin's acute effects are limited in part by LepR expression, LepR upregulation due to sustained increments in leptin levels (signaling energy or adipose excess) could unveil its latent anorexic and energy-expending actions.

Recent work suggests that the cellular actions of leptin on brain neurons may also be amplified with time via synergism with other neurotransmitters or neuromodulators. An example from our work: we investigated whether and how leptin in PVN increases SNA and blood pressure, despite a scarcity of LepR. Bilateral PVN injections increased lumbar SNA (LSNA), which innervates hindquarter skeletal muscle, as well as heart rate, blood pressure, and brown adipose tissue (BAT) SNA, but the increases developed slowly, reaching significance only after several minutes [12]. Interestingly, the slowly developing responses were associated with increased LepR expression (see above). These results were consistent with previous studies demonstrating that intracerebroventricular (ICV) leptin administration also slowly increased SNA and blood pressure [15,16] and that prior ICV administration of leptin amplified the hypertension elicited by a second leptin challenge [17]. Further studies revealed that the leptin-induced sympathoexcitatory response depended in part on glutamate, likely via synergism with leptin directly on PVN pre-sympathetic neurons, as previously shown in the hippocampus [18], as well as possibly via the activation of glutamatergic interneurons.

3. Physiological Significance of Increases in Leptin

3.1. Neurocircuitry by Which Leptin Increases SNA and Energy Expenditure (Figure 1)

Before considering the data for and against a physiological role for increases in leptin in long-term energy balance, it is necessary to outline the brain sites and pathways by which leptin not only inhibits food intake and increases energy expenditure (via both BAT SNA and the HPT axis), but also influences the function of the cardiovascular system.

It is well established that the ArcN is a central hub mediating the varied effects of leptin on food intake and energy expenditure. Within the ArcN, key neuronal subtypes include the orexigenic neuropeptide Y/Agouti-related protein (NPY/AgRP) neurons and their counterpart, anorexic pro-opiomelanocortin (POMC) neurons [4,5] (Figure 1). NPY neurons and the downstream neurocircuitry react quickly to stimulate food intake; however, POMC neurons and downstream sites are only slowly engaged, particularly in the context of the control of food intake [5,19]. Since hormones such as leptin excite POMC neurons, this slow responsiveness is again consistent with leptin's role in counteracting long-term accrual of excess body fat. In turn, ArcN NPY and POMC neurons course to multiple hypothalamic sites, most notably the PVN and DMH [19,20].

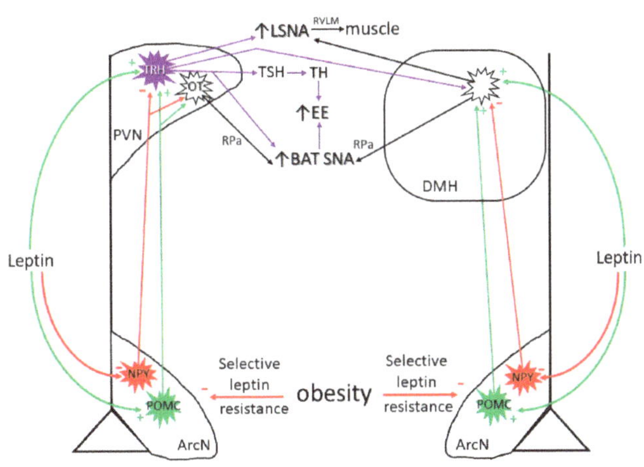

Figure 1. Hypothalamic neurocircuitry by which leptin increases SNA and activates the HPT axis. NPY: Neuropeptide Y; POMC: Pro-opiomelanocortin; ArcN: Arcuate nucleus; PVN: Paraventricular nucleus; TRH: Thyrotropin-releasing hormone; TSH: Thyroid-stimulating hormone; TH: Thyroid hormone; OT: Oxytocin; EE: Energy expenditure; BAT SNA: Brown adipose tissue sympathetic nerve activity; RVLM: Rostral ventrolateral medulla; RPa: Raphe pallidus; DMH: Dorsomedial hypothalamus. Green arrows: stimulatory; red arrows: inhibitory. See text for description and details.

Unlike the control of food intake, in which the ArcN is a major site of action, leptin binds to receptors in several hypothalamic sites to increase the activity of sympathetic nerves innervating many organs (kidney, muscle, gut, heart, BAT). Nevertheless, the ArcN again appears to be primary, as ArcN lesions prevented the effect of intravenous (IV) leptin to increase SNA [21] (Figure 1). Further evidence indicates that the PVN is also crucial in the downstream neurocircuitry (Figure 1). More specifically, the sympathoexcitatory response to ICV leptin (which presumably can access many hypothalamic sites [22]) was abolished by nonselective blockade of the PVN [16], also suggesting that the PVN is a site of confluence, regardless of where leptin is binding to initiate increases in SNA. Consistent with this model, ArcN NPY neuronal fibers that project to the PVN tonically inhibit presympathetic neurons reaching the brainstem cardiovascular center, the rostral ventrolateral medulla (RVLM) [23,24], as well as the brainstem center regulating BAT SNA, the raphe pallidus (RPa) [25] (Figure 1). In parallel, ArcN POMC neurons project to and excite presympathetic neurons that increase LSNA (which innervates skeletal muscle) [16,24] and BAT SNA [26] via release of alpha-melanocyte-stimulating hormone (α-MSH) (Figure 1). Importantly, reversal of tonic NPY sympathoinhibition is required before α-MSH can stimulate PVN presympathetic neurons [24].

The PVN contains at least four subsets of pre-sympathetic neurons that project to the brainstem or preganglionic neurons in the spinal cord: those expressing corticotropin releasing hormone (CRH), thyrotropin-releasing hormone (TRH), vasopressin, or oxytocin (OT) [12,27,28]. Increased leptin levels inhibit ArcN NPY neurons and stimulate POMC neurons, thereby increasing LSNA, splanchnic SNA (SSNA), and BAT SNA [12,24,29] via both TRH and OT presympathetic neurons [12,25,30,31] (Figure 1). Leptin's actions on ArcN NPY and POMC neurons that project to the PVN also inhibit food intake [32]. Finally, considerable information indicates that NPY neurons inhibit and POMC neurons stimulate the release of TRH from PVN neurons that instead progress to the median eminence, where thyroid-stimulating hormone (TRH) is released into the blood to stimulate the production of thyroid hormone (TH) [33,34] (Figure 1). Therefore, leptin can increase energy expenditure by both increasing TH levels as well as BAT SNA.

In addition to the ArcN, leptin is also capable of direct stimulatory actions on PVN TRH neurons, despite a paucity of LepR [12,33,34] (Figure 1). These actions likely rely on LepR receptor induction and synergism with glutamate [12], as described above. Direct leptin stimulation of TRH neurons not only leads to increased production of TH, but it also increases SNA to skeletal muscle (LSNA) and BAT [12], thereby reinforcing the effects of ArcN leptin to increase energy expenditure via these same pathways.

A third hierarchical site of leptin action is the DMH, not only via direct actions [12,35–37] but also possibly via indirect actions by suppression of ArcN NPY neurons [20,25] and stimulation of ArcN POMC neurons that project to the DMH [26,38] (Figure 1). As in the PVN, these actions lead to increases in LSNA and BAT SNA.

3.2. Diet-Induced Increases in Leptin

Given that leptin can induce the expression of its own receptor, are increments in leptin levels physiologically significant? Leptin-induced increases in LSNA could serve, over a longer time frame than insulin, to stimulate glucose uptake into muscle [39,40], whereas increases in BAT SNA generate heat and increase energy expenditure, both of which are appropriate actions in the face of increased body energy content, such as after eating. Thus, one physiological paradigm that has been extensively studied is diet-induced thermogenesis (DIT). It is important to emphasize, however, that while DIT was first hypothesized to play a role in long-term energy balance, the term DIT has more recently been used to describe the increases in body temperature and energy expenditure that occur as a result of eating a single meal. To distinguish between these two phenomena with widely differing time frames, we refer to the latter as "meal-induced thermogenesis" or MIT.

MIT results from the thermic effects of the digestion and absorption of food, largely of protein and carbohydrates; storage of excess food stuffs increases energy expenditure, but not heat (for reviews, see refs. [41–44]). In addition, while somewhat controversial due to the varying indirect approaches used to measure BAT activity, considerable evidence suggests that eating leads to BAT activation in both rodents and humans, in part mediated by increased BAT SNA (for reviews, see ref. [41,43,44]). Leptin increases BAT SNA and BAT thermogenesis [45] and leptin levels increase after eating; therefore, does leptin contribute to MIT? In humans, plasma leptin levels rise hours after eating [46,47], emphasizing its role in the long-term control of energy balance, but the rise occurs only after MIT has subsided, thus disputing a role of leptin in MIT. On the other hand, a recent study showed that re-feeding starved (48 h) rats increased body temperature and energy expenditure [48], and in parallel, leptin levels increased from low to normal. Further experiments revealed that the rise in body temperature, which was mediated by BAT activation (prevented by blockade of SNA or removal of BAT) but not an increase in energy expenditure, was dependent on leptin. Nevertheless, in this experimental paradigm, leptin levels increased from below normal back to normal, which again emphasizes the important physiological role of low to normal leptin plasma concentrations (although, notably, increases in leptin to twice normal levels evoked further increases in body temperature in this study). Thus, current evidence supports a major role for leptin in MIT or MIT-induced BAT activation only when its levels are low, as during fasting.

On the other hand, Rothwell and Stock's groundbreaking work suggested that leptin is critical to the increase in BAT-induced thermogenesis (DIT) that occurs in response to long-term overeating. In their experiments, rats fed a cafeteria diet to induce voluntary hyperphagia exhibited minimal body weight gain because of reduced feeding efficiency due to BAT activation, heat dissipation, and increased energy expenditure [49,50]. Evidence supporting a major role for leptin in increased heat production was that reduced feeding efficiency and DIT were not observed in *ob/ob* mice [51,52]. When humans were overfed for short periods, they also exhibited a compensatory increase in energy expenditure (independent of the increase in body mass) [5,53]. However, it is unknown whether this long-term DIT in overfed humans depends on leptin, although a case has been made for its existence [19,50]. On the other hand, a recent study suggested that another, unidentified, catabolic factor released from adipose tissue inhibited food intake and reduced feeding efficiency in mice overfed for days via gastric infusion [54]. Nevertheless, a role for leptin was not eliminated, since mice in which leptin levels were clamped below normal ate more after release from overfeeding than mice in which leptin levels were allowed to increase with fat mass. Therefore, current evidence implicates at least a partial role for leptin in DIT, but clearly more work is required; in particular, long-term challenge studies in humans need to be conducted.

4. Obesity-Induced Increases in Leptin

As indicated above, in the steady-state, energy (food) intake equals energy expended plus storage primarily in adipose tissue. Therefore, storage = food intake − (internal heat + external work). This relationship is the basis of the general consensus that obesity (excess storage) "is the consequence of small, cumulative imbalances between energy intake and expenditure" [5,55]. Nevertheless, some researchers have hypothesized that obesity could instead originate from derangements in food-partitioning or storage mechanisms ([56], see also ref. [5,57]). If so, then in order for obesity to occur, again there must also be derangements in the mechanisms that signal increased adiposity to the brain (such as leptin) and nullification of the subsequent, homeostatic, long-term control of food intake and/or expenditure that would tend to counteract increased adiposity. Therefore, in either case, there must be changes in these homeostatic mechanisms to ultimately allow small persistent increments in adipose to transpire in non-obese individuals. What are these changes? Many explanations have been suggested, including inherited, multi-modal genetic mutations or epigenetically modified genes, as well as environmental factors such as dietary composition, physical activity level, or social and hedonistic influences [5,58]. As an example of the latter, it is well known that many people gain weight during holiday periods [59] and that this excess weight can be sustained [60].

Regardless of how the obese state evolves, it is well established that excess adiposity yields the expected increases in plasma leptin levels. Nevertheless, the elevated leptin levels clearly fail to effectively suppress food intake or increase energy expenditure due to "leptin resistance" [5] (Figure 1). Thus, the excess adiposity comes to be considered normal. In other words, since increased leptin levels normally mitigate against the long-term accrual of body fat, there must be a factor that counteracts this normal response such that there is a shift in the threshold or "setpoint" of body weight maintenance to a higher leptin level. A hotly debated question is: why? One mechanism to explain the reduced effectiveness of elevated plasma leptin levels to counteract increased adiposity is that the transport of leptin from plasma to the brain is attenuated [61,62]. Second, while LepR-expressing hypothalamic neurons critical in the regulation of food intake and energy expenditure retain responsiveness to leptin in obese animals [10,63], responses were reduced with ICV administration of leptin, which bypassed the blood-brain barrier (BBB) transport mechanism [22,61,62]. Thus, leptin still bound and activated its receptor; however, the degree of activation and/or downstream signaling was less than predicted based on leptin levels alone. As a result, ArcN NPY levels were relatively increased and POMC levels were decreased [64]. What then attenuates the responsiveness of hypothalamic LepR to plasma leptin?

Obesity Suppresses Leptin's ArcN Anorexic Actions

As described above, one mechanism that normally tends to minimize leptin's inhibition of food intake and increase in energy expenditure is near maximal LepR occupation, although this limitation can be overcome with time by leptin's ability to induce the expression of its own receptor. Therefore, obesity could also reduce leptin responsiveness or shift the leptin setpoint by hindering leptin induction of its own receptor. In support of this view, while obesity has been shown to increase LepR expression in the ArcN [10,13], the administration of leptin to further increase plasma levels in obese mice failed to induce additional LepR expression, unlike in lean mice [13]. Thus, leptin-induced increased LepR expression appears to be restrained during obesity, thereby contributing to a reduction in leptin's actions.

With obesity, elevated levels may also suppress the ability of leptin to inhibit food intake via induction of cellular signaling negative feedback mechanisms, as described above, such as SOC3 and the tyrosine phosphatase, PTP1B, both of which are increased, specifically in the ArcN [10,65]. A recent study proposed that leptin's ability to mute its own actions via this feedback is a major contributor to leptin resetting [10,65] and that small decreases in leptin levels could actually unmask its anorexic actions. However, if so, why is weight loss in obese individuals, which is accompanied by a decrease in plasma leptin levels, so difficult to maintain (i.e., the fall in leptin levels causes hunger and reduced energy expenditure rather than reduced hunger; see below)?

In addition to limits in hypothalamic LepR expression and induction of cellular negative feedback signaling pathways, significant evidence indicates that obesity also induces hypothalamic inflammation, microglia activation, and cytokine production [64,66], which correlates with and contributes to ArcN leptin resistance [5]. The inflammation associated with obesity is a classic example of a low-grade systemic response affecting the whole body, brain included [67]. This low-grade, sustained over time, inflammatory response has also been termed metabolic inflammation or meta-inflammation [68]. We will now describe immune derangements, beginning first with peripheral inflammation and then the occurrence and mechanisms of central inflammation.

5. Obesity: A State of Inflammation

5.1. Obesity-Induced Systemic Inflammation

Adipose tissue is directly involved in immune function. Adipocytes and macrophages have many characteristics in common; indeed, pre-adipocytes can differentiate into macrophages [69]. White adipose tissue (WAT) has the ability to release more than 600 different bioactive molecules, including cytokines and chemokines, which are collectively known as adipokines [70–72] (Table 1).

Table 1. Obesity-induced cytokine responses.

Cytokine	Description	Changes with Obesity
Leptin	Pro-inflammatory adipokine released by fat cells. Stimulates T cells, macrophages, and neutrophils to release pro-inflammatory cytokines [73]. Leptin is essential for normal T-cell proliferation and its deficiency causes thymus atrophy and severe immune dysfunction [73,74]. Surprisingly, leptin improves survival during sepsis [75].	Levels directly related to fat stores; increase with obesity. WAT inflammation was strongly reduced when LepR was knocked out in leukocytes in DIO mice [76].
Adiponectin	An anti-inflammatory adipokine that modulates a number of metabolic processes [77].	Adiponectin circulating levels are usually inversely proportional to the level of visceral adiposity; therefore, obese individuals have very low levels of adiponectin [78]. The adiponectin–leptin ratio is often considered a functional marker of inflammation associated with obesity [77].
Resistin	Also known as adipose tissue-specific secretory factor, it plays a role in the pathogenesis of atherosclerosis by enhancing the synthesis of hepatic LDL [79].	Elevated with non-morbid obesity. Acts locally on leukocytes, located in the WAT, to induce the release of pro-inflammatory cytokines [80].
Tumor necrosis factor α (TNF)	TNF is a necessary and sufficient mediator of inflammation, acutely released by macrophages, T cells, and natural killer cells during infection [81]. TNF is also released by adipocytes [82].	Its plasma levels are generally high in obese individuals, especially in those presenting visceral obesity rather than subcutaneous obesity [83]. However, it is not clear whether its levels decrease after weight loss [84].
IL-6	A proinflammatory cytokine produced by immune, endothelial, and muscle cells as well as adipocytes [70]. Surprisingly, an anti-IL-6 antibody therapy, used for the treatment of rheumatoid arthritis, causes weight gain [85]	Plasma levels are correlated with BMI and especially with adipose tissue mass [86]. Surgery- induced weight loss is associated with a significant decrease in IL-6 levels [87].
Monocyte chemoattractant protein-1 (MCP-1)	Key chemokine that regulates migration and infiltration of monocytes and macrophages [88].	Circulating MCP-1 levels do not differ between lean and obese individuals [89]; however, levels of this chemokine increase selectively in the WAT of obese adults [90]. MCP-1 levels in plasma significantly decrease after Roux-en-Y Gastric Bypass (RYGB) surgery or a low calorie diet [70,91].
Interleukin 8 (IL-8)	Pro-inflammatory and chemoattractant cytokine [92].	Systemic levels are closely correlated with BMI, waist circumference, and other obesity-related parameters [93]. Weight loss is not always associated with a decrease in IL-8 plasma levels; in fact, low calorie diet-induced weight loss is associated with an increase in IL-8 levels [94], whereas weight loss induced by RYGB produces a decrease in IL-8 levels [95].
Interleukin 10 (IL-10)	Anti-inflammatory cytokine released by M2 macrophages, Th2 T cells, neutrophils, and adipocytes [96].	Systemic IL-10 levels are generally inversely correlated with BMI and body fat percentage [70,97]. Nevertheless, plasma IL-10 levels have been reported to be elevated in obese women; however, obese women with lower IL-10 levels were more prone to develop metabolic syndrome [98].

Table 1. *Cont.*

Cytokine	Description	Changes with Obesity
C-reactive protein (CRP)	Released by hepatocytes in response to trauma, infection, or injury [99].	CRP levels are significantly higher in obese individuals than in lean subjects [86] and decrease with diet or weight loss induced by surgical intervention [100,101], making it a good marker of meta-inflammation.
Transforming growth factor β (TGF-β)	Cytokine released by all leukocytes [102].	Regardless of the location of fat mass, obesity is associated with enhanced levels of TGF-β [103], while weight loss decreases TGF- β circulating levels [104].

The concept that obesity is associated with a low-grade inflammatory state stems from demonstrations that circulating levels of interleukin-6 (IL-6) and C-reactive protein (CRP), both pro-inflammatory markers, are elevated in obese animals and humans [105], while they decrease back to normal levels after weight loss [84]. In addition, obesity can be associated with an increase in the number of circulating neutrophils and other white blood cells (WBCs) but a specific decrease in cytotoxic T cells and NK cells [106,107]. The inflammation can be widespread: Fischer and colleagues showed that a high fat diet (HFD) for 20 weeks induced increases in body weight, glucose intolerance, and inflammation in the liver, perigonadal fat, and subcutaneous fat in mice. Most of these parameters returned to normal after 7 weeks of a low fat diet (LFD); however, inflammation persisted in the liver and perigonadal fat, which becomes critical in the context of sustained weight loss (see below) [108].

The low-grade systemic inflammatory status associated with obesity likely evolves from several sources [71]. As previously reviewed in detail [55,109], the number of adipocytes is set or locked in parallel with a low level of constant turnover in humans after the age of about 20. As a result, as lipid storage needs and adiposity increase, the excess lipid requires an enlargement of resident adipocytes. In turn, adipose cell growth necessitates a remodeling of the extracellular matrix (ECM) to support the larger fat cells, ultimately leading to fibrosis [109]. In addition, the increased adipocyte size causes hypoxia, oxidative stress [110], and the active secretion of cytokines. These adverse changes trigger infiltration of leukocytes into the adipose tissue [70,109]. The adaptive immune remodeling of WAT, in part induced via the recruitment of WBC by the chemokine, CxCl12 [111], parallels the increased expression of pro-inflammatory markers in adipose tissue [112]. Hypertrophic adipocytes can also eventually burst, thereby worsening the local inflammatory response [70]. WAT inflammation can therefore trigger a vicious cycle through which other organs become inflamed and the inflammation emanating from local sites yields increased systemic inflammation. Obesity can also induce an increase in gut permeability, which if exaggerated, might lead to endotoxemia [i.e., increased levels of lipopolysaccharide (LPS)] [113–115]. Finally, obesity is associated with an increase in SNA to the muscle and kidney (see below) [116,117]. Sympathetic nerves are a double-edge sword when we consider their role in the control of immune function: they promote anti-inflammatory effects during the course of a systemic infection [118,119], but they also have the ability to mediate pro-inflammatory effects when chronically overactivated [120,121], as is the case in obese people. Sympathetic nerve activation in WAT is also associated with enhanced non-esterified fatty acids and cytokine loads in the adipose microenvironment, including that of pancreatic fat, which likely impair beta-cell function and contribute to insulin resistance [122].

5.2. Obesity-Induced Neuroinflammation

Obesity triggers inflammatory responses in several areas of the CNS, including the hypothalamus, mediated by microglia and astroglia activation and proliferation. Importantly, hypothalamic inflammation first identified in preclinical studies has also been detected in obese humans [123,124]. Multiple mechanisms are likely involved. The first description of obesity-induced hypothalamic inflammation came from the work of De Souza and colleagues [125]. These authors showed that a high fat diet (HFD) for 4 months produced high levels of IL-1β, TNF, and IL-6 in the hypothalamus of rats, mediated through the JNK and NF-kB pathways. This was later confirmed in obese mice subjected to a much more restricted period of HFD (24–72 h) [123]. In the short term, saturated fatty acids (SFAs), especially long chain SFAs, bond to TLR-4 and triggered the classic pro-inflammatory physiological response [126].

The hypothalamic inflammatory response, therefore, can precede systemic inflammation and the establishment of obesity, suggesting a possible causal role for neuroinflammation in initiating a pathophysiological response that leads to obesity. Several studies support this scenario: (1) Specific and selective deletion of IKKB, a key protein involved in the intracellular inflammatory cascade response, in hypothalamic astrocytes reduced the susceptibility of mice to diet-induced obesity [127]. (2) In rats fed a HFD, ICV administration of IL-4, a key regulator in humoral and adaptive immunity, further increased HFD-mediated hypothalamic inflammation and caused even more excessive weight gain. If the rats were treated with an IKKB–NF-kB blocker, body weight and fat mass decreased [128]. (3) The inhibition of microglial expansion in the ArcN limited food intake with subsequent decreases in body weight and fat content in mice [129]. (4) One to three days of HFD increased the expression of the chemokine, CX3Cl1, in the hypothalamus of mice that were obesity-prone but not in those that were obesity-resistant. Inhibition of CX3Cl1 reduced the glucose intolerance and fat content of obesity-prone mice without affecting their body weight [130]. (5) Five days of HFD induced an overexpression of CXCl12 and its receptors, CXCR4 and CXCR7, both in the PVN and lateral hypothalamus (LH). Hypothalamic administration of CXCl12 in normal diet (ND)-fed animals recapitulated the effects of HFD consumption; it: (i) reduced novelty-induced locomotor activity, (ii) increased encephalin gene expression in the PVN, and (iii) induced an overconsumption of HFD rather than ND if the rats had a choice. Collectively these responses contributed to weight gain [131].

With chronic obesity, a second contributor to hypothalamic inflammation may be peripheral inflammation. Several adipokines, such as leptin, as well as cytokines, can cross the BBB to activate microglia [71,132]. Other potential, but to our knowledge currently unproven, scenarios include: (1) HFD rapidly alters the gut microbiome and increases gut permeability, leading to leakage of bacterial LPS into the blood [114,115] and activation of brain microglia [133,134]. Microglia activation may be direct, as LPS or its components has been shown to bind to brain endothelial cells and tanocytes and enter the brain, possibly via a lipoprotein transport mechanism [135]. Interestingly, LPS can also activate iNOS, which has been shown to trigger the transformation of ArcN perivascular macrophages into microglia-like immune cells, which move into the brain parenchyma [136]. Alternatively, SFA or gut bacteria (or their byproduct, LPS) may activate vagal afferents [137,138], which can increase hypothalamic inflammation [139]. Intriguingly, LPS, like obesity, has been shown to cause leptin resistance via induction of the negative cellular regulator, PTP-1B [140]. (2) Systemic inflammation can lead to disruption of the BBB, promoting leukocyte extravasation and increasing diffusion of solutes across the BBB, including the entry of pathogens and toxins in the brain parenchyma. This has the ability to trigger a vicious cycle responsible for inducing further inflammation [141].

5.3. Potential Actions of Leptin to Facilitate Inflammation with Obesity

The most direct evidence for a role of leptin in establishing the low-grade systemic inflammatory response associated with obesity is that WAT inflammation was strongly reduced when LepR were knocked out in mouse leukocytes [76]. Leptin also directly stimulates T cells, macrophages, and neutrophils [73] and is essential for normal T-cell proliferation; its deficiency caused thymus atrophy and severe immune dysfunction [73,74]. In addition to systemic proinflammatory actions, leptin may also enable central inflammation. For example, by binding to brain endothelial receptors, leptin promoted neutrophil migration to and into the brain following LPS administration [132,142]. Indeed, obesity, via increased leptin levels, exaggerated the neuroinflammatory response to LPS administration (for review, see ref. [71]). However, to our knowledge, a specific role for increased leptin levels in facilitating or mediating brain inflammation during obesity has not been established.

6. Obesity Induces Selective Leptin Resistance

The information summarized above suggests that, in obese individuals, elevated leptin levels fail to adequately suppress food intake or increase energy expenditure (i.e., leptin resistance) due to hypothalamic inflammation, maximal LepR occupancy, and decreased LepR responsiveness secondary to cellular negative feedback inhibition. However, much of the extensive research on this topic has been conducted without specific attention to potential sex differences. It is well-established that plasma leptin levels are higher in females than in males, either lean or obese. The pioneering studies of Clegg and colleagues [143] also documented that females are more responsive to the anorexic actions of leptin than males due to the sensitizing effects of estrogen. Nevertheless, the fact that women, like men, become obese despite elevated leptin levels suggests that resistance to the anorexic actions of leptin do develop, although preclinical studies suggest that a longer time on a HFD is required [144,145]. Clearly, more work is needed.

While, especially in males, the anorexic responses to leptin dissipate with obesity, leptin's ability to influence other systems remains intact or is even enhanced, giving rise to so called "selective leptin resistance." This selectivity is apparent on at least two levels: (1) within the ArcN, obesity suppresses the effect of leptin to inhibit food intake but not its control of other modalities; and (2) leptin's ability to suppress food intake and trigger appropriate signaling mechanisms that underlie the regulation of many modalities are muted in the ArcN but not in other leptin-receptive hypothalamic sites. As with the anorexic actions of leptin, sex differences have also been observed. We will address two of leptin's effects that elude the development of resistance: leptin-induced activation of cardiovascularly relevant SNA and the HPT axis, including what is currently known about sex differences.

6.1. Selective Leptin Resistance in the ArcN: Preserved Leptin-Induced Increases in SNA (in Obese Males)

The concept of selective leptin resistance was first identified in the context of leptin-induced activation of the sympathetic nervous system and its contribution to obesity-induced hypertension. As nicely summarized in their review [146], Mark and colleagues were the first to show that leptin's actions to suppress food intake and increase energy expenditure were suppressed in genetic mouse models of obesity and diet-induced obesity, but its ability to increase renal SNA (RSNA) was preserved or enhanced. Head and colleagues similarly reported that in HF diet-induced obese compared to lean rabbits, ICV administration of leptin induced larger increases in RSNA but smaller increases in c-fos in many hypothalamic regions, including the ArcN and PVN [22], again highlighting the selective preservation of leptin's sympathoexcitatory mechanisms despite more global cellular resistance.

As recently reviewed [24,147], several lines of evidence indicate that elevated leptin levels in obese males increase SNA and can, if it occurs, contribute to hypertension: (1) increases in MSNA correlate with leptin and adiposity levels in humans [24,148–151]; (2) nonspecific blockade of the ArcN or PVN decreased SNA and blood pressure in obese rodents [152,153]; and (3) selective blockade of LepR, including specifically in the ArcN, decreased SNA and blood pressure in obese animals [21,154,155]. Recent work has investigated whether changes in the function of ArcN NPY or POMC neurons in males with obesity underlie sensitization to the sympathoexcitatory effects of leptin and/or its metabolic partner, insulin. Briefly, the data suggest (see ref. [24] for a review) that tonic inhibition of PVN presympathetic neurons by NPY is absent in obese males, allowing sympathetic stimulation by leptin and insulin to progress unimpeded [156]. In addition, it appears that simultaneously enhanced signaling mechanisms in ArcN POMC neurons mediate the amplified sympathoexcitation to leptin and insulin [153].

One mechanism that has been hypothesized to mediate the select preservation of leptin-induced actions within the ArcN is that different leptin-induced signaling pathways are engaged in resistant versus sensitive neurons [146,157]. Another hypothesized mechanism involves the actions of the renin-angiotensin system (RAS) in the brain, since the sympathoexcitatory and hypertensive actions of leptin require functional AngII type 1 receptors (AT1aR), whereas the anorexic effects apparently do not [24,146] (Figure 2). Similar to leptin and insulin, AngII in the ArcN increased SNA and blood pressure [158,159] in male rats through binding to AT1aR; the downstream pathway depended upon inhibition of NPY and stimulation of POMC neurons that project to the PVN [159]. A detailed neuroanatomical assessment of the location of AT1aR in the ArcN of male rats [159,160] revealed that (1) approximately 10% of AT1aR were expressed on NPY neurons; (2) AT1aR were rarely found on POMC neurons (4%); (3) instead, AT1aR were expressed largely in neurons that co-expressed both tyrosine hydroxylase (TyH; dopaminergic) and VGat2 (GABAergic) (see also ref. [161]); (4) LepR positive neurons rarely co-expressed AT1aR. The distribution of AT1aR in the mouse ArcN differed in that AT1aR were more commonly found in NPY- and LepR-containing neurons, although the paucity of AT1aR in POMC neurons was similar (2%) [162]. Based on information in rats, it was hypothesized that ArcN AngII increases SNA through either direct or indirect (via stimulation of TH/GABAergic interneurons) inhibition of sympathoinhibitory ArcN-to-PVN NPY neurons. This NPY disinhibition allows other stimulatory factors to activate sympathoexcitatory POMC neurons, such as AngII, leptin, insulin, or even inflammation. Obesity activates the RAS both systemically and centrally [163–165]. Therefore, one hypothesis to explain the select sensitization of ArcN presympathetic neurons to leptin in obese males is that this very small cohort of sensitized neurons are those that co-express AT1aR with POMC and/or LepR.

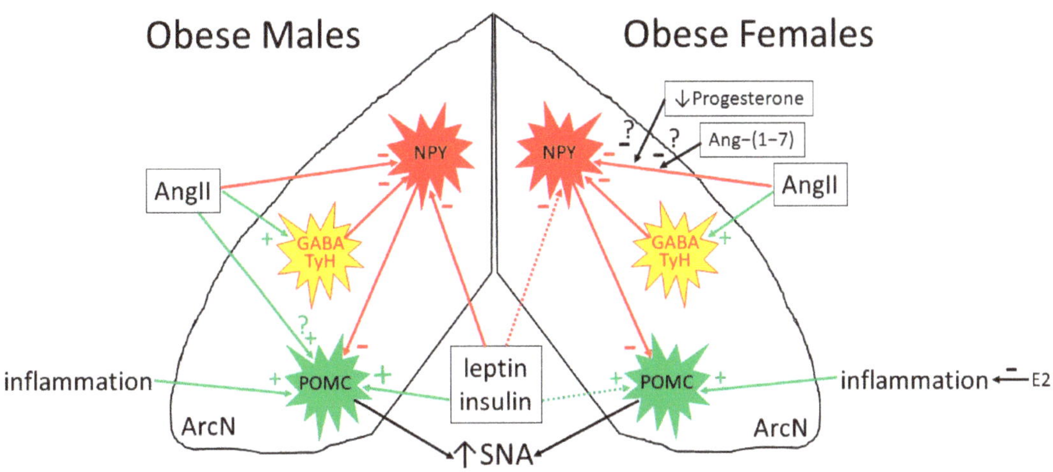

Figure 2. Hypothesized mechanisms underlying sex differences in the actions of elevated leptin and insulin levels to increase SNA during obesity. NPY: Neuropeptide Y; POMC: Pro-opiomelanocortin; ArcN: Arcuate nucleus; AngII: Angiotensin II; Ang-(1–7): Angiotensin 1–7; TyH: Tyrosine hydroxylase (marker of dopaminergic neurons); GABA: Gamma-aminobutyric acid; E2: Estrogen; SNA: Sympathetic nerve activity. Green arrows: stimulatory; red arrows: inhibitory. Dotted arrows indicate that NPY and POMC presympathetic neurons are resistant to the actions of leptin or insulin in obese females, whereas these actions are enhanced in obese males. See text and legend to Table 2 for details. Adapted in part from [24].

The situation in females is quite different. While increases in MSNA correlate with adiposity in men, a similar correlation is generally not observed in women (for reviews, see ref. [24,149,165]). At least two factors may be involved: (1) In lean males, leptin increases the activity of several sympathetic nerves; yet, in lean females, leptin's stimulation of some sympathetic nerves can vary with the reproductive cycle. More specifically, similar to leptin's anorexic actions in females [143], leptin-induced increases in LSNA and RSNA require proestrus (heightened) levels of estrogen, whereas increases in SSNA and HR do not vary during the estrus cycle and are similar between males and females [15]. Therefore, the dependence of leptin on elevated estrogen levels to increase lumbar (preclinical studies) or muscle (humans) SNA may lead to cycle-mediated variability that disrupts any correlation. (2) Strikingly, unlike in obese males in which the sympathoexcitatory responses to insulin or leptin can be enhanced (see above), insulin [156] and leptin (Shi and Brooks, unpublished data) fail to increase SNA in obese females; in other words, obese females come to exhibit leptin and insulin resistance rather than sensitization to its sympathoexcitatory actions. Investigations of the mechanism revealed that tonic PVN NPY sympathoinhibition is not reduced in obese females as it is in obese males, but it is unchanged [156]. More importantly, the tonic NPY sympathoinhibition was not inhibitable (by insulin), thereby imposing an irreversible blockade of sympathoexcitatory inputs.

In obese females, several possible (but unproven) mechanisms may resist the sympathoexcitory effects of leptin or insulin [24,163,166] (Figure 2): (1) POMC neurons were never found to co-express AT1aR, preventing this direct route of activation; (2) AT1aR expression is induced by progesterone, as during estrus or pregnancy (when ArcN AT1aR levels can become very high), but obesity tends to decrease progesterone levels; (3) obesity in females is not associated with central inflammation; and (4) obesity in females activates the anti-hypertensive arm of the RAS, Ang-(1-7), which could counteract AT1aR-mediated sympathoexcitation and hypertension, specifically in the ArcN.

6.2. Preserved or Enhanced Leptin Responsiveness in the PVN and DMH: Leptin Support of the HPT Axis and of BAT SNA with Obesity

In lean animals and humans, fasting suppresses the HPT axis and energy expenditure via a concurrent decrease in plasma leptin levels through reversal of leptin's tonic inhibition of ArcN NPY neurons and excitation of ArcN POMC neurons that project to PVN TRH neurons [33,34,167]. The direct effect of leptin on PVN TRH neurons is not involved, which is not surprising, given the marked paucity of LepR in the PVN compared to that in the ArcN. As described above, obesity renders ArcN NPY and POMC neurons resistant to leptin, including those that influence PVN TRH neurons. However, obese humans and diet-induced obese rats continue to demonstrate normal or even elevated TRH levels and activity of the HPT axis in both males and females [168–172]. At least in males, this enhanced HPT axis activity is instead mediated by the direct actions of leptin to stimulate PVN TRH neurons [168], perhaps through an upregulation of PVN LepR, although this has not been directly tested. With obesity, the direct action of leptin on PVN TRH presympathetic neurons that project to the RVLM may also contribute to hypertension, since knockdown of hypothalamic preproTRH normalized blood pressure in hypertensive diet-induced obese rats [173].

Finally, evidence supporting sustained activation of DMH leptin-receptive neurons was demonstrated in obese mice. More specifically, leptin activated BAT in DIO and *ob/ob* mice in the face of muted anorexic actions. Blockade of LepR in the DMH prevented the activation of BAT by leptin, implicating a role for this hypothalamic site in the sustained thermogenic effects of leptin during obesity [35].

A summary of the sites and mechanisms of select leptin resistance in males versus females is provided in Table 2.

Table 2. Selective Leptin Resistance: Summary of Sex differences.

	Males		Females	
	Lean	**Obese**	**Lean**	**Obese**
Plasma leptin levels	_	↑	↑	↑↑
Food intake with increased leptin	↓	_	↓↓ (dependent on estrogen)	_
LSNA with increased leptin	↑	↑↑	↑ (dependent on estrogen)	0
SSNA with increased leptin	↑	?	↑	?
Tonic PVN NPY sympathoinhibition	_	↓(0)	_	_
PVN POMC (α-MSH) sympathoexcitation	_	↑	_	↓
HPT axis	_	_	_	_
HPT axis: support mediated by ArcN leptin	_	↓	_	↓ (?)
HPT axis: support mediated by PVN leptin	0	↑	0 (?)	↑ (?)

Table 2. While the increased leptin levels that occur with obesity fail to suppress food intake in both sexes, its actions to increase SNA are increased in males (which can cause hypertension) but nullified in females (obese reproductively active females do not exhibit increased SNA). These sex differences appear to pivot on differential changes in the ArcN. More specifically, obese males exhibit sustained suppression of tonically inhibitory ArcN NPY neurons and activation of POMC neurons that project to the PVN. In addition, the data indirectly suggest that increased drive of PVN presympathetic TRH neurons due to increased expression of PVN LepR and leptin binding may also contribute to hypertension development. In contrast, obesity leads to leptin resistance in ArcN NPY and POMC presympathetic neurons in females. In both sexes, the activity of the HPT axis is maintained or increased. In males, there is evidence that despite the development of resistance to leptin in ArcN NPY and POMC neurons that control the HPT axis, TRH/TSH or thyroid hormone levels are normal or elevated due to increased stimulation of PVN TRH neurons by endogenous leptin, possibly in part because of elevated LepR expression. The symbol "_" denotes the response in lean males to which all other responses are compared. ↑: increased; ↑↑: increased more compared to other ↑ in rows; ↓: decreased; ↓↓: decreased more compared to other ↓ in row; ?: unknown or unclear.

6.3. The Role of Leptin in Weight Regain

Multiple studies, including those based on the show "The Biggest Loser," have clearly documented that the increased body weight of obese individuals is homeostatically defended such that it is nearly impossible to maintain weight loss [5,8,109,174]. While several factors fuel this recidivism [8,109], here we focus on the contribution of sustained leptin resetting. More specifically, elevated leptin levels fall with weight loss, which act inappropriately as a starvation signal to increase hunger and decrease energy expenditure in both men and women [8,174–176] due to activation of NPY neurons and inhibition of POMC neuron, in the ArcN (i.e., unremitting leptin resetting) [5,109]. Nevertheless, the activity of systems exhibiting sustained leptin sensitivity with obesity, such as the HPT axis and SNA to the vascular beds of cardiovascularly relevant organs, decreases with weight loss, leading to resolution of hypertension [177,178] and a decrease in TH levels. Thus, the decrease in energy expenditure (relative to body weight) is mediated in part by suppressed activity of BAT as well as of the HPT axis [174,179,180]. Evidence that irreversible leptin resetting and decreasing leptin and TH levels are a major contributor to the inability of formerly obese humans to maintain weight loss is that leptin and/or thyroxine administration increased the indices of SNA and also supported weight loss [181–183]. This sustained leptin resetting seems to dispute the hypothesized action of decreases in endogenous leptin (such as with weight loss) to enhance LepR responsiveness via decreases in negative signaling regulators or a role for maximal LepR occupation as major contributors to the leptin resetting with obesity. Therefore, the question then becomes: which mechanisms underlie sustained leptin resetting of the control of food intake in formerly obese individuals? One such mechanism pivots on the failure of obesogenic adipose inflammation to dissipate with a reduction in body weight and adipose tissue.

As described above, adipocytes enlarge with obesity, thereby promoting inflammation. However, as fat cells shrink due to a negative energy balance and weight loss in obese individuals, mechanical stress in the ECM ensues along with maintained fibrosis and accompanying sustained or even augmented adipose tissue inflammation [108,109,184–187], despite the resolution of systemic and muscle inflammation and improved glucose handling. Importantly, the sustained adipose inflammation is associated with sustained or increased hypothalamic inflammation, as deduced from the expression levels of cytokines and other inflammatory mediators in hypothalamic tissue blocks [188–190], as well as immunohistochemical assessment of iba-1, a microglial marker, in the ArcN [64,191].

Additional indirect evidence that unrelenting hypothalamic inflammation mediates sustained leptin resetting during obesity reversal is that the two treatments that most successfully support long-term weight loss and reverse leptin resetting also reverse hypothalamic inflammation: Roux-en-Y gastric bypass (RYGB) [192] and liraglutide [a glucagon-like-peptide 1 (GLP-1) agonist] treatment [64]. Unlike calorie restriction, which causes weight loss in association with increased ArcN orexigenic NPY/AgRP expression and decreased anorexigenic POMC expression, RYGB decreases body weight without changes in these neuromodulators [190,193–195]. In parallel, RYGB decreases hypothalamic indices of inflammation [190,192,196] and increases leptin sensitivity [192,194]. However, while RYGB can decrease adipose inflammation, this has not been a consistent finding [197–199]; therefore, are there other mechanisms that might explain how RYGB reverses hypothalamic inflammation? Current hypotheses stem from changes in gut function [195]. First, modification of the intestinal microbiome may reduce systemic metabolic endotoxicity, which is known to impact the hypothalamus in such varied conditions as cancer-induced cachexia and autonomic responses to bacterial toxemia [192]. Second, levels of GLP-1, a peptide released from the small intestine, are significantly increased after RYGB [195] and exert anti-inflammatory effects [200]. As mentioned above, liraglutade is a GLP-1 agonist currently being tested as an obesity reversal treatment [201,202]. Liraglutide not only inhibits or reverses ArcN microglial activation in obese subjects losing weight [64,191], but it also appears to suppress inhibitory PTP1B signaling, thereby enhancing leptin sensitivity [191,203] in part by direct actions on POMC and NPY neurons [204]. Collectively, these actions reverse leptin resetting and support sustained weight loss. Thus, sustained hypothalamic inflammation with weight loss secondary to calorie restriction correlates with sustained leptin resetting, but further work is required to mechanistically establish this link.

7. Summary and Future Research Directions

It is well-established that decreased plasma leptin levels, as with fasting, signal starvation and elicit appropriate physiological responses, such as increasing the drive to eat (and the intake of food, if available) and decreasing energy expenditure. These responses are mediated by suppression of the actions of leptin in the hypothalamus, most notably on ArcN orexigenic NPY neurons and anorexic POMC neurons. However, the main question addressed in this review is whether the effects of increased endogenous (or exogenous) leptin levels are also physiologically significant on the long-term control of energy balance and the sympathetic nervous system, despite conventional wisdom to the contrary. To summarize, we suggest the following:

While eating increases leptin levels, the rise is delayed relative to MIT; therefore, it appears that leptin is not a major player in meal-evoked thermogenesis, except in fasting individuals. On the other hand, DIT occurs over a much longer time frame and helps to limit weight gain despite chronic overeating. Current evidence suggests that increases in leptin levels contribute, but also that other unidentified hormonal mechanisms are involved.

Obesity also increases plasma leptin levels. While failing to decrease food intake due to ArcN leptin resistance or resetting (secondary to hypothalamic inflammation, maximal

LepR binding, and negative feedback mechanisms), increased leptin levels exert other significant effects. In both sexes, elevated leptin levels support the HPT axis, likely via increased actions in the PVN, possibly due to the induction of LepR expression in TRH neurons. Particularly in obese males, increased leptin levels increase SNA to the muscle and kidneys, thereby contributing to hypertension development. In sharp contrast, obese females exhibit leptin (and insulin) resistance and are less likely to develop hypertension secondary to sympathoexcitation. Obesity engenders a low-grade inflammatory state, which may also depend, in part, on increased leptin levels. Lastly, a major early but disappointing finding was that exogenous leptin administration failed to reverse obesity. Nevertheless, preliminary investigations suggest that leptin administration to individuals undergoing weight loss, when leptin levels fall due to sustained hypothalamic inflammation and leptin resetting, may help sustain weight loss. Agents that reverse hypothalamic inflammation, such as GLP-1, may also be beneficial.

Nevertheless, much remains to be discovered. Here, we suggest a few questions that, when answered, could yield high impact. (1) Do increased leptin levels contribute to obesity- or DIT-induced increases in energy expenditure (the HPT axis or BAT SNA) or hypertension via induction of its receptor in PVN or ArcN (if so, which cell type)? Does leptin induce its receptor in other brain areas? Leptin is a cytokine: what are the roles of LepR in microglia or astroglia to aid and abet obesity-induced hypothalamic inflammation? (2) What are the contributions of LPS, vagal afferent stimulation, and leptin in obesity-induced neuroinflammation? (3) Does select reversal of hypothalamic inflammation after weight loss reduce or eliminate weight regain? (4) What other catabolic hormones contribute to homeostatic BW regulation with sustained increases in food intake and/or decreases in energy expenditure? (5) How is adipose or systemic inflammation transmitted to the CNS? (6) Does anti-immune therapy with or without leptin treatment [55,205] or treatment with a GLP-1 agonist [201,202] help sustain weight loss? Minocycline has been successfully used to reverse or prevent hypothalamic microglial activation in rodents fed a high fat diet [206,207]; does this or a similar agent help sustain weight loss? (7) Finally, studies focusing on sex differences and their mechanisms are lacking. In this context, studies of the brain RAS and its role in energy balance, obesity, or obesity-induced increases in SNA and hypertension development are clearly needed.

Author Contributions: Conceptualization: V.L.B. and D.M.; Writing—Original Draft Preparation: V.L.B. and D.M.; Writing—Review & Editing: V.L.B. and D.M.; Visualization: V.L.B. All authors have read and agreed to the published version of the manuscript.

Funding: This work was funded in part by NIH Grant HL128181.

Institutional Review Board Statement: Not applicable.

Informed Consent Statement: Not applicable.

Data Availability Statement: Not applicable.

Conflicts of Interest: The authors declare no conflict of interest.

References

1. Taggart, N. Diet, activity and body-weight. A study of variations in a woman. *Br. J. Nutr.* **1962**, *16*, 223–235. [CrossRef]
2. Norberg, M.; Lindvall, K.; Jenkins, P.L.; Emmelin, M.; Lönnberg, G.; Nafziger, A.N. Self-rated health does not predict 10-year weight change among middle-aged adults in a longitudinal population study. *BMC Public Health* **2011**, *11*, 748. [CrossRef] [PubMed]
3. Chow, C.C.; Hall, K.D. Short and long-term energy intake patterns and their implications for human body weight regulation. *Physiol. Behav.* **2014**, *134*, 60–65. [CrossRef] [PubMed]
4. Friedman, J.M. Leptin and the endocrine control of energy balance. *Nat. Metab.* **2019**, *1*, 754–764. [CrossRef]
5. Schwartz, M.W.; Seeley, R.J.; Zeltser, L.M.; Drewnowski, A.; Ravussin, E.; Redman, L.M.; Leibel, R.L. Obesity Pathogenesis: An Endocrine Society Scientific Statement. *Endocr. Rev.* **2017**, *38*, 267–296. [CrossRef]
6. Leibel, R.L. The role of leptin in the control of body weight. *Nutr. Rev.* **2002**, *60*, S15–S19. [CrossRef]
7. Aronne, L.J.; Hall, K.D.; Jakicic, J.M.; Leibel, R.L.; Lowe, M.R.; Rosenbaum, M.; Klein, S. Describing the Weight-Reduced State: Physiology, Behavior, and Interventions. *Obesity* **2021**, *29* (Suppl. S1), S9–S24. [CrossRef] [PubMed]

8. Berthoud, H.R.; Seeley, R.J.; Roberts, S.B. Physiology of Energy Intake in the Weight-Reduced State. *Obesity* **2021**, *29* (Suppl. S1), S25–S30. [CrossRef] [PubMed]
9. Flier, J.S. Starvation in the Midst of Plenty: Reflections on the History and Biology of Insulin and Leptin. *Endocr. Rev.* **2019**, *40*, 1–16. [CrossRef]
10. Zhao, S.; Kusminski, C.M.; Elmquist, J.K.; Scherer, P.E. Leptin: Less Is More. *Diabetes* **2020**, *69*, 823–829. [CrossRef]
11. Halaas, J.L.; Boozer, C.; Blair-West, J.; Fidahusein, N.; Denton, D.A.; Friedman, J.M. Physiological response to long-term peripheral and central leptin infusion in lean and obese mice. *Proc. Natl. Acad. Sci. USA* **1997**, *94*, 8878–8883. [CrossRef] [PubMed]
12. Shi, Z.; Pelletier, N.E.; Wong, J.; Li, B.; Sdrulla, A.D.; Madden, C.J.; Marks, D.L.; Brooks, V.L. Leptin increases sympathetic nerve activity via induction of its own receptor in the paraventricular nucleus. *Elife* **2020**, *9*, e55357. [CrossRef] [PubMed]
13. Mitchell, S.E.; Nogueiras, R.; Morris, A.; Tovar, S.; Grant, C.; Cruickshank, M.; Rayner, D.V.; Dieguez, C.; Williams, L.M. Leptin receptor gene expression and number in the brain are regulated by leptin level and nutritional status. *J. Physiol.* **2009**, *587*, 3573–3585. [CrossRef] [PubMed]
14. Tang, C.H.; Lu, D.Y.; Yang, R.S.; Tsai, H.Y.; Kao, M.C.; Fu, W.M.; Chen, Y.F. Leptin-induced IL-6 production is mediated by leptin receptor, insulin receptor substrate-1, phosphatidylinositol 3-kinase, Akt, NF-kappaB, and p300 pathway in microglia. *J. Immunol.* **2007**, *179*, 1292–1302. [CrossRef]
15. Shi, Z.; Brooks, V.L. Leptin differentially increases sympathetic nerve activity and its baroreflex regulation in female rats: Role of oestrogen. *J. Physiol.* **2015**, *593*, 1633–1647. [CrossRef]
16. Shi, Z.; Li, B.; Brooks, V.L. Role of the Paraventricular Nucleus of the Hypothalamus in the Sympathoexcitatory Effects of Leptin. *Hypertension* **2015**, *66*, 1034–1041. [CrossRef]
17. Han, C.; Wu, W.; Ale, A.; Kim, M.S.; Cai, D. Central Leptin and Tumor Necrosis Factor-α (TNFα) in Diurnal Control of Blood Pressure and Hypertension. *J. Biol. Chem.* **2016**, *291*, 15131–15142. [CrossRef]
18. Moult, P.R.; Cross, A.; Santos, S.D.; Carvalho, A.L.; Lindsay, Y.; Connolly, C.N.; Irving, A.J.; Leslie, N.R.; Harvey, J. Leptin regulates AMPA receptor trafficking via PTEN inhibition. *J. Neurosci.* **2010**, *30*, 4088–4101. [CrossRef]
19. Münzberg, H.; Singh, P.; Heymsfield, S.B.; Yu, S.; Morrison, C.D. Recent advances in understanding the role of leptin in energy homeostasis. *F1000Research* **2020**, *9*, 451. [CrossRef]
20. Shi, Z.; Madden, C.J.; Brooks, V.L. Arcuate neuropeptide Y inhibits sympathetic nerve activity via multiple neuropathways. *J. Clin. Investig.* **2017**, *127*, 2868–2880. [CrossRef]
21. Harlan, S.M.; Morgan, D.A.; Agassandian, K.; Guo, D.F.; Cassell, M.D.; Sigmund, C.D.; Mark, A.L.; Rahmouni, K. Ablation of the leptin receptor in the hypothalamic arcuate nucleus abrogates leptin-induced sympathetic activation. *Circ. Res.* **2011**, *108*, 808–812. [CrossRef] [PubMed]
22. Prior, L.J.; Eikelis, N.; Armitage, J.A.; Davern, P.J.; Burke, S.L.; Montani, J.P.; Barzel, B.; Head, G.A. Exposure to a high-fat diet alters leptin sensitivity and elevates renal sympathetic nerve activity and arterial pressure in rabbits. *Hypertension* **2010**, *55*, 862–868. [CrossRef] [PubMed]
23. Cassaglia, P.A.; Shi, Z.; Li, B.; Reis, W.L.; Clute-Reinig, N.M.; Stern, J.E.; Brooks, V.L. Neuropeptide Y acts in the paraventricular nucleus to suppress sympathetic nerve activity and its baroreflex regulation. *J. Physiol.* **2014**, *592*, 1655–1675. [CrossRef]
24. Shi, Z.; Wong, J.; Brooks, V.L. Obesity: Sex and sympathetics. *Biol. Sex Differ.* **2020**, *11*, 10. [CrossRef] [PubMed]
25. Shi, Z.; Bonillas, A.C.; Wong, J.; Padilla, S.L.; Brooks, V.L. Neuropeptide Y suppresses thermogenic and cardiovascular sympathetic nerve activity via Y1 receptors in the paraventricular nucleus and dorsomedial hypothalamus. *J. Neuroendocrinol.* **2021**, *33*, e13006. [CrossRef] [PubMed]
26. Chitravanshi, V.C.; Kawabe, K.; Sapru, H.N. Stimulation of the hypothalamic arcuate nucleus increases brown adipose tissue nerve activity via hypothalamic paraventricular and dorsomedial nuclei. *Am. J. Physiol. Heart Circ. Physiol.* **2016**, *311*, H433–H444. [CrossRef]
27. Coote, J.H. A role for the paraventricular nucleus of the hypothalamus in the autonomic control of heart and kidney. *Exp. Physiol.* **2005**, *90*, 169–173. [CrossRef]
28. Elsaafien, K.; Kirchner, M.K.; Mohammed, M.; Eikenberry, S.A.; West, C.; Scott, K.A.; de Kloet, A.D.; Stern, J.E.; Krause, E.G. Identification of Novel Cross-Talk between the Neuroendocrine and Autonomic Stress Axes Controlling Blood Pressure. *J. Neurosci.* **2021**, *41*, 4641–4657. [CrossRef]
29. Pandit, R.; Beerens, S.; Adan, R.A.H. Role of leptin in energy expenditure: The hypothalamic perspective. *Am. J. Physiol. Regul. Integr. Comp. Physiol.* **2017**, *312*, R938–R947. [CrossRef]
30. Doslikova, B.; Tchir, D.; McKinty, A.; Zhu, X.; Marks, D.L.; Baracos, V.E.; Colmers, W.F. Convergent neuronal projections from paraventricular nucleus, parabrachial nucleus, and brainstem onto gastrocnemius muscle, white and brown adipose tissue in male rats. *J. Comp. Neurol.* **2019**, *527*, 2826–2842. [CrossRef]
31. Oldfield, B.J.; Giles, M.E.; Watson, A.; Anderson, C.; Colvill, L.M.; McKinley, M.J. The neurochemical characterisation of hypothalamic pathways projecting polysynaptically to brown adipose tissue in the rat. *Neuroscience* **2002**, *110*, 515–526. [CrossRef] [PubMed]
32. Kerem, L.; Lawson, E.A. The Effects of Oxytocin on Appetite Regulation, Food Intake and Metabolism in Humans. *Int. J. Mol. Sci.* **2021**, *22*, 7737. [CrossRef] [PubMed]
33. Hollenberg, A.N. The role of the thyrotropin-releasing hormone (TRH) neuron as a metabolic sensor. *Thyroid* **2008**, *18*, 131–139. [CrossRef] [PubMed]

34. Nillni, E.A. Regulation of the hypothalamic thyrotropin releasing hormone (TRH) neuron by neuronal and peripheral inputs. *Front. Neuroendocrinol.* **2010**, *31*, 134–156. [CrossRef]
35. Enriori, P.J.; Sinnayah, P.; Simonds, S.E.; Garcia Rudaz, C.; Cowley, M.A. Leptin action in the dorsomedial hypothalamus increases sympathetic tone to brown adipose tissue in spite of systemic leptin resistance. *J. Neurosci.* **2011**, *31*, 12189–12197. [CrossRef]
36. Zhang, Y.; Kerman, I.A.; Laque, A.; Nguyen, P.; Faouzi, M.; Louis, G.W.; Jones, J.C.; Rhodes, C.; Münzberg, H. Leptin-receptor-expressing neurons in the dorsomedial hypothalamus and median preoptic area regulate sympathetic brown adipose tissue circuits. *J. Neurosci.* **2011**, *31*, 1873–1884. [CrossRef]
37. Marsh, A.J.; Fontes, M.A.; Killinger, S.; Pawlak, D.B.; Polson, J.W.; Dampney, R.A. Cardiovascular responses evoked by leptin acting on neurons in the ventromedial and dorsomedial hypothalamus. *Hypertension* **2003**, *42*, 488–493. [CrossRef]
38. Lim, K.; Barzel, B.; Burke, S.L.; Armitage, J.A.; Head, G.A. Origin of Aberrant Blood Pressure and Sympathetic Regulation in Diet-Induced Obesity. *Hypertension* **2016**, *68*, 491–500. [CrossRef]
39. Seoane-Collazo, P.; Fernø, J.; Gonzalez, F.; Diéguez, C.; Leis, R.; Nogueiras, R.; López, M. Hypothalamic-autonomic control of energy homeostasis. *Endocrine* **2015**, *50*, 276–291. [CrossRef]
40. Shiuchi, T.; Toda, C.; Okamoto, S.; Coutinho, E.A.; Saito, K.; Miura, S.; Ezaki, O.; Minokoshi, Y. Induction of glucose uptake in skeletal muscle by central leptin is mediated by muscle β(2)-adrenergic receptor but not by AMPK. *Sci. Rep.* **2017**, *7*, 15141. [CrossRef]
41. Saito, M.; Matsushita, M.; Yoneshiro, T.; Okamatsu-Ogura, Y. Brown Adipose Tissue, Diet-Induced Thermogenesis, and Thermogenic Food Ingredients: From Mice to Men. *Front. Endocrinol.* **2020**, *11*, 222. [CrossRef] [PubMed]
42. Din, M.U.; Saari, T.; Raiko, J.; Kudomi, N.; Maurer, S.F.; Lahesmaa, M.; Fromme, T.; Amri, E.Z.; Klingenspor, M.; Solin, O.; et al. Postprandial Oxidative Metabolism of Human Brown Fat Indicates Thermogenesis. *Cell Metab.* **2018**, *28*, 207–216.e3. [CrossRef]
43. Ho, K.K.Y. Diet-induced thermogenesis: Fake friend or foe? *J. Endocrinol.* **2018**, *238*, R185–R191. [CrossRef]
44. Chan, P.C.; Hsieh, P.S. The Role and Regulatory Mechanism of Brown Adipose Tissue Activation in Diet-Induced Thermogenesis in Health and Diseases. *Int. J. Mol. Sci.* **2022**, *23*, 9448. [CrossRef] [PubMed]
45. Seoane-Collazo, P.; Martínez-Sánchez, N.; Milbank, E.; Contreras, C. Incendiary Leptin. *Nutrients* **2020**, *12*, 472. [CrossRef]
46. Romon, M.; Lebel, P.; Velly, C.; Marecaux, N.; Fruchart, J.C.; Dallongeville, J. Leptin response to carbohydrate or fat meal and association with subsequent satiety and energy intake. *Am. J. Physiol.* **1999**, *277*, E855–E861. [CrossRef]
47. Elimam, A.; Marcus, C. Meal timing, fasting and glucocorticoids interplay in serum leptin concentrations and diurnal profile. *Eur. J. Endocrinol.* **2002**, *147*, 181–188. [CrossRef]
48. Perry, R.J.; Lyu, K.; Rabin-Court, A.; Dong, J.; Li, X.; Yang, Y.; Qing, H.; Wang, A.; Yang, X.; Shulman, G.I. Leptin mediates postprandial increases in body temperature through hypothalamus-adrenal medulla-adipose tissue crosstalk. *J. Clin. Investig.* **2020**, *130*, 2001–2016. [CrossRef] [PubMed]
49. Rothwell, N.J.; Stock, M.J. A role for brown adipose tissue in diet-induced thermogenesis. *Nature* **1979**, *281*, 31–35. [CrossRef]
50. Stock, M.J. Gluttony and thermogenesis revisited. *Int. J. Obes. Relat. Metab. Disord.* **1999**, *23*, 1105–1117. [CrossRef]
51. Trayhurn, P.; Arch, J.R.S. Is energy expenditure reduced in obese mice with mutations in the leptin/leptin receptor genes? *J. Nutr. Sci.* **2020**, *9*, e23. [CrossRef] [PubMed]
52. Trayhurn, P.; Jones, P.M.; McGuckin, M.M.; Goodbody, A.E. Effects of overfeeding on energy balance and brown fat thermogenesis in obese (ob/ob) mice. *Nature* **1982**, *295*, 323–325. [CrossRef] [PubMed]
53. Leibel, R.L.; Rosenbaum, M.; Hirsch, J. Changes in energy expenditure resulting from altered body weight. *N. Engl. J. Med.* **1995**, *332*, 621–628. [CrossRef]
54. Ravussin, Y.; Edwin, E.; Gallop, M.; Xu, L.; Bartolomé, A.; Kraakman, M.J.; LeDuc, C.A.; Ferrante, A.W., Jr. Evidence for a Non-leptin System that Defends against Weight Gain in Overfeeding. *Cell Metab.* **2018**, *28*, 289–299.e5. [CrossRef]
55. Van Baak, M.A.; Mariman, E.C.M. Mechanisms of weight regain after weight loss—The role of adipose tissue. *Nat. Rev. Endocrinol.* **2019**, *15*, 274–287. [CrossRef]
56. Ludwig, D.S.; Aronne, L.J.; Astrup, A.; de Cabo, R.; Cantley, L.C.; Friedman, M.I.; Heymsfield, S.B.; Johnson, J.D.; King, J.C.; Krauss, R.M.; et al. The carbohydrate-insulin model: A physiological perspective on the obesity pandemic. *Am. J. Clin. Nutr.* **2021**, *114*, 1873–1885. [CrossRef]
57. Hall, K.D.; Farooqi, I.S.; Friedman, J.M.; Klein, S.; Loos, R.J.F.; Mangelsdorf, D.J.; O'Rahilly, S.; Ravussin, E.; Redman, L.M.; Ryan, D.H.; et al. The energy balance model of obesity: Beyond calories in, calories out. *Am. J. Clin. Nutr.* **2022**, *115*, 1243–1254. [CrossRef] [PubMed]
58. Greenway, F.L. Physiological adaptations to weight loss and factors favouring weight regain. *Int. J. Obes.* **2015**, *39*, 1188–1196. [CrossRef]
59. Zorbas, C.; Reeve, E.; Naughton, S.; Batis, C.; Whelan, J.; Waqa, G.; Bell, C. The Relationship between Feasting Periods and Weight Gain: A Systematic Scoping Review. *Curr. Obes. Rep.* **2020**, *9*, 39–62. [CrossRef]
60. Yanovski, J.A.; Yanovski, S.Z.; Sovik, K.N.; Nguyen, T.T.; O'Neil, P.M.; Sebring, N.G. A prospective study of holiday weight gain. *N. Engl. J. Med.* **2000**, *342*, 861–867. [CrossRef]
61. Myers, M.G.; Cowley, M.A.; Münzberg, H. Mechanisms of leptin action and leptin resistance. *Annu. Rev. Physiol.* **2008**, *70*, 537–556. [CrossRef] [PubMed]
62. Münzberg, H.; Björnholm, M.; Bates, S.H.; Myers, M.G., Jr. Leptin receptor action and mechanisms of leptin resistance. *Cell Mol. Life Sci.* **2005**, *62*, 642–652. [CrossRef] [PubMed]

63. Balland, E.; Chen, W.; Tiganis, T.; Cowley, M.A. Persistent Leptin Signaling in the Arcuate Nucleus Impairs Hypothalamic Insulin Signaling and Glucose Homeostasis in Obese Mice. *Neuroendocrinology* **2019**, *109*, 374–390. [CrossRef] [PubMed]
64. Liao, T.; Zhang, S.L.; Yuan, X.; Mo, W.Q.; Wei, F.; Zhao, S.N.; Yang, W.; Liu, H.; Rong, X. Liraglutide Lowers Body Weight Set Point in DIO Rats and its Relationship with Hypothalamic Microglia Activation. *Obesity* **2020**, *28*, 122–131. [CrossRef] [PubMed]
65. Zhao, S.; Zhu, Y.; Schultz, R.D.; Li, N.; He, Z.; Zhang, Z.; Caron, A.; Zhu, Q.; Sun, K.; Xiong, W.; et al. Partial Leptin Reduction as an Insulin Sensitization and Weight Loss Strategy. *Cell Metab.* **2019**, *30*, 706–719.e706. [CrossRef]
66. Dorfman, M.D.; Thaler, J.P. Hypothalamic inflammation and gliosis in obesity. *Curr. Opin. Endocrinol. Diabetes Obes.* **2015**, *22*, 325–330. [CrossRef]
67. Bastard, J.P.; Maachi, M.; Lagathu, C.; Kim, M.J.; Caron, M.; Vidal, H.; Capeau, J.; Feve, B. Recent advances in the relationship between obesity, inflammation, and insulin resistance. *Eur. Cytokine Netw.* **2006**, *17*, 4–12.
68. Gregor, M.F.; Hotamisligil, G.S. Inflammatory mechanisms in obesity. *Annu. Rev. Immunol.* **2011**, *29*, 415–445. [CrossRef]
69. Charrière, G.; Cousin, B.; Arnaud, E.; André, M.; Bacou, F.; Penicaud, L.; Casteilla, L. Preadipocyte conversion to macrophage. Evidence of plasticity. *J. Biol. Chem.* **2003**, *278*, 9850–9855. [CrossRef]
70. Phillips, C.L.; Grayson, B.E. The immune remodel: Weight loss-mediated inflammatory changes to obesity. *Exp. Biol. Med.* **2020**, *245*, 109–121. [CrossRef]
71. Aguilar-Valles, A.; Inoue, W.; Rummel, C.; Luheshi, G.N. Obesity, adipokines and neuroinflammation. *Neuropharmacology* **2015**, *96*, 124–134. [CrossRef]
72. Trayhurn, P.; Wood, I.S. Adipokines: Inflammation and the pleiotropic role of white adipose tissue. *Br. J. Nutr.* **2004**, *92*, 347–355. [CrossRef] [PubMed]
73. La Cava, A.; Matarese, G. The weight of leptin in immunity. *Nat. Rev. Immunol.* **2004**, *4*, 371–379. [CrossRef] [PubMed]
74. Howard, J.K.; Lord, G.M.; Matarese, G.; Vendetti, S.; Ghatei, M.A.; Ritter, M.A.; Lechler, R.I.; Bloom, S.R. Leptin protects mice from starvation-induced lymphoid atrophy and increases thymic cellularity in ob/ob mice. *J. Clin. Investig.* **1999**, *104*, 1051–1059. [CrossRef]
75. Siegl, D.; Annecke, T.; Johnson, B.L., 3rd; Schlag, C.; Martignoni, A.; Huber, N.; Conzen, P.; Caldwell, C.C.; Tschöp, J. Obesity-induced hyperleptinemia improves survival and immune response in a murine model of sepsis. *Anesthesiology* **2014**, *121*, 98–114. [CrossRef]
76. Dib, L.H.; Ortega, M.T.; Fleming, S.D.; Chapes, S.K.; Melgarejo, T. Bone marrow leptin signaling mediates obesity-associated adipose tissue inflammation in male mice. *Endocrinology* **2014**, *155*, 40–46. [CrossRef]
77. Frühbeck, G.; Catalán, V.; Rodríguez, A.; Ramírez, B.; Becerril, S.; Salvador, J.; Colina, I.; Gómez-Ambrosi, J. Adiponectin-leptin Ratio is a Functional Biomarker of Adipose Tissue Inflammation. *Nutrients* **2019**, *11*, 454. [CrossRef]
78. Fantuzzi, G. Adiponectin and inflammation: Consensus and controversy. *J. Allergy Clin. Immunol.* **2008**, *121*, 326–330. [CrossRef] [PubMed]
79. Pang, S.S.; Le, Y.Y. Role of resistin in inflammation and inflammation-related diseases. *Cell Mol. Immunol.* **2006**, *3*, 29–34.
80. Vendrell, J.; Broch, M.; Vilarrasa, N.; Molina, A.; Gómez, J.M.; Gutiérrez, C.; Simón, I.; Soler, J.; Richart, C. Resistin, adiponectin, ghrelin, leptin, and proinflammatory cytokines: Relationships in obesity. *Obes. Res.* **2004**, *12*, 962–971. [CrossRef]
81. Bradley, J.R. TNF-mediated inflammatory disease. *J. Pathol.* **2008**, *214*, 149–160. [CrossRef] [PubMed]
82. Pereira, S.S.; Alvarez-Leite, J.I. Low-Grade Inflammation, Obesity, and Diabetes. *Curr. Obes. Rep.* **2014**, *3*, 422–431. [CrossRef] [PubMed]
83. Tsigos, C.; Kyrou, I.; Chala, E.; Tsapogas, P.; Stavridis, J.C.; Raptis, S.A.; Katsilambros, N. Circulating tumor necrosis factor alpha concentrations are higher in abdominal versus peripheral obesity. *Metabolism* **1999**, *48*, 1332–1335. [CrossRef] [PubMed]
84. Rao, S.R. Inflammatory markers and bariatric surgery: A meta-analysis. *Inflamm. Res.* **2012**, *61*, 789–807. [CrossRef]
85. Younis, S.; Rosner, I.; Rimar, D.; Boulman, N.; Rozenbaum, M.; Odeh, M.; Slobodin, G. Interleukin 6 blockade-associated weight gain with abdominal enlargement in a patient with rheumatoid arthritis. *J. Clin. Rheumatol.* **2013**, *19*, 48–49. [CrossRef]
86. Park, H.S.; Park, J.Y.; Yu, R. Relationship of obesity and visceral adiposity with serum concentrations of CRP, TNF-alpha and IL-6. *Diabetes Res. Clin. Pract.* **2005**, *69*, 29–35. [CrossRef]
87. Askarpour, M.; Khani, D.; Sheikhi, A.; Ghaedi, E.; Alizadeh, S. Effect of Bariatric Surgery on Serum Inflammatory Factors of Obese Patients: A Systematic Review and Meta-Analysis. *Obes. Surg.* **2019**, *29*, 2631–2647. [CrossRef]
88. Deshmane, S.L.; Kremlev, S.; Amini, S.; Sawaya, B.E. Monocyte chemoattractant protein-1 (MCP-1): An overview. *J. Interferon Cytokine Res.* **2009**, *29*, 313–326. [CrossRef]
89. Dahlman, I.; Kaaman, M.; Olsson, T.; Tan, G.D.; Bickerton, A.S.; Wåhlén, K.; Andersson, J.; Nordström, E.A.; Blomqvist, L.; Sjögren, A.; et al. A unique role of monocyte chemoattractant protein 1 among chemokines in adipose tissue of obese subjects. *J. Clin. Endocrinol. Metab.* **2005**, *90*, 5834–5840. [CrossRef]
90. Tourniaire, F.; Romier-Crouzet, B.; Lee, J.H.; Marcotorchino, J.; Gouranton, E.; Salles, J.; Malezet, C.; Astier, J.; Darmon, P.; Blouin, E.; et al. Chemokine Expression in Inflamed Adipose Tissue Is Mainly Mediated by NF-κB. *PLoS ONE* **2013**, *8*, e66515. [CrossRef]
91. Dalmas, E.; Rouault, C.; Abdennour, M.; Rovere, C.; Rizkalla, S.; Bar-Hen, A.; Nahon, J.L.; Bouillot, J.L.; Guerre-Millo, M.; Clément, K.; et al. Variations in circulating inflammatory factors are related to changes in calorie and carbohydrate intakes early in the course of surgery-induced weight reduction. *Am. J. Clin. Nutr.* **2011**, *94*, 450–458. [CrossRef] [PubMed]
92. Baggiolini, M.; Clark-Lewis, I. Interleukin-8, a chemotactic and inflammatory cytokine. *FEBS Lett.* **1992**, *307*, 97–101. [CrossRef] [PubMed]

93. Kim, C.S.; Park, H.S.; Kawada, T.; Kim, J.H.; Lim, D.; Hubbard, N.E.; Kwon, B.S.; Erickson, K.L.; Yu, R. Circulating levels of MCP-1 and IL-8 are elevated in human obese subjects and associated with obesity-related parameters. *Int. J. Obes.* **2006**, *30*, 1347–1355. [CrossRef] [PubMed]
94. Bruun, J.M.; Verdich, C.; Toubro, S.; Astrup, A.; Richelsen, B. Association between measures of insulin sensitivity and circulating levels of interleukin-8, interleukin-6 and tumor necrosis factor-alpha. Effect of weight loss in obese men. *Eur. J. Endocrinol.* **2003**, *148*, 535–542. [CrossRef]
95. Alvehus, M.; Simonyte, K.; Andersson, T.; Söderström, I.; Burén, J.; Rask, E.; Mattsson, C.; Olsson, T. Adipose tissue IL-8 is increased in normal weight women after menopause and reduced after gastric bypass surgery in obese women. *Clin. Endocrinol.* **2012**, *77*, 684–690. [CrossRef] [PubMed]
96. Mosser, D.M.; Zhang, X. Interleukin-10: New perspectives on an old cytokine. *Immunol. Rev.* **2008**, *226*, 205–218. [CrossRef]
97. Gotoh, K.; Fujiwara, K.; Anai, M.; Okamoto, M.; Masaki, T.; Kakuma, T.; Shibata, H. Role of spleen-derived IL-10 in prevention of systemic low-grade inflammation by obesity [Review]. *Endocr. J.* **2017**, *64*, 375–378. [CrossRef]
98. Esposito, K.; Pontillo, A.; Giugliano, F.; Giugliano, G.; Marfella, R.; Nicoletti, G.; Giugliano, D. Association of low interleukin-10 levels with the metabolic syndrome in obese women. *J. Clin. Endocrinol. Metab.* **2003**, *88*, 1055–1058. [CrossRef]
99. Mold, C.; Gewurz, H.; Du Clos, T.W. Regulation of complement activation by C-reactive protein. *Immunopharmacology* **1999**, *42*, 23–30. [CrossRef]
100. Holloszy, J.O.; Fontana, L. Caloric restriction in humans. *Exp. Gerontol.* **2007**, *42*, 709–712. [CrossRef]
101. Ress, C.; Tschoner, A.; Engl, J.; Klaus, A.; Tilg, H.; Ebenbichler, C.F.; Patsch, J.R.; Kaser, S. Effect of bariatric surgery on circulating chemerin levels. *Eur. J. Clin. Investig.* **2010**, *40*, 277–280. [CrossRef]
102. Letterio, J.J.; Roberts, A.B. Regulation of immune responses by TGF-beta. *Annu. Rev. Immunol.* **1998**, *16*, 137–161. [CrossRef] [PubMed]
103. Alessi, M.C.; Bastelica, D.; Morange, P.; Berthet, B.; Leduc, I.; Verdier, M.; Geel, O.; Juhan-Vague, I. Plasminogen activator inhibitor 1, transforming growth factor-beta1, and BMI are closely associated in human adipose tissue during morbid obesity. *Diabetes* **2000**, *49*, 1374–1380. [CrossRef] [PubMed]
104. Cottam, D.R.; Mattar, S.G.; Barinas-Mitchell, E.; Eid, G.; Kuller, L.; Kelley, D.E.; Schauer, P.R. The chronic inflammatory hypothesis for the morbidity associated with morbid obesity: Implications and effects of weight loss. *Obes. Surg.* **2004**, *14*, 589–600. [CrossRef] [PubMed]
105. Festa, A.; D'Agostino, R., Jr.; Williams, K.; Karter, A.J.; Mayer-Davis, E.J.; Tracy, R.P.; Haffner, S.M. The relation of body fat mass and distribution to markers of chronic inflammation. *Int. J. Obes. Relat. Metab. Disord.* **2001**, *25*, 1407–1415. [CrossRef]
106. Dixon, J.B.; O'Brien, P.E. Obesity and the white blood cell count: Changes with sustained weight loss. *Obes. Surg.* **2006**, *16*, 251–257. [CrossRef]
107. Lynch, L.A.; O'Connell, J.M.; Kwasnik, A.K.; Cawood, T.J.; O'Farrelly, C.; O'Shea, D.B. Are natural killer cells protecting the metabolically healthy obese patient? *Obesity* **2009**, *17*, 601–605. [CrossRef]
108. Fischer, I.P.; Irmler, M.; Meyer, C.W.; Sachs, S.J.; Neff, F.; Hrabě de Angelis, M.; Beckers, J.; Tschöp, M.H.; Hofmann, S.M.; Ussar, S. A history of obesity leaves an inflammatory fingerprint in liver and adipose tissue. *Int. J. Obes.* **2018**, *42*, 507–517. [CrossRef] [PubMed]
109. MacLean, P.S.; Higgins, J.A.; Giles, E.D.; Sherk, V.D.; Jackman, M.R. The role for adipose tissue in weight regain after weight loss. *Obes. Rev.* **2015**, *16* (Suppl. S1), 45–54. [CrossRef]
110. Wood, I.S.; de Heredia, F.P.; Wang, B.; Trayhurn, P. Cellular hypoxia and adipose tissue dysfunction in obesity. *Proc. Nutr. Soc.* **2009**, *68*, 370–377. [CrossRef]
111. Kim, D.; Kim, J.; Yoon, J.H.; Ghim, J.; Yea, K.; Song, P.; Park, S.; Lee, A.; Hong, C.P.; Jang, M.S.; et al. CXCL12 secreted from adipose tissue recruits macrophages and induces insulin resistance in mice. *Diabetologia* **2014**, *57*, 1456–1465. [CrossRef] [PubMed]
112. Schenk, S.; Saberi, M.; Olefsky, J.M. Insulin sensitivity: Modulation by nutrients and inflammation. *J. Clin. Investig.* **2008**, *118*, 2992–3002. [CrossRef] [PubMed]
113. Basu, S.; Haghiac, M.; Surace, P.; Challier, J.C.; Guerre-Millo, M.; Singh, K.; Waters, T.; Minium, J.; Presley, L.; Catalano, P.M.; et al. Pregravid obesity associates with increased maternal endotoxemia and metabolic inflammation. *Obesity* **2011**, *19*, 476–482. [CrossRef] [PubMed]
114. Zhang, P.; Yu, Y.; Qin, Y.; Zhou, Y.; Tang, R.; Wang, Q.; Li, X.; Wang, H.; Weston-Green, K.; Huang, X.F.; et al. Alterations to the microbiota-colon-brain axis in high-fat-diet-induced obese mice compared to diet-resistant mice. *J. Nutr. Biochem.* **2019**, *65*, 54–65. [CrossRef]
115. Chompre, G.; Sambolin, L.; Cruz, M.L.; Sanchez, R.; Rodriguez, Y.; Rodríguez-Santiago, R.E.; Yamamura, Y.; Appleyard, C.B. A one month high fat diet disrupts the gut microbiome and integrity of the colon inducing adiposity and behavioral despair in male Sprague Dawley rats. *Heliyon* **2022**, *8*, e11194. [CrossRef]
116. Canale, M.P.; Manca di Villahermosa, S.; Martino, G.; Rovella, V.; Noce, A.; De Lorenzo, A.; Di Daniele, N. Obesity-related metabolic syndrome: Mechanisms of sympathetic overactivity. *Int. J. Endocrinol.* **2013**, *2013*, 865965. [CrossRef]
117. Esler, M.; Straznicky, N.; Eikelis, N.; Masuo, K.; Lambert, G.; Lambert, E. Mechanisms of sympathetic activation in obesity-related hypertension. *Hypertension* **2006**, *48*, 787–796. [CrossRef]

118. Occhinegro, A.; Wong, C.Y.; Chua, B.Y.; Jackson, D.C.; McKinley, M.J.; McAllen, R.M.; Martelli, D. The endogenous inflammatory reflex inhibits the inflammatory response to different immune challenges in mice. *Brain Behav. Immun.* **2021**, *97*, 371–375. [CrossRef]
119. McKinley, M.J.; Martelli, D.; Trevizan-Baú, P.; McAllen, R.M. Divergent splanchnic sympathetic efferent nerve pathways regulate interleukin-10 and tumour necrosis factor-α responses to endotoxaemia. *J. Physiol.* **2022**, *600*, 4521–4536. [CrossRef]
120. van der Heijden, C.; Groh, L.; Keating, S.T.; Kaffa, C.; Noz, M.P.; Kersten, S.; van Herwaarden, A.E.; Hoischen, A.; Joosten, L.A.B.; Timmers, H.; et al. Catecholamines Induce Trained Immunity in Monocytes In Vitro and In Vivo. *Circ. Res.* **2020**, *127*, 269–283. [CrossRef]
121. Straub, R.H.; Dufner, B.; Rauch, L. Proinflammatory α-Adrenergic Neuronal Regulation of Splenic IFN-γ, IL-6, and TGF-β of Mice from Day 15 onwards in Arthritis. *Neuroimmunomodulation* **2020**, *27*, 58–68. [CrossRef]
122. Oquendo, M.B.; Lorza-Gil, E.; Juarez-Lopez, D.; Wagner, R.; Birkenfeld, A.L.; Ullrich, S.; Gerst, F. Effects of adrenergic-stimulated lipolysis and cytokine production on in vitro mouse adipose tissue-islet interactions. *Sci. Rep.* **2022**, *12*, 15831. [CrossRef] [PubMed]
123. Thaler, J.P.; Yi, C.X.; Schur, E.A.; Guyenet, S.J.; Hwang, B.H.; Dietrich, M.O.; Zhao, X.; Sarruf, D.A.; Izgur, V.; Maravilla, K.R.; et al. Obesity is associated with hypothalamic injury in rodents and humans. *J. Clin. Investig.* **2012**, *122*, 153–162. [CrossRef] [PubMed]
124. Kreutzer, C.; Peters, S.; Schulte, D.M.; Fangmann, D.; Türk, K.; Wolff, S.; van Eimeren, T.; Ahrens, M.; Beckmann, J.; Schafmayer, C.; et al. Hypothalamic Inflammation in Human Obesity Is Mediated by Environmental and Genetic Factors. *Diabetes* **2017**, *66*, 2407–2415. [CrossRef] [PubMed]
125. De Souza, C.T.; Araujo, E.P.; Bordin, S.; Ashimine, R.; Zollner, R.L.; Boschero, A.C.; Saad, M.J.; Velloso, L.A. Consumption of a fat-rich diet activates a proinflammatory response and induces insulin resistance in the hypothalamus. *Endocrinology* **2005**, *146*, 4192–4199. [CrossRef]
126. Le Thuc, O.; Stobbe, K.; Cansell, C.; Nahon, J.L.; Blondeau, N.; Rovère, C. Hypothalamic Inflammation and Energy Balance Disruptions: Spotlight on Chemokines. *Front. Endocrinol.* **2017**, *8*, 197. [CrossRef]
127. Douglass, J.D.; Dorfman, M.D.; Fasnacht, R.; Shaffer, L.D.; Thaler, J.P. Astrocyte IKKβ/NF-κB signaling is required for diet-induced obesity and hypothalamic inflammation. *Mol. Metab.* **2017**, *6*, 366–373. [CrossRef]
128. Benzler, J.; Ganjam, G.K.; Pretz, D.; Oelkrug, R.; Koch, C.E.; Legler, K.; Stöhr, S.; Culmsee, C.; Williams, L.M.; Tups, A. Central inhibition of IKKβ/NF-κB signaling attenuates high-fat diet-induced obesity and glucose intolerance. *Diabetes* **2015**, *64*, 2015–2027. [CrossRef] [PubMed]
129. André, C.; Guzman-Quevedo, O.; Rey, C.; Rémus-Borel, J.; Clark, S.; Castellanos-Jankiewicz, A.; Ladeveze, E.; Leste-Lasserre, T.; Nadjar, A.; Abrous, D.N.; et al. Inhibiting Microglia Expansion Prevents Diet-Induced Hypothalamic and Peripheral Inflammation. *Diabetes* **2017**, *66*, 908–919. [CrossRef]
130. Morari, J.; Anhe, G.F.; Nascimento, L.F.; de Moura, R.F.; Razolli, D.; Solon, C.; Guadagnini, D.; Souza, G.; Mattos, A.H.; Tobar, N.; et al. Fractalkine (CX3CL1) is involved in the early activation of hypothalamic inflammation in experimental obesity. *Diabetes* **2014**, *63*, 3770–3784. [CrossRef]
131. Poon, K.; Barson, J.R.; Ho, H.T.; Leibowitz, S.F. Relationship of the Chemokine, CXCL12, to Effects of Dietary Fat on Feeding-Related Behaviors and Hypothalamic Neuropeptide Systems. *Front. Behav. Neurosci.* **2016**, *10*, 51. [CrossRef] [PubMed]
132. Pflieger, F.J.; Hernandez, J.; Schweighöfer, H.; Herden, C.; Rosengarten, B.; Rummel, C. The role of neutrophil granulocytes in immune-to-brain communication. *Temperature* **2018**, *5*, 296–307. [CrossRef] [PubMed]
133. Brandi, E.; Torres-Garcia, L.; Svanbergsson, A.; Haikal, C.; Liu, D.; Li, W.; Li, J.Y. Brain region-specific microglial and astrocytic activation in response to systemic lipopolysaccharides exposure. *Front. Aging Neurosci.* **2022**, *14*, 910988. [CrossRef] [PubMed]
134. Brochu, S.; Olivier, M.; Rivest, S. Neuronal activity and transcription of proinflammatory cytokines, IkappaBalpha, and iNOS in the mouse brain during acute endotoxemia and chronic infection with Trypanosoma brucei brucei. *J. Neurosci. Res.* **1999**, *57*, 801–816. [CrossRef]
135. Vargas-Caraveo, A.; Sayd, A.; Maus, S.R.; Caso, J.R.; Madrigal, J.L.M.; García-Bueno, B.; Leza, J.C. Lipopolysaccharide enters the rat brain by a lipoprotein-mediated transport mechanism in physiological conditions. *Sci. Rep.* **2017**, *7*, 13113. [CrossRef]
136. Lee, C.H.; Kim, H.J.; Lee, Y.S.; Kang, G.M.; Lim, H.S.; Lee, S.H.; Song, D.K.; Kwon, O.; Hwang, I.; Son, M.; et al. Hypothalamic Macrophage Inducible Nitric Oxide Synthase Mediates Obesity-Associated Hypothalamic Inflammation. *Cell Rep.* **2018**, *25*, 934–946.e935. [CrossRef]
137. Waise, T.M.Z.; Toshinai, K.; Naznin, F.; NamKoong, C.; Md Moin, A.S.; Sakoda, H.; Nakazato, M. One-day high-fat diet induces inflammation in the nodose ganglion and hypothalamus of mice. *Biochem. Biophys. Res. Commun.* **2015**, *464*, 1157–1162. [CrossRef]
138. de La Serre, C.B.; de Lartigue, G.; Raybould, H.E. Chronic exposure to low dose bacterial lipopolysaccharide inhibits leptin signaling in vagal afferent neurons. *Physiol. Behav.* **2015**, *139*, 188–194. [CrossRef]
139. Hosoi, T.; Okuma, Y.; Nomura, Y. Electrical stimulation of afferent vagus nerve induces IL-1beta expression in the brain and activates HPA axis. *Am. J. Physiol. Regul. Integr. Comp. Physiol.* **2000**, *279*, R141–R147. [CrossRef]
140. Borges Bde, C.; Rorato, R.C.; Uchoa, E.T.; Marangon, P.B.; Elias, C.F.; Antunes-Rodrigues, J.; Elias, L.L. Protein tyrosine phosphatase-1B contributes to LPS-induced leptin resistance in male rats. *Am. J. Physiol. Endocrinol. Metab.* **2015**, *308*, E40–E50. [CrossRef]
141. Van Dyken, P.; Lacoste, B. Impact of Metabolic Syndrome on Neuroinflammation and the Blood-Brain Barrier. *Front. Neurosci.* **2018**, *12*, 930. [CrossRef]

142. Rummel, C.; Inoue, W.; Poole, S.; Luheshi, G.N. Leptin regulates leukocyte recruitment into the brain following systemic LPS-induced inflammation. *Mol. Psychiatry* **2010**, *15*, 523–534. [CrossRef] [PubMed]
143. Clegg, D.J.; Brown, L.M.; Woods, S.C.; Benoit, S.C. Gonadal hormones determine sensitivity to central leptin and insulin. *Diabetes* **2006**, *55*, 978–987. [CrossRef] [PubMed]
144. Côté, I.; Green, S.M.; Toklu, H.Z.; Morgan, D.; Carter, C.S.; Tümer, N.; Scarpace, P.J. Differential physiological responses to central leptin overexpression in male and female rats. *J. Neuroendocrinol.* **2017**, *29*, jne12552. [CrossRef] [PubMed]
145. Harris, R.B.; Bowen, H.M.; Mitchell, T.D. Leptin resistance in mice is determined by gender and duration of exposure to high-fat diet. *Physiol. Behav.* **2003**, *78*, 543–555. [CrossRef]
146. Mark, A.L. Selective leptin resistance revisited. *Am. J. Physiol. Regul. Integr. Comp. Physiol.* **2013**, *305*, R566–R581. [CrossRef]
147. Bell, B.B.; Rahmouni, K. Leptin as a Mediator of Obesity-Induced Hypertension. *Curr. Obes. Rep.* **2016**, *5*, 397–404. [CrossRef]
148. Lambert, E.; Sari, C.I.; Dawood, T.; Nguyen, J.; McGrane, M.; Eikelis, N.; Chopra, R.; Wong, C.; Chatzivlastou, K.; Head, G.; et al. Sympathetic nervous system activity is associated with obesity-induced subclinical organ damage in young adults. *Hypertension* **2010**, *56*, 351–358. [CrossRef]
149. Fu, Q. Sex differences in sympathetic activity in obesity and its related hypertension. *Ann. N. Y. Acad. Sci.* **2019**, *1454*, 31–41. [CrossRef]
150. Lambert, E.; Straznicky, N.; Eikelis, N.; Esler, M.; Dawood, T.; Masuo, K.; Schlaich, M.; Lambert, G. Gender differences in sympathetic nervous activity: Influence of body mass and blood pressure. *J. Hypertens.* **2007**, *25*, 1411–1419. [CrossRef]
151. Tank, J.; Heusser, K.; Diedrich, A.; Hering, D.; Luft, F.C.; Busjahn, A.; Narkiewicz, K.; Jordan, J. Influences of gender on the interaction between sympathetic nerve traffic and central adiposity. *J. Clin. Endocrinol. Metab.* **2008**, *93*, 4974–4978. [CrossRef] [PubMed]
152. Chen, F.; Cham, J.L.; Badoer, E. High-fat feeding alters the cardiovascular role of the hypothalamic paraventricular nucleus. *Am. J. Physiol. Regul. Integr. Comp. Physiol.* **2010**, *298*, R799–R807. [CrossRef] [PubMed]
153. Shi, Z.; Zhao, D.; Cassaglia, P.A.; Brooks, V.L. Sites and sources of sympathoexcitation in obese male rats: Role of brain insulin. *Am. J. Physiol. Regul. Integr. Comp. Physiol.* **2020**, *318*, R634–R648. [CrossRef] [PubMed]
154. Lim, K.; Burke, S.L.; Head, G.A. Obesity-related hypertension and the role of insulin and leptin in high-fat-fed rabbits. *Hypertension* **2013**, *61*, 628–634. [CrossRef]
155. Simonds, S.E.; Pryor, J.T.; Ravussin, E.; Greenway, F.L.; Dileone, R.; Allen, A.M.; Bassi, J.; Elmquist, J.K.; Keogh, J.M.; Henning, E.; et al. Leptin mediates the increase in blood pressure associated with obesity. *Cell* **2014**, *159*, 1404–1416. [CrossRef]
156. Shi, Z.; Cassaglia, P.A.; Pelletier, N.E.; Brooks, V.L. Sex differences in the sympathoexcitatory response to insulin in obese rats: Role of neuropeptide Y. *J. Physiol.* **2019**, *597*, 1757–1775. [CrossRef]
157. Evans, M.C.; Lord, R.A.; Anderson, G.M. Multiple Leptin Signalling Pathways in the Control of Metabolism and Fertility: A Means to Different Ends? *Int. J. Mol. Sci.* **2021**, *22*, 9210. [CrossRef]
158. Arakawa, H.; Chitravanshi, V.C.; Sapru, H.N. The hypothalamic arcuate nucleus: A new site of cardiovascular action of angiotensin-(1-12) and angiotensin II. *Am. J. Physiol. Heart Circ. Physiol.* **2011**, *300*, H951–H960. [CrossRef]
159. Shi, Z.; Stornetta, D.S.; Stornetta, R.L.; Brooks, V.L. Arcuate Angiotensin II Increases Arterial Pressure via Coordinated Increases in Sympathetic Nerve Activity and Vasopressin Secretion. *eNeuro* **2022**, *9*, 1–23. [CrossRef]
160. Shi, Z.; Stornetta, R.L.; Stornetta, D.S.; Abbott, S.B.G.; Brooks, V.L. The arcuate nucleus: A site of synergism between Angiotensin II and leptin to increase sympathetic nerve activity and blood pressure in rats. *Neurosci. Lett.* **2022**, *785*, 136773. [CrossRef]
161. Jöhren, O.; Sanvitto, G.L.; Egidy, G.; Saavedra, J.M. Angiotensin II AT1A receptor mRNA expression is induced by estrogen-progesterone in dopaminergic neurons of the female rat arcuate nucleus. *J. Neurosci.* **1997**, *17*, 8283–8292. [CrossRef]
162. Claflin, K.E.; Sandgren, J.A.; Lambertz, A.M.; Weidemann, B.J.; Littlejohn, N.K.; Burnett, C.M.; Pearson, N.A.; Morgan, D.A.; Gibson-Corley, K.N.; Rahmouni, K.; et al. Angiotensin AT1A receptors on leptin receptor-expressing cells control resting metabolism. *J. Clin. Investig.* **2017**, *127*, 1414–1424. [CrossRef]
163. Mehay, D.; Silberman, Y.; Arnold, A.C. The Arcuate Nucleus of the Hypothalamus and Metabolic Regulation: An Emerging Role for Renin-Angiotensin Pathways. *Int. J. Mol. Sci.* **2021**, *22*, 7050. [CrossRef]
164. Cassis, L.A.; Police, S.B.; Yiannikouris, F.; Thatcher, S.E. Local adipose tissue renin-angiotensin system. *Curr. Hypertens. Rep.* **2008**, *10*, 93–98. [CrossRef]
165. Brooks, V.L.; Shi, Z.; Holwerda, S.W.; Fadel, P.J. Obesity-induced increases in sympathetic nerve activity: Sex matters. *Auton. Neurosci.* **2015**, *187*, 18–26. [CrossRef]
166. Morselli, E.; Frank, A.P.; Palmer, B.F.; Rodriguez-Navas, C.; Criollo, A.; Clegg, D.J. A sexually dimorphic hypothalamic response to chronic high-fat diet consumption. *Int. J. Obes.* **2016**, *40*, 206–209. [CrossRef]
167. Joseph-Bravo, P.; Jaimes-Hoy, L.; Uribe, R.M.; Charli, J.L. 60 YEARS OF NEUROENDOCRINOLOGY: TRH, the first hypophysiotropic releasing hormone isolated: Control of the pituitary-thyroid axis. *J. Endocrinol.* **2015**, *226*, T85–T100. [CrossRef]
168. Perello, M.; Cakir, I.; Cyr, N.E.; Romero, A.; Stuart, R.C.; Chiappini, F.; Hollenberg, A.N.; Nillni, E.A. Maintenance of the thyroid axis during diet-induced obesity in rodents is controlled at the central level. *Am. J. Physiol. Endocrinol. Metab.* **2010**, *299*, E976–E989. [CrossRef]
169. Araujo, R.L.; Andrade, B.M.; Padrón, A.S.; Gaidhu, M.P.; Perry, R.L.; Carvalho, D.P.; Ceddia, R.B. High-fat diet increases thyrotropin and oxygen consumption without altering circulating 3,5,3′-triiodothyronine (T3) and thyroxine in rats: The role of iodothyronine deiodinases, reverse T3 production, and whole-body fat oxidation. *Endocrinology* **2010**, *151*, 3460–3469. [CrossRef]

170. Reinehr, T. Obesity and thyroid function. *Mol. Cell Endocrinol.* **2010**, *316*, 165–171. [CrossRef]
171. Iacobellis, G.; Ribaudo, M.C.; Zappaterreno, A.; Iannucci, C.V.; Leonetti, F. Relationship of thyroid function with body mass index, leptin, insulin sensitivity and adiponectin in euthyroid obese women. *Clin. Endocrinol.* **2005**, *62*, 487–491. [CrossRef] [PubMed]
172. Sari, R.; Balci, M.K.; Altunbas, H.; Karayalcin, U. The effect of body weight and weight loss on thyroid volume and function in obese women. *Clin. Endocrinol.* **2003**, *59*, 258–262. [CrossRef] [PubMed]
173. Landa, M.S.; García, S.I.; Schuman, M.L.; Burgueño, A.; Alvarez, A.L.; Saravia, F.E.; Gemma, C.; Pirola, C.J. Knocking down the diencephalic thyrotropin-releasing hormone precursor gene normalizes obesity-induced hypertension in the rat. *Am. J. Physiol. Endocrinol. Metab.* **2007**, *292*, E1388–E1394. [CrossRef] [PubMed]
174. Ravussin, E.; Smith, S.R.; Ferrante, A.W., Jr. Physiology of Energy Expenditure in the Weight-Reduced State. *Obesity* **2021**, *29* (Suppl. S1), S31–S38. [CrossRef] [PubMed]
175. Schwartz, A.; Doucet, E. Relative changes in resting energy expenditure during weight loss: A systematic review. *Obes. Rev.* **2010**, *11*, 531–547. [CrossRef] [PubMed]
176. Busetto, L.; Bettini, S.; Makaronidis, J.; Roberts, C.A.; Halford, J.C.G.; Batterham, R.L. Mechanisms of weight regain. *Eur. J. Intern. Med.* **2021**, *93*, 3–7. [CrossRef]
177. Fantin, F.; Giani, A.; Zoico, E.; Rossi, A.P.; Mazzali, G.; Zamboni, M. Weight Loss and Hypertension in Obese Subjects. *Nutrients* **2019**, *11*, 1667. [CrossRef]
178. Cohen, J.B. Hypertension in Obesity and the Impact of Weight Loss. *Curr. Cardiol. Rep.* **2017**, *19*, 98. [CrossRef]
179. Knuth, N.D.; Johannsen, D.L.; Tamboli, R.A.; Marks-Shulman, P.A.; Huizenga, R.; Chen, K.Y.; Abumrad, N.N.; Ravussin, E.; Hall, K.D. Metabolic adaptation following massive weight loss is related to the degree of energy imbalance and changes in circulating leptin. *Obesity* **2014**, *22*, 2563–2569. [CrossRef]
180. Rosenbaum, M.; Hirsch, J.; Murphy, E.; Leibel, R.L. Effects of changes in body weight on carbohydrate metabolism, catecholamine excretion, and thyroid function. *Am. J. Clin. Nutr.* **2000**, *71*, 1421–1432. [CrossRef]
181. Kissileff, H.R.; Thornton, J.C.; Torres, M.I.; Pavlovich, K.; Mayer, L.S.; Kalari, V.; Leibel, R.L.; Rosenbaum, M. Leptin reverses declines in satiation in weight-reduced obese humans. *Am. J. Clin. Nutr.* **2012**, *95*, 309–317. [CrossRef] [PubMed]
182. Rosenbaum, M.; Goldsmith, R.; Bloomfield, D.; Magnano, A.; Weimer, L.; Heymsfield, S.; Gallagher, D.; Mayer, L.; Murphy, E.; Leibel, R.L. Low-dose leptin reverses skeletal muscle, autonomic, and neuroendocrine adaptations to maintenance of reduced weight. *J. Clin. Investig.* **2005**, *115*, 3579–3586. [CrossRef] [PubMed]
183. Rosenbaum, M.; Goldsmith, R.L.; Haddad, F.; Baldwin, K.M.; Smiley, R.; Gallagher, D.; Leibel, R.L. Triiodothyronine and leptin repletion in humans similarly reverse weight-loss-induced changes in skeletal muscle. *Am. J. Physiol. Endocrinol. Metab.* **2018**, *315*, E771–E779. [CrossRef] [PubMed]
184. Cottam, M.A.; Caslin, H.L.; Winn, N.C.; Hasty, A.H. Multiomics reveals persistence of obesity-associated immune cell phenotypes in adipose tissue during weight loss and weight regain in mice. *Nat. Commun.* **2022**, *13*, 2950. [CrossRef] [PubMed]
185. Zamarron, B.F.; Mergian, T.A.; Cho, K.W.; Martinez-Santibanez, G.; Luan, D.; Singer, K.; DelProposto, J.L.; Geletka, L.M.; Muir, L.A.; Lumeng, C.N. Macrophage Proliferation Sustains Adipose Tissue Inflammation in Formerly Obese Mice. *Diabetes* **2017**, *66*, 392–406. [CrossRef] [PubMed]
186. Shirakawa, K.; Endo, J.; Katsumata, Y.; Yamamoto, T.; Kataoka, M.; Isobe, S.; Yoshida, N.; Fukuda, K.; Sano, M. Negative legacy of obesity. *PLoS ONE* **2017**, *12*, e0186303. [CrossRef]
187. Schmitz, J.; Evers, N.; Awazawa, M.; Nicholls, H.T.; Brönneke, H.S.; Dietrich, A.; Mauer, J.; Blüher, M.; Brüning, J.C. Obesogenic memory can confer long-term increases in adipose tissue but not liver inflammation and insulin resistance after weight loss. *Mol. Metab.* **2016**, *5*, 328–339. [CrossRef]
188. Wang, X.; Ge, A.; Cheng, M.; Guo, F.; Zhao, M.; Zhou, X.; Liu, L.; Yang, N. Increased hypothalamic inflammation associated with the susceptibility to obesity in rats exposed to high-fat diet. *Exp. Diabetes Res.* **2012**, *2012*, 847246. [CrossRef]
189. de Castro, J.M.; Stein, D.J.; Medeiros, H.R.; de Oliveira, C.; Torres, I.L.S. Nicotinamide Riboside Neutralizes Hypothalamic Inflammation and Increases Weight Loss Without Altering Muscle Mass in Obese Rats Under Calorie Restriction: A Preliminary Investigation. *Front. Nutr.* **2021**, *8*, 648893. [CrossRef]
190. Patkar, P.P.; Hao, Z.; Mumphrey, M.B.; Townsend, R.L.; Berthoud, H.R.; Shin, A.C. Unlike calorie restriction, Roux-en-Y gastric bypass surgery does not increase hypothalamic AgRP and NPY in mice on a high-fat diet. *Int. J. Obes.* **2019**, *43*, 2143–2150. [CrossRef]
191. Barreto-Vianna, A.R.; Aguila, M.B.; Mandarim-de-Lacerda, C.A. Effects of liraglutide in hypothalamic arcuate nucleus of obese mice. *Obesity* **2016**, *24*, 626–633. [CrossRef] [PubMed]
192. Chen, J.; Haase, N.; Haange, S.B.; Sucher, R.; Münzker, J.; Jäger, E.; Schischke, K.; Seyfried, F.; von Bergen, M.; Hankir, M.K.; et al. Roux-en-Y gastric bypass contributes to weight loss-independent improvement in hypothalamic inflammation and leptin sensitivity through gut-microglia-neuron-crosstalk. *Mol. Metab.* **2021**, *48*, 101214. [CrossRef]
193. Frank, A.P.; Zechner, J.F.; Clegg, D.J. Gastric Bypass Surgery but not Caloric Restriction Improves Reproductive Function in Obese Mice. *Obes. Surg.* **2016**, *26*, 467–473. [CrossRef] [PubMed]
194. Liu, J.Y.; Mu, S.; Zhang, S.P.; Guo, W.; Li, Q.F.; Xiao, X.Q.; Zhang, J.; Wang, Z.H. Roux-en-Y gastric bypass surgery suppresses hypothalamic PTP1B protein level and alleviates leptin resistance in obese rats. *Exp. Ther. Med.* **2017**, *14*, 2536–2542. [CrossRef]
195. Manning, S.; Pucci, A.; Batterham, R.L. Roux-en-Y gastric bypass: Effects on feeding behavior and underlying mechanisms. *J. Clin. Investig.* **2015**, *125*, 939–948. [CrossRef] [PubMed]

196. Hankir, M.K.; Rullmann, M.; Seyfried, F.; Preusser, S.; Poppitz, S.; Heba, S.; Gousias, K.; Hoyer, J.; Schütz, T.; Dietrich, A.; et al. Roux-en-Y gastric bypass surgery progressively alters radiologic measures of hypothalamic inflammation in obese patients. *JCI Insight* **2019**, *4*, e131329. [CrossRef]
197. Kratz, M.; Hagman, D.K.; Kuzma, J.N.; Foster-Schubert, K.E.; Chan, C.P.; Stewart, S.; van Yserloo, B.; Westbrook, E.O.; Arterburn, D.E.; Flum, D.R.; et al. Improvements in glycemic control after gastric bypass occur despite persistent adipose tissue inflammation. *Obesity* **2016**, *24*, 1438–1445. [CrossRef]
198. Billeter, A.T.; Vittas, S.; Israel, B.; Scheurlen, K.M.; Hidmark, A.; Fleming, T.H.; Kopf, S.; Büchler, M.W.; Müller-Stich, B.P. Gastric bypass simultaneously improves adipose tissue function and insulin-dependent type 2 diabetes mellitus. *Langenbecks Arch. Surg.* **2017**, *402*, 901–910. [CrossRef]
199. Neff, K.J.; O'Donohoe, P.K.; le Roux, C.W. Anti-inflammatory effects of gastric bypass surgery and their association with improvement in metabolic profile. *Expert Rev. Endocrinol. Metab.* **2015**, *10*, 435–446. [CrossRef]
200. Bendotti, G.; Montefusco, L.; Lunati, M.E.; Usuelli, V.; Pastore, I.; Lazzaroni, E.; Assi, E.; Seelam, A.J.; El Essawy, B.; Jang, J.; et al. The anti-inflammatory and immunological properties of GLP-1 Receptor Agonists. *Pharmacol. Res.* **2022**, *182*, 106320. [CrossRef]
201. O'Neil, P.M.; Birkenfeld, A.L.; McGowan, B.; Mosenzon, O.; Pedersen, S.D.; Wharton, S.; Carson, C.G.; Jepsen, C.H.; Kabisch, M.; Wilding, J.P.H. Efficacy and safety of semaglutide compared with liraglutide and placebo for weight loss in patients with obesity: A randomised, double-blind, placebo and active controlled, dose-ranging, phase 2 trial. *Lancet* **2018**, *392*, 637–649. [CrossRef] [PubMed]
202. Phillips, A.; Clements, J.N. Clinical review of subcutaneous semaglutide for obesity. *J. Clin. Pharm. Ther.* **2022**, *47*, 184–193. [CrossRef] [PubMed]
203. Kanoski, S.E.; Ong, Z.Y.; Fortin, S.M.; Schlessinger, E.S.; Grill, H.J. Liraglutide, leptin and their combined effects on feeding: Additive intake reduction through common intracellular signalling mechanisms. *Diabetes Obes. Metab.* **2015**, *17*, 285–293. [CrossRef] [PubMed]
204. He, Z.; Gao, Y.; Lieu, L.; Afrin, S.; Cao, J.; Michael, N.J.; Dong, Y.; Sun, J.; Guo, H.; Williams, K.W. Direct and indirect effects of liraglutide on hypothalamic POMC and NPY/AgRP neurons—Implications for energy balance and glucose control. *Mol. Metab.* **2019**, *28*, 120–134. [CrossRef]
205. Tam, C.S.; Lecoultre, V.; Ravussin, E. Novel strategy for the use of leptin for obesity therapy. *Expert Opin. Biol. Ther.* **2011**, *11*, 1677–1685. [CrossRef]
206. Coker, C.R.; White, M.; Singal, A.; Bingaman, S.S.; Paul, A.; Arnold, A.C.; Silberman, Y. Minocycline Reduces Hypothalamic Microglia Activation and Improves Metabolic Dysfunction in High Fat Diet-Induced Obese Mice. *Front. Physiol.* **2022**, *13*, 933706. [CrossRef]
207. Vaughn, A.C.; Cooper, E.M.; DiLorenzo, P.M.; O'Loughlin, L.J.; Konkel, M.E.; Peters, J.H.; Hajnal, A.; Sen, T.; Lee, S.H.; de La Serre, C.B.; et al. Energy-dense diet triggers changes in gut microbiota, reorganization of gut-brain vagal communication and increases body fat accumulation. *Acta Neurobiol. Exp.* **2017**, *77*, 18–30. [CrossRef]

International Journal of
Molecular Sciences

MDPI

Review

Chronic Pain-Associated Cardiovascular Disease: The Role of Sympathetic Nerve Activity

Christian A. Reynolds [1,2,*] and Zeljka Minic [1,2,*]

1 Department of Emergency Medicine, Wayne State University School of Medicine, 540 E Canfield St., Detroit, MI 48201, USA
2 Department of Biotechnology, University of Rijeka, 51000 Rijeka, Croatia
* Correspondence: careynol@med.wayne.edu (C.A.R.); zminic@med.wayne.edu (Z.M.); Tel.: +1-313-577-8648 (C.A.R.); Fax: +1-313-993-7703 (C.A.R.)

Abstract: Chronic pain affects many people world-wide, and this number is continuously increasing. There is a clear link between chronic pain and the development of cardiovascular disease through activation of the sympathetic nervous system. The purpose of this review is to provide evidence from the literature that highlights the direct relationship between sympathetic nervous system dysfunction and chronic pain. We hypothesize that maladaptive changes within a common neural network regulating the sympathetic nervous system and pain perception contribute to sympathetic overactivation and cardiovascular disease in the setting of chronic pain. We review clinical evidence and highlight the basic neurocircuitry linking the sympathetic and nociceptive networks and the overlap between the neural networks controlling the two.

Keywords: sympathetic nerve activity; chronic pain; cardiovascular disease

Citation: Reynolds, C.A.; Minic, Z. Chronic Pain-Associated Cardiovascular Disease: The Role of Sympathetic Nerve Activity. *Int. J. Mol. Sci.* **2023**, *24*, 5378. https://doi.org/10.3390/ijms24065378

Academic Editor: Yutang Wang

Received: 15 December 2022
Revised: 26 February 2023
Accepted: 9 March 2023
Published: 11 March 2023

1. Introduction

Chronic pain syndromes affect 20–30% of the world's population [1] and there is a clear link between chronic pain and the development of cardiovascular disease, which exists across a spectrum of chronic pain syndromes, including back or pelvic pain, neuropathic pain, and fibromyalgia. A meta-analysis that included 25 large observational studies of patients across the spectrum of chronic pain syndromes, found a significant association between chronic pain and cardiovascular disease [2]. A second meta-analysis found a similar relationship between chronic musculoskeletal pain and cardiovascular disease [3]. Multiple longitudinal studies have also reported an association between chronic pain and the development of cardiovascular disease [4–16], which (for many of the studies) remained significant after adjustment for cardiovascular risk factors [4,9,11–16]. The largest of these studies recently reported on the relationship between chronic pain and cardiovascular disease among 475,171 participants in the UK Biobank [16]. The results of this large study indicate that participants with chronic localized pain and chronic widespread pain had a significantly increased risk for future incidence of myocardial infarction, heart failure, stroke, cardiovascular mortality, and composite cardiovascular disease. The study also provides direct evidence of a dose–response relationship between chronic pain severity and cardiovascular morbidity. As summarized in Table 1, multiple studies have identified increased risk among chronic pain sufferers for myocardial infarction, angina, arrhythmia, coronary artery disease, hypertension, stroke, heart failure, and cardiovascular mortality.

Table 1. Chronic pain-associated cardiovascular outcomes.

Cardiovascular Outcome	References
Myocardial Infarction	[10,14,16–22]
Angina	[14,17,18,22–24]
Arrhythmia	[20,25]
Coronary Artery Disease	[14,22,25–27]
Hypertension	[20,25,28–30]
Stroke	[10,16,17,19–21,29]
Heart failure	[16,20]
Cardiovascular mortality	[8,16,21,22,30–33]

Scientific evidence linking pain to the cardiovascular system dates to the early 1900s when Sir Charles Sherrington observed that experimental pain triggers acute cardiovascular responses [34,35]. Over the decades, a wealth of literature reporting on pharmacological, neuroanatomical, electrophysiological, and behavioral data indicated that a highly conserved neural network regulates both the sympathetic nervous system (SNS) and pain perception. We hypothesize that chronic pain-related changes within these neural networks contribute to increased cardiovascular disease in chronic pain sufferers. In addition to reviewing available clinical evidence, we emphasize the basic neurocircuitry driving cardiovascular disease in chronic pain syndromes and highlight the causative role of the SNS in precipitating chronic pain-associated cardiovascular comorbidities. While SNS overactivation in chronic pain patients may be partially explained by the direct effects of nociceptive stimulation on sympathetic preganglionic neurons [36–38], we highlight the central neural network regulating sympathetic nerve activity (SNA) and pain perception, which we propose contributes to SNS overactivation in the setting of chronic pain.

2. Sympathetic Nervous System and Cardiovascular Disease

The SNS is critical for general cardiovascular homeostasis [39] as it exerts direct actions on heart rate and cardiac contractility as well as venous capacitance, arteriolar resistance, and blood volume (via sodium and water balance in the kidneys). However, excessive SNS activation increases cardiovascular morbidity and mortality [39,40]. Acute overactivation of the SNS can manifest as adverse cardiac events including: ventricular arrhythmias, myocardial infarction, atrial fibrillation, stroke, and Takotsubo cardiomyopathy [40]. Chronic overactivation of the SNS leads to hypertension, ischemic heart disease, heart failure, and renal failure [39].

In chronic pain, cardiovascular indices that indicate SNS overactivation, including increased blood pressure and heart rate, have been well documented [41–45]. Additionally, SNS overactivation contributes to the process of atherosclerosis by inducing platelet activation [46–48] and promoting mechanical injury to the vascular endothelial cells because of increased blood pressure and flow velocity. At the level of the heart, the ensuing atherosclerosis manifests with coronary artery disease and can trigger myocardial infarction. Furthermore, chronic, and excessive sympathetic drive to the heart (i) limits myocardial oxygen delivery via coronary vasoconstriction, while also (ii) increasing myocardial oxygen demand because of increased energy utilization. Ultimately, this combination of reduced oxygen delivery and increased oxygen demand leads to myocardial ischemia, which can result in angina, arrhythmia, or even heart failure [49].

3. Basic Neurocircuitry of the Sympathetic Nervous System

Sympathetic preganglionic neurons are cholinergic neurons located within the intermediolateral (IML) cell column of the spinal cord. These neurons innervate pre- and para-vertebral ganglia where they synapse with adrenergic postganglionic neurons. One exception is the adrenal gland, which is directly innervated by sympathetic preganglionic neurons and drives the systemic release of epinephrine. The regulation of sympathetic preganglionic neuron-firing is influenced by various interconnected neuronal networks, including: (i) the central autonomic network providing descending projections from the brain, and (ii) the intraspinal network consisting of propriospinal neurons and spinal interneurons. The level of sympathetic activity targeting the effector organs is ultimately determined by the balance of excitatory and inhibitory inputs to sympathetic preganglionic neurons in the IML.

The central autonomic network serves an essential physiological role in the coordinated, real-time adaptations of SNS in response to external and internal stimuli. Brain regions, which comprise the central autonomic network include: (i) brainstem centers (both medulla and midbrain), (ii) diencephalon, and (iii) cortical sites; all of which work in a coordinated manner to influence sympathetic tone and cardiovascular function [50]. Arguably, to date, most is known about the role of brainstem centers in controlling SNA. It was over 100 years ago when the rostral ventrolateral medulla (RVLM) was first identified as a primary regulator of sympathetic tone [51,52] as bilateral lesioning or pharmacological blockade of the ventrolateral medulla produced decreases in sympathetic activity and arterial hypotension [51–53]. Since then, many studies have substantiated this initial observation and highlighted the RVLM as one of the most important sources of descending excitatory drive to sympathetic preganglionic neurons located in the IML [54–58]. Additionally, the RVLM and adjacent brainstem centers are involved in baroreflex processing, which is critical for quick reflex-mediated changes in SNA (Figure 1). Afferent neurons innervating the carotid sinus and aortic arch carry afferent, baroreceptor-related information, to the medulla, via the vagal and glossopharyngeal nerves. During baroreceptor loading, as occurs during increases in blood pressure, the nucleus of the solitary tract (NTS) receives excitatory glutamatergic input from these afferent neurons and activates inhibitory GABAergic neurons within the caudal ventrolateral medulla (CVLM). The CVLM, in turn, inhibits RVLM neurons and decreases the excitatory drive to the sympathetic preganglionic neurons located in the IML. It is important to note that in addition to decreasing excitatory RVLM drive, baroreceptor loading increases descending inhibitory drive to sympathetic preganglionic neurons located in the IML [59–61]. Such descending inhibitory projections may arise from nuclei of the CVLM, the reticular formation, or the locus coeruleus (LC), which also receive direct projections from barosensitive NTS neurons. While the central neural networks involved in driving descending inhibition of IML neurons are not well defined, in response to baroreceptor loading, SNA can be reduced to levels similar to those observed following ganglionic blockade (Figure 2, panel A), which is well below the levels recorded following bilateral pharmacological inhibition of RVLM [62] or spinal cord transection (Figure 2, panel B) [63–66]. These data suggest that: (i) intraspinal networks alone can maintain baseline SNA tone and blood pressure and that (ii) the inhibition of SNA in response to baroreceptor loading (phenylephrine infusion) originates from within the supraspinal centers, possibly the LC or other brainstem centers.

In addition to reflex regulation, cognitive processing and emotion regulation are important drivers of sympathetic activity involving the central autonomic network. Classical cardiovascular responses to emotional stress (fight and flight response or defense reaction) are produced by disinhibition of the paraventricular and dorsomedial nuclei of the hypothalamus and drive sympathoexcitatory responses and increases in blood pressure (Figure 3) [67–69]. Similarly, increases in SNA are observed during strenuous cognitive tasks, i.e., mental stress and they seem to be gender specific [70,71]. To sustain increases in sympathetic activity in response to emotional stress, the medullary baroreflex circuitry is reset or overridden by the higher brain structures, which helps to prepare the body for action [72–75]. Similarly, baroreflex resetting occurs during perceived mental stress and highlights the important role of perception in controlling SNS [76]. Such baroreflex resetting is believed to involve direct cortico-medullary projections converging in the NTS to attenuate the medullary baroreflex processing [77,78]. Emotional and mental stress-induced increases in sympathetic activity involve various cortical brain regions some of which have been implicated in tonic control of sympathetic activity at rest [79–82]. For example, electrical stimulation of discrete areas of the insular cortex can markedly increase blood pressure by approximately 50 mmHg and increase heart rate by about 40 bpm [83]. In addition to the insular cortex, other important cortical regions are the anterior cingulate cortex and prefrontal cortex. These regions form an integrative network where the insular and anterior cingulate cortex send projections to the hypothalamus and the amygdala [80]. The amygdala as a central feedback regulator sends inhibitory projections to the NTS, thereby deactivating the medullary baroreflex mechanism and activating the hypothalamic nuclei, which drive sympathoexcitatory responses associated with stress reaction [84].

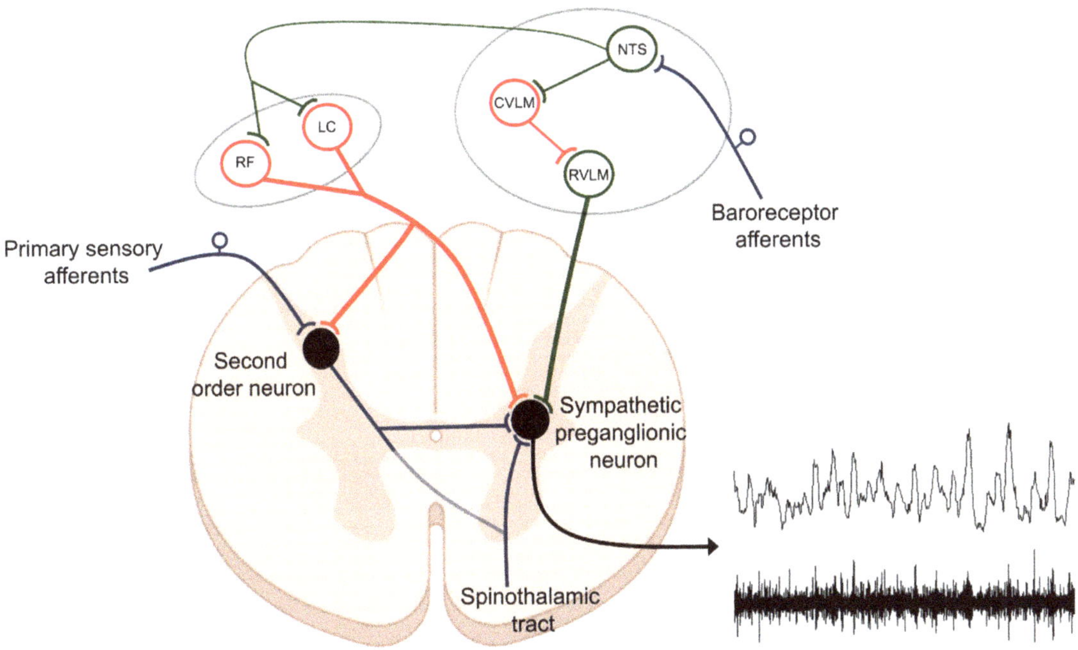

Figure 1. Control of sympathetic preganglionic neurons. Efferent sympathetic nerve activity (SNA) is determined by the net product of excitatory and inhibitory projections to sympathetic preganglionic neurons located within the intermediolateral (IML) cell column of the thoracic spinal cord. The rostral ventrolateral medulla (RVLM) is one of the main sources of descending excitatory drive (green). Additionally, second-order sensory neurons, relaying information from primary sensory afferents, are an important intraspinal source of excitatory drive to sympathetic preganglionic neurons. Descending inhibitory projections to sympathetic preganglionic neurons (red) arise from multiple brain regions including from the locus coeruleus (LC) and the reticular formation (RF). Much of the descending excitatory and inhibitory drive to sympathetic preganglionic neurons is regulated by neurons located within the nucleus of the solitary tract (NTS), which is the primary integrative site for baroreceptor and chemoreceptor afferent fibers which drive autonomic reflexes. Included above is an example of raw (lower tracing) and integrated (upper tracing) SNA recorded from postganglionic renal sympathetic nerve fibers in a rat as described in [63] (lower right).

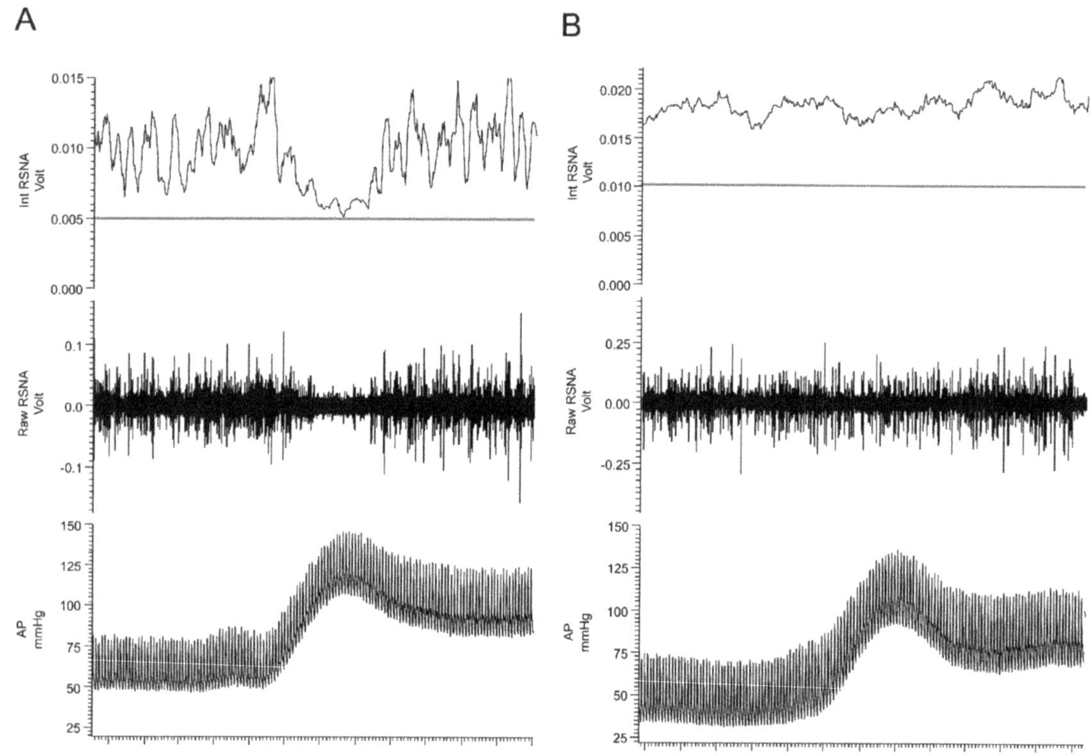

Figure 2. Baroreceptor loading strongly inhibits sympathetic nerve activity. Responses in arterial pressure (AP) and renal sympathetic nerve activity (RSNA) to intravenous administration of phenylephrine (40 μg/kg) in (**A**) spinal intact and (**B**) T4 spinal-transected rats. The red line in the top panel represents the level of recorded activity (zero) at the end of the experiment following ganglionic blockade (hexamethonium, 20 mg/kg). See text for additional information.

In addition to the descending excitatory and inhibitory projections controlling sympathetic activity outlined above, sympathetic preganglionic neurons within the IML receive extensive input from intraspinal networks [85,86]. Spinal circuits can contribute both to baseline and reflex responses in sympathetic activity. Resting sympathetic activity and blood pressure are maintained (albeit at a lower level), following cervical or thoracic spinal cord transection and they can be further reduced by ganglionic blockade [63,64]. Although tonically low, sympathetic activity can be markedly increased following the activation of spinal sympathetic pathways. The spinal sympathetic reflex circuitry is similar in composition to the spinal motor reflex circuits that drive muscle contraction in response to muscle stretch or cutaneous stimulation, and like spinal motor reflexes, the spinal sympathetic reflexes are exaggerated by spinal cord transection [87–91]. The spinal motoneurons are controlled by the descending corticothalamic projections, which exhibit a tonic inhibitory drive to the motoneurons [92,93]. Similarly, we speculate that descending inhibitory projections arising from or above the level of the brainstem, modulate sympathetic drive. Tonic sympathoinhibitory projections are important regulators of overall sympathetic activity and they can arise from various central structures including, the CVLM [94,95], raphe nuclei [96], and the spinal cord [97]. These structures have been shown to inhibit sympathetic activity tonically [95,98,99]. This inhibition likely involves both direct bulbospinal projections [100–102], and projections to the RVLM [94,95,98]. Additional baroreceptor-independent inhibitory projections descend via the dorsolateral funiculus and reduce spinal sympathetic reflex activation [103,104]. Finally, the inhibitory influences on the sympathetic preganglionic neurons can be exerted by the spinal interneurons located within and around the IML. To that end, electrical stimulation of the central autonomic area of the spinal cord triggered inhibitory postsynaptic potentials within the IML, and these potentials were blocked by the GABA receptor blockade, suggesting that spinal GABA-ergic interneurons modulate sympathetic preganglionic transmission and influence tonic sympathetic activity [105].

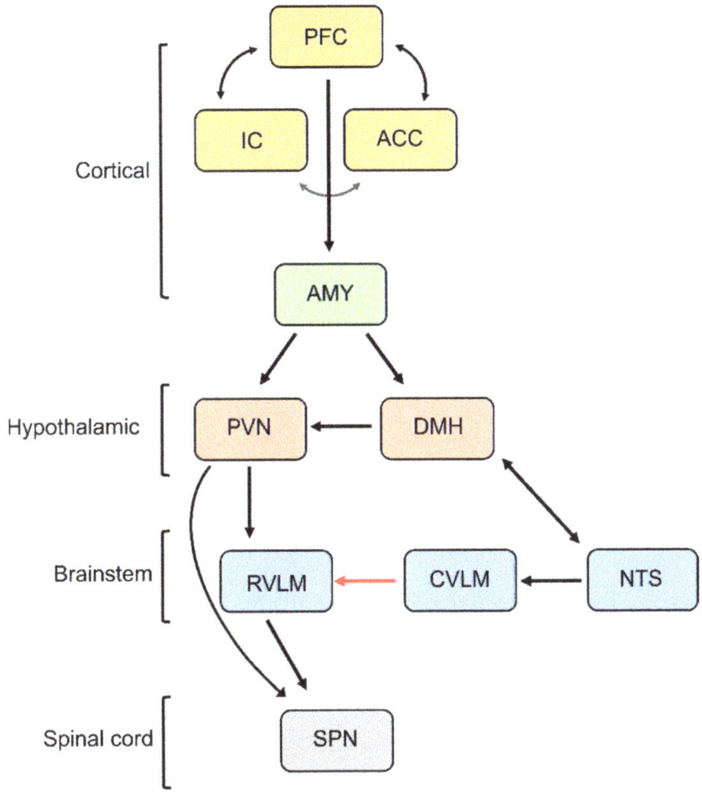

Figure 3. Central autonomic network. Depicted is the neural circuitry controlling SNA during emotional stress. PFC, prefrontal cortex; IC, insular cortex; ACC, anterior cingulate cortex; AMY, amygdala; DMH, dorsomedial hypothalamus; PVN, paraventricular nucleus; RVLM, rostral ventrolateral medulla; CVLM, caudal ventrolateral medulla; NTS, nucleus of the solitary tract; SPN, sympathetic preganglionic neurons. See text for additional information.

4. Basic Neurocircuitry of Nociception

In this section, we try to briefly summarize functional neural circuitry involved in the detection, transmission, and central integration of nociceptive signals. Based on the presence of thermal, chemical, or mechanical stimuli, the signal is detected by specialized receptors (nociceptors) expressed on peripheral terminals of small thinly myelinated or unmyelinated A-delta and C fibers of primary afferent neurons, respectively [106,107]. These afferents carry nociceptive information into laminae I-III of the dorsal horn of the spinal cord where they undergo complex processing. The excitatory and inhibitory interneuronal circuitry integrates incoming information and transmits it to projecting neurons for relay to the brain. Nociceptive information travels to the brain via the contralateral spinothalamic tract synapsing within the ventromedial and mediodorsal nuclei of the thalamus [108]. The nociceptive information also reaches the medulla and brainstem centers via the spinoreticular and spinomesencephalic tracts and the hypothalamus via the spinohypothalamic tract [109]. For the processing of nociceptive signals, especially important is the NTS, which integrates ascending noxious and non-noxious information and provides an interface between sensory and autonomic outflows. The NTS has dense projections with other sub-cortical brain regions such as periaqueductal gray, nucleus raphe magnus, and the hypothalamic nuclei, all of which are also involved in autonomic processing [108]. Finally, nociceptive information is subjected to processing at the cortical level where pain is conceptualized through localization and intensity discrimination. Regarding its localization, somatosensory cortices I and II are mostly responsible for determining the position of nociceptive stimulus while anterior cingulate cortex is involved in assigning an affective component to the pain [110,111]. Furthermore, the insular cortex can encode the intensity, and localization of the painful stimulus as it relates to the generation of moods and feelings [112]. It is the strong relationship between the anterior cingulate cortex and the insula that forms a link between pain, autonomic, and motor systems through the formation of motivation and emotion related to pain.

Interestingly, numerous monosynaptic descending projections have been described in the literature that simultaneously project to second-order neurons in the dorsal horn and to sympathetic preganglionic neurons in the IML. Table 2 highlights brain regions with monosynaptic descending projections to both the dorsal horn and the IML, and the effect of these projections on nociception and SNA. Inhibitory and excitatory projections can arise from all three integrative levels: (i) the brainstem, (ii) the hypothalamus, and (iii) the cortex, with most regions having a common excitatory and/or inhibitory effect on SNA and pain. Importantly, the NTS and LC, elicit strong inhibition of both second-order sensory neurons and sympathetic preganglionic neurons, which reduce nociceptive processing and SNA, respectively.

Table 2. Overview of monosynaptic pathways descending to the spinal cord that influence both second-order sensory neurons and sympathetic preganglionic neurons.

Structure	Effect(s) on DH Neurons	Effect(s) on IML Neurons	References
Hypothalamus			
Paraventricular nucleus	Inhibition	Inhibition/Excitation	[113–116]
Arcuate nucleus	Inhibition/Facilitation	Inhibition/Excitation	[117–119]
Parabrachial nucleus	Inhibition	Excitation	[120–122]
Brainstem			
Nucleus of the solitary tract	Inhibition	Inhibition	[123–126]
Raphe magnus/pallidus	Inhibition/Facilitation	Inhibition/Excitation	[127–130]
Rostroventromedial medulla	Facilitation	Excitation	[131]
Locus coeruleus	Inhibition	Inhibition	[132–134]
Medullary reticular formation	Inhibition/Facilitation	Inhibition/Excitation	[135,136]
Cerebral cortex			
Frontal/parietal	Inhibition	Inhibition/Excitation	[137–142]

DH, dorsal horn; IML intermediolateral cell column.

5. The Relationship between Blood Pressure and Acute Pain Sensitivity

In pain-free individuals, there is a linear, inverse (negative) relationship between blood pressure and acute pain sensitivity [143–153]. Importantly, the diminished acute pain sensitivity association with elevated resting blood pressure involves the medullary baroreflex circuitry, reducing both SNA and pain processing in response to increases in systemic blood pressure. In animals, baroreceptor stimulation induces antinociception [154–156], and surgical denervation of baroreceptor afferents produces hyperalgesia [157–159]. In humans, stimulation of baroreceptors (e.g., via application of external suction to the carotid artery) reduces acute pain sensitivity [160–163]. Consistent with the notion that baroreceptor loading is greatest during systole and lowest during diastole, pain sensitivity to ultra-rapid electrical pain stimuli was found to be lowest during systole and the greatest during diastole [164,165]. Altogether, these studies indicate that in addition to reducing SNA, baroreceptor stimulation triggers descending pain inhibitory activity, which contributes to the inverse relationship between resting blood pressure and acute pain.

In the setting of chronic pain, the relationship between blood pressure and pain sensitivity is dramatically altered. In contrast to the significant inverse (negative) relationship observed in pain-free individuals, chronic pain sufferers exhibit a direct (positive) relationship between resting blood pressure and acute pain [42,43,144]. Changes to the blood pressure-pain sensitivity relationship in the chronic pain setting directly contribute to SNS overactivation. A large retrospective study identified an increased prevalence of hypertension in chronic back pain patients and pain intensity was a significant predictor of hypertensive status [166]. This suggests that chronic pain may drive SNS overactivation and thereby promote cardiovascular morbidity and mortality.

6. Chronic Pain as a Driver of SNS Overactivation

As introduced above, spinal sympathetic reflex arcs result in the direct activation of sympathetic preganglionic neurons within the IML by ascending sensory/nociceptive pathways. This likely represents the most straightforward neural mechanism by which chronic pain can drive SNS overactivation. Within the dorsal horn, sensory information is directed to: (i) interneurons, which terminate within the same spinal segment, (ii) propriospinal neurons, which terminate within a different spinal segment, and (iii) projecting neurons, which terminate within supraspinal structures. Activation of sympathetic preganglionic neurons occurs either via second-order interneurons, second-order propriospinal neurons, or second-order spinothalamic projecting neurons, which have extensive axon collaterals within the thoracic spinal cord [167]. Additional interneurons (third order) terminating within the intermediolateral cell column of the spinal cord are likely critical for sensory-sympathetic coupling, as second-order neurons (arising from the dorsal horn) rarely make direct connections with cell bodies of sympathetic preganglionic neurons [168]. Thus, in the setting of chronic pain, SNS overactivation may result directly from excess spinal nociceptive input via a spinal sympathetic reflex arc.

A second mechanism by which chronic pain may increase sympathetic activity is secondary to chronic cognitive or emotional stress. Excessive sympathetic stress responses are associated with numerous human disorders, including post-traumatic stress disorder, cerebral palsy, traumatic brain injury, autism spectrum disorders, bipolar disorder, epilepsy, and Charcot–Marie–Tooth disease, among others [169–172], and can manifest with acute myocardial injury and/or increased susceptibility to sudden cardiac death [173–175]. Cardiovascular consequences of stress are similar in all mammals [67,176,177], and chronic or repetitive stress leading to increases in sympathetic activity is one proven cause of hypertension and heart failure [176,178]. Pain, especially when present in a chronic form, is a source of discomfort and substantial emotional suffering. In chronic orofacial pain patients, sympathetic responses to mental stressors are significantly greater than responses in pain-free, age and sex-matched controls [179]. Similarly, patients with chronic low back pain or chronic arthritic pain display a significantly higher baseline and stress-induced sympathetic (electrodermal) arousal than what is observed in pain-free controls [45,180,181].

7. Maladaptive Changes in the Neural Circuitry Leading to Sympathetic Overactivation and Chronic Pain

Our understanding of neuroplastic changes associated with chronic pain has been largely made possible by advancements in in vivo imaging using magnetic resonance imaging (MRI) techniques. Enhanced spatial resolution and contrast enable detailed structural morphology studies to be conducted in patients with chronic pain. Functional magnetic resonance imaging (fMRI) studies in humans have demonstrated the coordinated activation of several brain areas in response to noxious somatic and visceral stimuli, including the thalamus, anterior cingulate cortex, insular cortex, primary and secondary sensory cortices, prefrontal cortex, basal ganglia, cerebellum, and amygdala. This network of brain regions involved in both sensory discriminative and emotional-affective aspects of pain is termed the "pain matrix" [182]. Studies utilizing resting state fMRI revealed changes in network properties in chronic pain when compared to healthy controls. Specifically, atrophy within the dorsolateral prefrontal cortex (dlPFC) and increased activity of the medial prefrontal cortex (mPFC) have been observed in chronic back pain sufferers [183]. Similarly, a general shift from nociceptive to emotional circuits has been observed in chronic back pain [184], which may corroborate the involvement of emotional circuits in driving sympathetic overactivation in chronic pain. As mentioned above, these brain structures are also involved in the control of sympathetic activity both at baseline and during emotional stress. Thus, SNS overactivation among chronic pain sufferers likely involves an exaggerated excitatory drive to sympathetic preganglionic neurons as part of a chronic emotional stress response.

Chronic pain is associated with plastic changes at every level of the neural axis including primary sensory endings, as well as spinal and supraspinal sites [185]. The changes within the supraspinal centers are particularly important as they may contribute to altered nociceptive processing and perception of pain. Given the anatomical and functional overlap between the central nociceptive and central autonomic networks, the aberrant changes that occur within the supraspinal centers may simultaneously manifest with dysregulated sympathetic control and chronic pain. Recent studies have found that chronic pain sufferers irrespective of the origin of pain, exhibit common brain signatures associated with the loss of gray matter in cortical and subcortical structures such as the prefrontal and insular cortex, orbitofrontal cortex, and pons [185]. Using connectivity analyses of resting state fMRI, chronic migraine sufferers exhibited reduced gray matter volume and reduced cortical thickness in brain regions involved in the affective processing of pain including dlPFC [186]. The cortical thickness was inversely associated with the intensity and duration of pain suggesting a potential causative relationship between structural changes and the clinical phenotype of chronic pain. Additionally, decreased connectivity of the prefrontal cortex and increased connectivity with the insular cortex were found to be correlated with the intensity of chronic pain [187]. Both the insular cortex and prefrontal cortex represent areas of the brain involved in the emotional processing that led to an increased drive to the central autonomic network. The baseline activity of dlPFC is elevated in chronic pain sufferers [188] and direct transcranial stimulation of dlPFC in healthy individuals was found to increase SNA and blood pressure [137]. Additionally, specific neuronal ensembles within the mPFC appear to be critical for processing nociceptive information and regulating pain chronicity [189]. These data suggest the reorganization of the cortical centers associated with chronic pain states may also promote sympathetic overactivation and the development of cardiovascular disease via increased descending excitatory drive to sympathetic preganglionic neurons in the spinal cord.

Lastly, given that a common baroreceptor-sensitive, descending, inhibitory pathway regulates the spinal transmission of afferent nociceptive information and efferent sympathetic activity, it is conceivable that alterations within this central inhibitory network would simultaneously manifest with increased pain and increased sympathetic activity. Central noradrenergic pathways, particularly those mediated by alpha-2 adrenergic receptors ($\alpha 2$), are a crucial component of both the descending pain inhibitory system [190–192] and the descending pathway regulating SNS [60,193]. The contribution of $\alpha 2$ receptors to descending nociceptive inhibition is well documented [194–199] and the activation of central $\alpha 2$ receptors is important for the inhibition of sympathetic nerve activity in response to baroreceptor loading [59,61,193,200]. Multiple central structures, including the locus coeruleus (LC), are sources of descending pain-sympathetic modulation via $\alpha 2$ receptor activation [190,201–204], and likely contribute to descending inhibition of both pain and SNS. Therefore, alterations in this central neural network, which inhibits both spinal sympathetic preganglionic neurons and spinal transmission from afferent nociceptive neurons, would contribute to a concomitant increase in both pain and SNS activity.

Within the NTS there are numerous glutamatergic and GABAergic terminals arising from interneurons or from cortical and hypothalamic nuclei, which facilitate or inhibit excitation of the NTS, (e.g., during baroreceptor loading). Notably, the amygdala, the bed nucleus of the stria terminalis, and the PVN all project to the NTS [205,206]. Projections from cortical and hypothalamic centers to the NTS are known to inhibit excitatory neurotransmission of barosensitive neurons and participate in the upward resetting of the baroreflex [207]. Thus, it is tempting to speculate that chronic maladaptive changes within the central neural network, resulting in enhanced inhibition of barosensitive neurons within the NTS, serve as another plausible mechanism by which both chronic pain and increased sympathetic activity would simultaneously manifest.

8. Conclusions

There are many points along the neuraxis where nociceptive signals interact with SNS processing and alterations in these neural networks in the setting of chronic pain likely contribute to the SNS overactivation and development of cardiovascular disease. It remains unclear what neural processes contribute to SNS overactivation in the setting of chronic pain. Recent findings in human models have identified the structures within the prefrontal cortex that are involved in pain chronicity [189] and regulate sympathetic nerve activity and cardiovascular function [137]. These new findings indicate there are structural connections between chronic pain and cardiovascular function. Additionally, we highlight other direct and indirect neural networks that may play a role in driving chronic pain-associated cardiovascular disease. Future studies are needed to understand this relationship in more detail.

Author Contributions: C.A.R.: concept and design, writing, editing, Z.M.: concept and design, writing, editing. All authors have read and agreed to the published version of the manuscript.

Funding: This work was supported by NIH–5R01DK121812.

Conflicts of Interest: The authors declare no conflict of interest.

Abbreviations

α-2	Alpha-2 adrenergic receptors
ACC	Anterior cingulate cortex
DH	Dorsal horn
dlPFC	Dorsolateral prefrontal cortex
DMH	Dorsomedial hypothalamus
CVLM	Caudal ventrolateral medulla
fMRI	Functional magnetic resonance imaging
IC	Insular cortex
IML	Intermediolateral cell column
LC	Locus coeruleus
MRI	Magnetic resonance imaging
mPFC	Medial prefrontal cortex
GABA	γ- aminobutyric acid
NTS	Nucleus of the solitary tract
PFC	Prefrontal cortex
PVN	Paraventricular nucleus
RF	Reticular formation
RVLM	Rostral ventrolateral medulla
SNA	Sympathetic nerve activity
SNS	Sympathetic nerve system
SPN	Sympathetic preganglionic neurons

References

1. Global burden of 369 diseases and injuries in 204 countries and territories, 1990–2019: A systematic analysis for the Global Burden of Disease Study 2019. *Lancet* **2020**, *396*, 1204–1222. [CrossRef]
2. Fayaz, A.; Ayis, S.; Panesar, S.S.; Langford, R.M.; Donaldson, L.J. Assessing the relationship between chronic pain and cardiovascular disease: A systematic review and meta-analysis. *Scand. J. Pain* **2016**, *13*, 76–90. [CrossRef]

3. Oliveira, C.B.; Maher, C.G.; Franco, M.R.; Kamper, S.J.; Williams, C.M.; Silva, F.G.; Pinto, R.Z. Co-occurrence of Chronic Musculoskeletal Pain and Cardiovascular Diseases: A Systematic Review with Meta-analysis. *Pain Med.* **2020**, *21*, 1106–1121. [CrossRef]
4. Macfarlane, G.J.; Barnish, M.S.; Jones, G.T. Persons with chronic widespread pain experience excess mortality: Longitudinal results from UK Biobank and meta-analysis. *Ann. Rheum. Dis.* **2017**, *76*, 1815–1822. [CrossRef]
5. Holmberg, T.; Davidsen, M.; Thygesen, L.C.; Krøll, M.J.; Tolstrup, J.S. Mortality among persons experiencing musculoskeletal pain: A prospective study among Danish men and women. *BMC Musculoskelet. Disord.* **2020**, *21*, 666. [CrossRef] [PubMed]
6. Tesarz, J.; Eich, W.; Baumeister, D.; Kohlmann, T.; D'Agostino, R.; Schuster, A.K. Widespread pain is a risk factor for cardiovascular mortality: Results from the Framingham Heart Study. *Eur. Heart J.* **2019**, *40*, 1609–1617. [CrossRef] [PubMed]
7. McBeth, J.; Symmons, D.; Silman, A.; Allison, T.; Webb, R.; Brammah, T.; Macfarlane, G. Musculoskeletal pain is associated with a long-term increased risk of cancer and cardiovascular-related mortality. *Rheumatology* **2009**, *48*, 74–77. [CrossRef]
8. Andersson, H.I. Increased mortality among individuals with chronic widespread pain relates to lifestyle factors: A prospective population-based study. *Disabil. Rehabil.* **2009**, *31*, 1980–1987. [CrossRef]
9. Rodríguez-Sánchez, I.; Ortolá, R.; Graciani, A.; Martínez-Gómez, D.; Banegas, J.R.; Rodríguez-Artalejo, F.; García-Esquinas, E. Pain characteristics, cardiovascular risk factors, and cardiovascular disease. *J. Gerontol. Ser. A* **2022**, *77*, 204–213. [CrossRef]
10. Lindgren, H.; Bergman, S. Chronic musculoskeletal pain predicted hospitalisation due to serious medical conditions in a 10 year follow up study. *BMC Musculoskelet. Disord.* **2010**, *11*, 127. [CrossRef] [PubMed]
11. Kendzerska, T.; Jüni, P.; King, L.K.; Croxford, R.; Stanaitis, I.; Hawker, G.A. The longitudinal relationship between hand, hip and knee osteoarthritis and cardiovascular events: A population-based cohort study. *Osteoarthr. Cartil.* **2017**, *25*, 1771–1780. [CrossRef]
12. Atiquzzaman, M.; Karim, M.E.; Kopec, J.; Wong, H.; Anis, A.H. Role of nonsteroidal antiinflammatory drugs in the association between osteoarthritis and cardiovascular diseases: A longitudinal study. *Arthritis Rheumatol.* **2019**, *71*, 1835–1843. [CrossRef]
13. Rahman, M.M.; Kopec, J.A.; Anis, A.H.; Cibere, J.; Goldsmith, C.H. Risk of cardiovascular disease in patients with osteoarthritis: A prospective longitudinal study. *Arthritis Care Res.* **2013**, *65*, 1951–1958. [CrossRef] [PubMed]
14. Su, C.-H.; Chen, J.-H.; Lan, J.-L.; Wang, Y.-C.; Tseng, C.-H.; Hsu, C.-Y.; Huang, L. Increased risk of coronary heart disease in patients with primary fibromyalgia and those with concomitant comorbidity—A Taiwanese population-based cohort study. *PLoS ONE* **2015**, *10*, e0137137. [CrossRef]
15. Kluzek, S.; Sanchez-Santos, M.; Leyland, K.; Judge, A.; Spector, T.; Hart, D.; Cooper, C.; Newton, J.; Arden, N. Painful knee but not hand osteoarthritis is an independent predictor of mortality over 23 years follow-up of a population-based cohort of middle-aged women. *Ann. Rheum. Dis.* **2016**, *75*, 1749–1756. [CrossRef] [PubMed]
16. Rönnegård, A.-S.; Nowak, C.; Äng, B.; Ärnlöv, J. The association between short-term, chronic localized and chronic widespread pain and risk for cardiovascular disease in the UK Biobank. *Eur. J. Prev. Cardiol.* **2022**, *29*, 1994–2002. [CrossRef] [PubMed]
17. Ha, I.-H.; Lee, J.; Kim, M.-R.; Kim, H.; Shin, J.-S. The association between the history of cardiovascular diseases and chronic low back pain in South Koreans: A cross-sectional study. *PLoS ONE* **2014**, *9*, e93671. [CrossRef] [PubMed]
18. Parsons, S.; McBeth, J.; Macfarlane, G.; Hannaford, P.; Symmons, D. Self-reported pain severity is associated with a history of coronary heart disease. *Eur. J. Pain* **2015**, *19*, 167–175. [CrossRef]
19. Von Korff, M.; Crane, P.; Lane, M.; Miglioretti, D.L.; Simon, G.; Saunders, K.; Stang, P.; Brandenburg, N.; Kessler, R. Chronic spinal pain and physical–mental comorbidity in the United States: Results from the national comorbidity survey replication. *Pain* **2005**, *113*, 331–339. [CrossRef]
20. Keller, J.J.; Chen, Y.K.; Lin, H.C. Comorbidities of bladder pain syndrome/interstitial cystitis: A population-based study. *BJU Int.* **2012**, *110*, E903–E909. [CrossRef]
21. Dreyer, L.; Kendall, S.; Danneskiold-Samsøe, B.; Bartels, E.M.; Bliddal, H. Mortality in a cohort of Danish patients with fibromyalgia: Increased frequency of suicide. *Arthritis Rheum.* **2010**, *62*, 3101–3108. [CrossRef] [PubMed]
22. Zhu, K.; Devine, A.; Dick, I.M.; Prince, R.L. Association of back pain frequency with mortality, coronary heart events, mobility, and quality of life in elderly women. *Spine* **2007**, *32*, 2012–2018. [CrossRef] [PubMed]
23. Conway, L.C.; Smith, B.H.; Hocking, L.J.; McGilchrist, M.M.; Dominiczak, A.F.; Morris, A.; Porteous, D.J.; Goebel, A.; Goodson, N.J. The prevalence of rose angina is increased in people reporting chronic pain: Results from a cross-sectional general population study. *Rheumatology* **2014**, *53*, i68. [CrossRef]
24. Svensson, H.-O.; Vedin, A.; Wilhelmsson, C.; Andersson, G.B. Low-back pain in relation to other diseases and cardiovascular risk factors. *Spine* **1983**, *8*, 277–285. [CrossRef]
25. Pontari, M.A.; McNaughton-Collins, M.; O'leary, M.P.; Calhoun, E.A.; Jang, T.; Kusek, J.W.; Landis, J.R.; Knauss, J.; Litwin, M.S.; Group, C.S. A case-control study of risk factors in men with chronic pelvic pain syndrome. *BJU Int.* **2005**, *96*, 559–565. [CrossRef]
26. Ablin, J.N.; Beilinson, N.; Aloush, V.; Elkayam, O.; Finkelstein, A. Association between fibromyalgia and coronary heart disease and coronary catheterization. *Clin. Cardiol. Int. Index. Peer-Rev. J. Adv. Treat. Cardiovasc. Dis.* **2009**, *32*, E7–E11. [CrossRef] [PubMed]
27. Tsai, P.-S.; Fan, Y.-C.; Huang, C.-J. Fibromyalgia is associated with coronary heart disease: A population-based cohort study. *Reg. Anesth. Pain Med.* **2015**, *40*, 37–42. [CrossRef]
28. Mäkela, M.; Heliövaara, M.; Sievers, K.; Impivaara, O.; Knekt, P.; Aromaa, A. Prevalence, determinants, and consequences of chronic neck pain in Finland. *Am. J. Epidemiol.* **1991**, *134*, 1356–1367. [CrossRef]

29. Ohayon, M.M.; Stingl, J.C. Prevalence and comorbidity of chronic pain in the German general population. *J. Psychiatr. Res.* **2012**, *46*, 444–450. [CrossRef] [PubMed]
30. Ryan, C.; McDonough, S.; Kirwan, J.; Leveille, S.; Martin, D. An investigation of association between chronic musculoskeletal pain and cardiovascular disease in the Health Survey for England (2008). *Eur. J. Pain* **2014**, *18*, 740–750. [CrossRef]
31. Heliövaara, M.; Mäkelä, M.; Aromaa, A.; Impivaara, O.; Knekt, P.; Reunanen, A. Low back pain and subsequent cardiovascular mortality. *Spine* **1995**, *20*, 2109–2111. [CrossRef] [PubMed]
32. Smith, B.H.; Elliott, A.M.; Hannaford, P.C. Pain and subsequent mortality and cancer among women in the Royal College of General Practitioners Oral Contraception Study. *Br. J. Gen. Pract.* **2003**, *53*, 45–46. [PubMed]
33. Torrance, N.; Elliott, A.M.; Lee, A.J.; Smith, B.H. Severe chronic pain is associated with increased 10 year mortality. A cohort record linkage study. *Eur. J. Pain* **2010**, *14*, 380–386. [CrossRef]
34. Sherrington, C.S. Experiments on the value of vascular and visceral factors for the genesis of emotion. *Proc. R. Soc. Lond.* **1900**, *66*, 390–403.
35. Sherrington, C.S. Address on the spinal animal. *Med. Chir. Trans.* **1899**, *82*, 449. [CrossRef]
36. Reis, D.J.; Ruggiero, D.A.; Morrison, S.F. The CI area of the rostral ventrolateral medulla oblongata: A critical brainstem region for control of resting and reflex integration of arterial pressure. *Am. J. Hypertens.* **1989**, *2*, 363S–374S. [CrossRef] [PubMed]
37. Stornetta, R.L.; Morrison, S.F.; Ruggiero, D.A.; Reis, D.J. Neurons of rostral ventrolateral medulla mediate somatic pressor reflex. *Am. J. Physiol.-Regul. Integr. Comp. Physiol.* **1989**, *256*, R448–R462. [CrossRef]
38. Basbaum, A.I.; Fields, H.L. Endogenous pain control systems: Brainstem spinal pathways and endorphin circuitry. *Annu. Rev. Neurosci.* **1984**, *7*, 309–338. [CrossRef]
39. Malpas, S.C. Sympathetic nervous system overactivity and its role in the development of cardiovascular disease. *Physiol. Rev.* **2010**, *90*, 513–557. [CrossRef]
40. Rozanski, A.; Blumenthal, J.A.; Kaplan, J. Impact of psychological factors on the pathogenesis of cardiovascular disease and implications for therapy. *Circulation* **1999**, *99*, 2192–2217. [CrossRef]
41. Maixner, W.; Sigurdsson, A.; Fillingim, R.; Lundeen, T.; Booker, D. Regulation of acute and chronic orofacial pain. In *Orofacial Pain and Temopromandibular Disorders*; Raven Press: New York, NY, USA, 1995; pp. 85–102.
42. Bruehl, S.; Chung, O.Y.; Ward, P.; Johnson, B.; McCubbin, J.A. The relationship between resting blood pressure and acute pain sensitivity in healthy normotensives and chronic back pain sufferers: The effects of opioid blockade. *Pain* **2002**, *100*, 191–201. [CrossRef]
43. Bruehl, S.; Burns, J.W.; McCubbin, J.A. Altered cardiovascular/pain regulatory relationships in chronic pain. *Int. J. Behav. Med.* **1998**, *5*, 63–75. [CrossRef] [PubMed]
44. Brody, S.; Angrilli, A.; Weiss, U.; Birbaumer, N.; Mini, A.; Veit, R.; Rau, H. Somatotosensory evoked potentials during baroreceptor stimulation in chronic low back pain patients and normal controls. *Int. J. Psychophysiol.* **1997**, *25*, 201–210. [CrossRef]
45. Perry, F.; Heller, P.H.; Kamiya, J.; Levine, J.D. Altered autonomic function in patients with arthritis or with chronic myofascial pain. *Pain* **1989**, *39*, 77–84. [CrossRef] [PubMed]
46. Hjemdahl, P.; Larsson, P.; Wallen, N. Effects of stress and beta-blockade on platelet function. *Circulation* **1991**, *84*, VI44–VI61. [PubMed]
47. Ardlie, N.; Glew, G.; Schwartz, C. Influence of catecholamines on nucleotide-induced platelet aggregation. *Nature* **1966**, *212*, 415–417. [CrossRef]
48. O'brien, J. Some effects of adrenaline and anti-adrenaline compounds on platelets in vitro and in vivo. *Nature* **1963**, *200*, 763–764. [CrossRef]
49. Remme, W. The sympathetic nervous system and ischaemic heart disease. *Eur. Heart J.* **1998**, *19*, F62–F71.
50. Benarroch, E.E. The central autonomic network: Functional organization, dysfunction, and perspective. In *Proceedings of Mayo Clinic Proceedings*; Elsevier: Amsterdam, The Netherlands, 1993; pp. 988–1001.
51. Calaresu, F.; Yardley, C. Medullary basal sympathetic tone. *Annu. Rev. Physiol.* **1988**, *50*, 511–524. [CrossRef]
52. Guertzenstein, P.; Silver, A. Fall in blood pressure produced from discrete regions of the ventral surface of the medulla by glycine and lesions. *J. Physiol.* **1974**, *242*, 489–503. [CrossRef]
53. Stein, R.; Weaver, L.; Yardley, C. Ventrolateral medullary neurones: Effects on magnitude and rhythm of discharge of mesenteric and renal nerves in cats. *J. Physiol.* **1989**, *408*, 571–586. [CrossRef]
54. Guyenet, P.G.; Haselton, J.R.; Sun, M.-K. Sympathoexcitatory neurons of the rostroventrolateral medulla and the origin of the sympathetic vasomotor tone. *Prog. Brain Res.* **1989**, *81*, 105–116. [PubMed]
55. Sakima, A.; Yamazato, M.; Sesoko, S.; Muratani, H.; Fukiyama, K. Cardiovascular and sympathetic effects of L-glutamate and glycine injected into the rostral ventrolateral medulla of conscious rats. *Hypertens. Res.* **2000**, *23*, 633–641. [CrossRef]
56. Willette, R.; Barcas, P.; Krieger, A.; Sapru, H.N. Vasopressor and depressor areas in the rat medulla: Identification by microinjection of L-glutamate. *Neuropharmacology* **1983**, *22*, 1071–1079. [CrossRef] [PubMed]
57. Dampney, R.; Moon, E.A. Role of ventrolateral medulla in vasomotor response to cerebral ischemia. *Am. J. Physiol.-Heart Circ. Physiol.* **1980**, *239*, H349–H358. [CrossRef] [PubMed]
58. Sun, M.-K.; Reis, D.J. Central neural mechanisms mediating excitation of sympathetic neurons by hypoxia. *Prog. Neurobiol.* **1994**, *44*, 197–219. [CrossRef]

59. Kubo, T.; Goshima, Y.; Hata, H.; Misu, Y. Evidence that endogenous catecholamines are involved inα2-adrenoceptor-mediated modulation of the aortic baroreceptor reflex in the nucleus tractus solitarii of the rat. *Brain Res.* **1990**, *526*, 313–317. [CrossRef]
60. Lawrence, A.J.; Jarrott, B. Neurochemical modulation of cardiovascular control in the nucleus tractus solitarius. *Prog. Neurobiol.* **1996**, *48*, 21–53. [CrossRef]
61. Murase, S.; Inui, K.; Nosaka, S. Baroreceptor inhibition of the locus coeruleus noradrenergic neurons. *Neuroscience* **1994**, *61*, 635–643. [CrossRef]
62. Morrison, S.F. RVLM and raphe differentially regulate sympathetic outflows to splanchnic and brown adipose tissue. *Am. J. Physiol.* **1999**, *276*, R962–R973. [CrossRef] [PubMed]
63. Reynolds, C.A.; O'Leary, D.S.; Ly, C.; Smith, S.A.; Minic, Z. Development of a decerebrate model for investigating mechanisms mediating viscero-sympathetic reflexes in the spinalized rat. *Am. J. Physiol.-Heart Circ. Physiol.* **2019**, *316*, H1332–H1340. [CrossRef] [PubMed]
64. Mayorov, D.N.; Adams, M.A.; Krassioukov, A.V. Telemetric blood pressure monitoring in conscious rats before and after compression injury of spinal cord. *J. Neurotrauma* **2001**, *18*, 727. [CrossRef] [PubMed]
65. Osborn, J.W.; Taylor, R.F.; Schramm, L.P. Determinants of arterial pressure after chronic spinal transection in rats. *Am. J. Physiol.-Regul. Integr. Comp. Physiol.* **1989**, *256*, R666–R673. [CrossRef] [PubMed]
66. Hayes, K.; Yardley, C.; Weaver, L. Evidence for descending tonic inhibition specifically affecting sympathetic pathways to the kidney in rats. *J. Physiol.* **1991**, *434*, 295–306. [CrossRef] [PubMed]
67. Fontes, M.; Tagawa, T.; Polson, J.; Cavanagh, S.-J.; Dampney, R. Descending pathways mediating cardiovascular response from dorsomedial hypothalamic nucleus. *Am. J. Physiol.-Heart Circ. Physiol.* **2001**, *280*, H2891–H2901. [CrossRef]
68. Cao, W.H.; Fan, W.; Morrison, S.F. Medullary pathways mediating specific sympathetic responses to activation of dorsomedial hypothalamus. *Neuroscience* **2004**, *126*, 229–240. [CrossRef]
69. Martin, D.S.; Segura, T.; Haywood, J.R. Cardiovascular responses to bicuculline in the paraventricular nucleus of the rat. *Hypertension* **1991**, *18*, 48–55. [CrossRef] [PubMed]
70. Carter, J.R.; Kupiers, N.T.; Ray, C.A. Neurovascular responses to mental stress. *J. Physiol.* **2005**, *564*, 321–327. [CrossRef]
71. Carter, J.R.; Ray, C.A. Sympathetic neural responses to mental stress: Responders, nonresponders and sex differences. *Am. J. Physiol.-Heart Circ. Physiol.* **2009**, *296*, H847–H853. [CrossRef]
72. Coote, J.H.; Dodds, W.N. The baroreceptor reflex and the cardiovascular changes associated with sustained muscular contraction in the cat. *Pflügers Arch.* **1976**, *363*, 167–173. [CrossRef]
73. Sheriff, D.; O'Leary, D.S.; Scher, A.M.; Rowell, L.B. Baroreflex attenuates pressor response to graded muscle ischemia in exercising dogs. *Am. J. Physiol.-Heart Circ. Physiol.* **1990**, *258*, H305–H310. [CrossRef] [PubMed]
74. Hatton, D.C.; Brooks, V.; Qi, Y.; McCarron, D.A. Cardiovascular response to stress: Baroreflex resetting and hemodynamics. *Am. J. Physiol. -Regul. Integr. Comp. Physiol.* **1997**, *272*, R1588–R1594. [CrossRef] [PubMed]
75. Raven, P.B.; Young, B.E.; Fadel, P.J. Arterial baroreflex resetting during exercise in humans: Underlying signaling mechanisms. *Exerc. Sport Sci. Rev.* **2019**, *47*, 129–141. [CrossRef]
76. Durocher, J.J.; Klein, J.C.; Carter, J.R. Attenuation of sympathetic baroreflex sensitivity during the onset of acute mental stress in humans. *Am. J. Physiol.-Heart Circ. Physiol.* **2011**, *300*, H1788–H1793. [CrossRef] [PubMed]
77. Terreberry, R.R.; Neafsey, E.J. Rat medial frontal cortex: A visceral motor region with a direct projection to the solitary nucleus. *Brain Res.* **1983**, *278*, 245–249. [CrossRef] [PubMed]
78. Terreberry, R.R.; Neafsey, E.J. The rat medial frontal cortex projects directly to autonomic regions of the brainstem. *Brain Res. Bull.* **1987**, *19*, 639–649. [CrossRef] [PubMed]
79. Macefield, V.G.; Henderson, L.A. "Real-time" imaging of cortical and subcortical sites of cardiovascular control: Concurrent recordings of sympathetic nerve activity and fMRI in awake subjects. *J. Neurophysiol.* **2016**, *116*, 1199–1207. [CrossRef]
80. Shoemaker, J.K.; Wong, S.W.; Cechetto, D.F. Cortical circuitry associated with reflex cardiovascular control in humans: Does the cortical autonomic network "speak" or "listen" during cardiovascular arousal. *Anat. Rec. Adv. Integr. Anat. Evol. Biol.* **2012**, *295*, 1375–1384. [CrossRef]
81. MacDonald, A.W.; Cohen, J.D.; Stenger, V.A.; Carter, C.S. Dissociating the role of the dorsolateral prefrontal and anterior cingulate cortex in cognitive control. *Science* **2000**, *288*, 1835–1838. [CrossRef]
82. Fechir, M.; Gamer, M.; Blasius, I.; Bauermann, T.; Breimhorst, M.; Schlindwein, P.; Schlereth, T.; Birklein, F. Functional imaging of sympathetic activation during mental stress. *Neuroimage* **2010**, *50*, 847–854. [CrossRef]
83. Ruggiero, D.A.; Mraovitch, S.; Granata, A.R.; Anwar, M.; Reis, D.J. A role of insular cortex in cardiovascular function. *J. Comp. Neurol.* **1987**, *257*, 189–207. [CrossRef] [PubMed]
84. Saha, S. Role of the central nucleus of the amygdala in the control of blood pressure: Descending pathways to medullary cardiovascular nuclei. *Clin. Exp. Pharmacol. Physiol.* **2005**, *32*, 450–456. [CrossRef]
85. Chau, D.; Johns, D.G.; Schramm, L.P. Ongoing and stimulus-evoked activity of sympathetically correlated neurons in the intermediate zone and dorsal horn of acutely spinalized rats. *J. Neurophysiol.* **2000**, *83*, 2699–2707. [CrossRef]
86. Barman, S.M.; Gebber, G.L. Spinal interneurons with sympathetic nerve-related activity. *Am. J. Physiol.-Regul. Integr. Comp. Physiol.* **1984**, *247*, R761–R767. [CrossRef] [PubMed]
87. Sherrington, C.; Laslett, E. Observations on some spinal reflexes and the interconnection of spinal segments. *J. Physiol.* **1903**, *29*, 58. [CrossRef] [PubMed]

88. Miller, F.R. Viscero-motor reflexes. II. *Am. J. Physiol.-Leg. Content* **1924**, *71*, 84–89. [CrossRef]
89. Miller, F.R.; Waud, R. Viscero-motor reflexes. IV. *Am. J. Physiol.-Leg. Content* **1925**, *73*, 329–340. [CrossRef]
90. Downman, C.; McSwiney, B. Reflexes elicited by visceral stimulation in the acute spinal animal. *J. Physiol.* **1946**, *105*, 80. [CrossRef]
91. Minic, Z.; O'Leary, D.S.; Reynolds, C.A. Spinal reflex control of arterial blood pressure: The role of TRP channels and their endogenous eicosanoid modulators. *Front. Physiol.* **2022**, *13*, 838175. [CrossRef]
92. Jankowska, E.; Padel, Y.; Tanaka, R. Disynaptic inhibition of spinal motoneurones from the motor cortex in the monkey. *J. Physiol.* **1976**, *258*, 467–487. [CrossRef]
93. Triggs, W.J.; Macdonell, R.A.; Cros, D.; Chiappa, K.H.; Shahani, B.T.; Day, B.J. Motor inhibition and excitation are independent effects of magnetic cortical stimulation. *Ann. Neurol. Off. J. Am. Neurol. Assoc. Child Neurol. Soc.* **1992**, *32*, 345–351. [CrossRef]
94. Agarwal, S.; Gelsema, A.; Calaresu, F. Neurons in rostral VLM are inhibited by chemical stimulation of caudal VLM in rats. *Am. J. Physiol. -Regul. Integr. Comp. Physiol.* **1989**, *257*, R265–R270. [CrossRef]
95. Willette, R.; Punnen, S.; Krieger, A.; Sapru, H. Interdependence of rostral and caudal ventrolateral medullary areas in the control of blood pressure. *Brain Res.* **1984**, *321*, 169–174. [CrossRef]
96. McCall, R.B. GABA-mediated inhibition of sympathoexcitatory neurons by midline medullary stimulation. *Am. J. Physiol.-Regul. Integr. Comp. Physiol.* **1988**, *255*, R605–R615. [CrossRef]
97. Llewellyn-Smith, I.J. GABA in the control of sympathetic preganglionic neurons. *Clin. Exp. Pharmacol. Physiol.* **2002**, *29*, 507–513. [CrossRef]
98. McCall, R.B.; Harris, L.T. Sympathetic alterations after midline medullary raphe lesions. *Am. J. Physiol.-Regul. Integr. Comp. Physiol.* **1987**, *253*, R91–R100. [CrossRef]
99. Pilowsky, P.; West, M.; Chalmers, J. Renal sympathetic nerve responses to stimulation, inhibition and destruction of the ventrolateral medulla in the rabbit. *Neurosci. Lett.* **1985**, *60*, 51–55. [CrossRef]
100. Loewy, A. Raphe pallidus and raphe obscurus projections to the intermediolateral cell column in the rat. *Brain Res.* **1981**, *222*, 129–133. [CrossRef]
101. Loewy, A.; McKellar, S. Serotonergic projections from the ventral medulla to the intermediolateral cell column in the rat. *Brain Res.* **1981**, *211*, 146–152. [CrossRef]
102. Fleetwood-Walker, S.M.; Coote, J.H.; Gilbey, M.P. Identification of spinally projecting neurones in the A1 catecholamine cell group of the ventrolateral medulla. *Brain Res.* **1983**, *273*, 25–33. [CrossRef]
103. Dembowsky, K.; Czachurski, J.; Amendt, K.; Seller, H. Tonic descending inhibition of the spinal somato-sympathetic reflex from the lower brain stem. *J. Auton. Nerv. Syst.* **1980**, *2*, 157–182. [CrossRef]
104. Müller, U.W.; Dembowsky, K.; Czachurski, J.; Seller, H. Tonic descending inhibition of the spinal cardio-sympathetic reflex in the cat. *J. Auton. Nerv. Syst.* **1988**, *23*, 111–123. [CrossRef] [PubMed]
105. Deuchars, S.A.; Milligan, C.J.; Stornetta, R.L.; Deuchars, J. GABAergic neurons in the central region of the spinal cord: A novel substrate for sympathetic inhibition. *J. Neurosci.* **2005**, *25*, 1063–1070. [CrossRef]
106. Bennett, G.J. Update on the neurophysiology of pain transmission and modulation: Focus on the NMDA-receptor. *J. Pain Symptom Manag.* **2000**, *19*, 2–6. [CrossRef]
107. Sandkühler, J.; Bromm, B.; Gebhart, G.F. *Nervous System Plasticity and Chronic Pain*; Elsevier Science Limited: Amsterdam, The Netherlands, 2000; Volume 129.
108. Craig, A.D. Pain mechanisms: Labeled lines versus convergence in central processing. *Annu. Rev. Neurosci.* **2003**, *26*, 1–30. [CrossRef] [PubMed]
109. Brooks, J.; Tracey, I. From nociception to pain perception: Imaging the spinal and supraspinal pathways. *J. Anat.* **2005**, *207*, 19–33. [CrossRef] [PubMed]
110. Bushnell, M.C.; Duncan, G.H.; Hofbauer, R.K.; Ha, B.; Chen, J.I.; Carrier, B. Pain perception: Is there a role for primary somatosensory cortex? *Proc. Natl. Acad. Sci. USA* **1999**, *96*, 7705–7709. [CrossRef]
111. Rainville, P.; Duncan, G.H.; Price, D.D.; Carrier, B.; Bushnell, M.C. Pain affect encoded in human anterior cingulate but not somatosensory cortex. *Science* **1997**, *277*, 968–971. [CrossRef] [PubMed]
112. Coghill, R.C.; Sang, C.N.; Maisog, J.M.; Iadarola, M.J. Pain intensity processing within the human brain: A bilateral, distributed mechanism. *J. Neurophysiol.* **1999**, *82*, 1934–1943. [CrossRef]
113. Strack, A.; Sawyer, W.; Hughes, J.; Platt, K.; Loewy, A. A general pattern of CNS innervation of the sympathetic outflow demonstrated by transneuronal pseudorabies viral infections. *Brain Res.* **1989**, *491*, 156–162. [CrossRef]
114. Yamashita, H.; Inenaga, K.; Koizumi, K. Possible projections from regions of paraventricular and supraoptic nuclei to the spinal cord: Electrophysiological studies. *Brain Res.* **1984**, *296*, 373–378. [CrossRef]
115. Badoer, E. Hypothalamic paraventricular nucleus and cardiovascular regulation. *Clin. Exp. Pharmacol. Physiol.* **2001**, *28*, 95–99. [CrossRef]
116. Ramchandra, R.; Hood, S.G.; Frithiof, R.; McKinley, M.J.; May, C.N. The role of the paraventricular nucleus of the hypothalamus in the regulation of cardiac and renal sympathetic nerve activity in conscious normal and heart failure sheep. *J. Physiol.* **2013**, *591*, 93–107. [CrossRef]
117. Cechetto, D.F.; Saper, C.B. Neurochemical organization of the hypothalamic projection to the spinal cord in the rat. *J. Comp. Neurol.* **1988**, *272*, 579–604. [CrossRef]

118. Elias, C.F.; Lee, C.; Kelly, J.; Aschkenasi, C.; Ahima, R.S.; Couceyro, P.R.; Kuhar, M.J.; Saper, C.B.; Elmquist, J.K. Leptin activates hypothalamic CART neurons projecting to the spinal cord. *Neuron* **1998**, *21*, 1375–1385. [CrossRef]
119. Sapru, H.N. Role of the hypothalamic arcuate nucleus in cardiovascular regulation. *Auton. Neurosci.* **2013**, *175*, 38–50. [CrossRef]
120. Fulwiler, C.E.; Saper, C.B. Subnuclear organization of the efferent connections of the parabrachial nucleus in the rat. *Brain Res. Rev.* **1984**, *7*, 229–259. [CrossRef]
121. Saleh, T.M.; Connell, B.J.; Cribb, A.E. Estrogen in the parabrachial nucleus attenuates the sympathoexcitation following MCAO in male rats. *Brain Res.* **2005**, *1066*, 187–195. [CrossRef]
122. Hayward, L.F.; Felder, R.B. Lateral parabrachial nucleus modulates baroreflex regulation of sympathetic nerve activity. *Am. J. Physiol.-Regul. Integr. Comp. Physiol.* **1998**, *274*, R1274–R1282. [CrossRef]
123. Mtui, E.P.; Anwar, M.; Gomez, R.; Reis, D.J.; Ruggiero, D.A. Projections from the nucleus tractus solitarii to the spinal cord. *J. Comp. Neurol.* **1993**, *337*, 231–252. [CrossRef]
124. Stornetta, R.L.; Guyenet, P.G. Distribution of glutamic acid decarboxylase mRNA-containing neurons in rat medulla projecting to thoracic spinal cord in relation to monoaminergic brainstem neurons. *J. Comp. Neurol.* **1999**, *407*, 367–380. [CrossRef]
125. Loewy, A.; Burton, H. Nuclei of the solitary tract: Efferent projections to the lower brain stem and spinal cord of the cat. *J. Comp. Neurol.* **1978**, *181*, 421–449. [CrossRef] [PubMed]
126. Madden, C. Consumption of a high fat diet inhibits sympathetic outflow to brown adipose tissue (BAT) via vagal afferent activation of neurons in the Nucleus Tractus Solitarius (NTS). *Auton. Neurosci. Basic Clin.* **2015**, *192*, 13. [CrossRef]
127. Haselton, J.R.; Winters, R.W.; Liskowsky, D.R.; Haselton, C.L.; McCabe, P.M.; Schneiderman, N. Anatomical and functional connections of neurons of the rostral medullary raphe of the rabbit. *Brain Res.* **1988**, *453*, 176–182. [CrossRef]
128. Arami, M.K.; Komaki, A.; Gharibzadeh, S. Contribution of nucleus raphe magnus to thermoregulation. *Physiol. Pharmacol.* **2020**, *24*, 165–173. [CrossRef]
129. Leung, C.G.; Mason, P. Spectral analysis of arterial blood pressure and raphe magnus neuronal activity in anesthetized rats. *Am. J. Physiol. -Regul. Integr. Comp. Physiol.* **1996**, *271*, R483–R489. [CrossRef]
130. Blessing, W.; Nalivaiko, E. Raphe magnus/pallidus neurons regulate tail but not mesenteric arterial blood flow in rats. *Neuroscience* **2001**, *105*, 923–929. [CrossRef]
131. Babic, T.; Ciriello, J. Medullary and spinal cord projections from cardiovascular responsive sites in the rostral ventromedial medulla. *J. Comp. Neurol.* **2004**, *469*, 391–412. [CrossRef]
132. Elam, M.; Svensson, T.H.; Thoren, P. Differentiated cardiovascular afferent regulation of locus coeruleus neurons and sympathetic nerves. *Brain Res.* **1985**, *358*, 77–84. [CrossRef]
133. Bruinstroop, E.; Cano, G.; Vanderhorst, V.G.; Cavalcante, J.C.; Wirth, J.; Sena-Esteves, M.; Saper, C.B. Spinal projections of the A5, A6 (locus coeruleus), and A7 noradrenergic cell groups in rats. *J. Comp. Neurol.* **2012**, *520*, 1985–2001. [CrossRef]
134. Miyawaki, T.; Kawamura, H.; Komatsu, K.; Yasugi, T. Chemical stimulation of the locus coeruleus: Inhibitory effects on hemodynamics and renal sympathetic nerve activity. *Brain Res.* **1991**, *568*, 101–108. [CrossRef] [PubMed]
135. Aicher, S.A.; Reis, D.J.; Nicolae, R.; Milner, T.A. Monosynaptic projections from the medullary gigantocellular reticular formation to sympathetic preganglionic neurons in the thoracic spinal cord. *J. Comp. Neurol.* **1995**, *363*, 563–580. [CrossRef] [PubMed]
136. Korkola, M.L.; Weaver, L.C. Role of dorsal medullary reticular formation in maintenance of vasomotor tone in rats. *J. Auton. Nerv. Syst.* **1994**, *46*, 161–169. [CrossRef] [PubMed]
137. Sesa-Ashton, G.; Wong, R.; McCarthy, B.; Datta, S.; Henderson, L.A.; Dawood, T.; Macefield, V.G. Stimulation of the dorsolateral prefrontal cortex modulates muscle sympathetic nerve activity and blood pressure in humans. *Cereb. Cortex Commun.* **2022**, *3*, tgac017. [CrossRef]
138. James, C.; Macefield, V.G.; Henderson, L.A. Real-time imaging of cortical and subcortical control of muscle sympathetic nerve activity in awake human subjects. *Neuroimage* **2013**, *70*, 59–65. [CrossRef]
139. Van Eden, C.G.; Buijs, R.M. Functional neuroanatomy of the prefrontal cortex: Autonomic interactions. *Prog. Brain Res.* **2000**, *126*, 49–62.
140. Miller, M. The origin of corticospinal projection neurons in rat. *Exp. Brain Res.* **1987**, *67*, 339–351. [CrossRef]
141. Hurley, K.M.; Herbert, H.; Moga, M.M.; Saper, C.B. Efferent projections of the infralimbic cortex of the rat. *J. Comp. Neurol.* **1991**, *308*, 249–276. [CrossRef]
142. Kuroda, R.; Kawao, N.; Yoshimura, H.; Umeda, W.; Takemura, M.; Shigenaga, Y.; Kawabata, A. Secondary somatosensory cortex stimulation facilitates the antinociceptive effect of the NO synthase inhibitor through suppression of spinal nociceptive neurons in the rat. *Brain Res.* **2001**, *903*, 110–116. [CrossRef] [PubMed]
143. Bruehl, S.; Carlson, C.R.; McCubbin, J.A. The relationship between pain sensitivity and blood pressure in normotensives. *Pain* **1992**, *48*, 463–467. [CrossRef]
144. Bruehl, S.; Chung, O.Y. Interactions between the cardiovascular and pain regulatory systems: An updated review of mechanisms and possible alterations in chronic pain. *Neurosci. Biobehav. Rev.* **2004**, *28*, 395–414. [CrossRef]
145. McCubbin, J.A.; Bruehl, S. Do endogenous opioids mediate the relationship between blood pressure and pain sensitivity in normotensives? *Pain* **1994**, *57*, 63–67. [CrossRef]
146. Fillingim, R.B.; Maixner, W. The influence of resting blood pressure and gender on pain responses. *Psychosom. Med.* **1996**, *58*, 326–332. [CrossRef]

147. Pfleeger, M.; Straneva, P.A.; Fillingim, R.B.; Maixner, W.; Girdler, S.S. Menstrual cycle, blood pressure and ischemic pain sensitivity in women: A preliminary investigation. *Int. J. Psychophysiol.* **1997**, *27*, 161–166. [CrossRef]
148. Fillingim, R.B.; Maixner, W.; Bunting, S.; Silva, S. Resting blood pressure and thermal pain responses among females: Effects on pain unpleasantness but not pain intensity. *Int. J. Psychophysiol.* **1998**, *30*, 313–318. [CrossRef]
149. Myers, C.D.; Robinson, M.E.; Riley III, J.L.; Sheffield, D. Sex, gender, and blood pressure: Contributions to experimental pain report. *Psychosom. Med.* **2001**, *63*, 545–550. [CrossRef]
150. Al'Absi, M.; Buchanan, T.; Lovallo, W.R. Pain perception and cardiovascular responses in men with positive parental history for hypertension. *Psychophysiology* **1996**, *33*, 655–661. [CrossRef]
151. Al'Absi, M.; Buchanan, T.W.; Marrero, A.; Lovallo, W.R. Sex differences in pain perception and cardiovascular responses in persons with parental history for hypertension. *Pain* **1999**, *83*, 331–338. [CrossRef]
152. al'Absi, M.; Petersen, K.L.; Wittmers, L.E. Blood pressure but not parental history for hypertension predicts pain perception in women. *Pain* **2000**, *88*, 61–68. [CrossRef]
153. al'Absi, M.; Petersen, K.L.; Wittmers, L.E. Adrenocortical and hemodynamic predictors of pain perception in men and women. *Pain* **2002**, *96*, 197–204. [CrossRef]
154. Randich, A.; Maixner, W. Interactions between cardiovascular and pain regulatory systems. *Neurosci. Biobehav. Rev.* **1984**, *8*, 343–367. [CrossRef] [PubMed]
155. Takeda, M.; Tanimoto, T.; Ojima, K.; Matsumoto, S. Suppressive effect of vagal afferents on the activity of the trigeminal spinal neurons related to the jaw-opening reflex in rats: Involvement of the endogenous opioid system. *Brain Res. Bull.* **1998**, *47*, 49–56. [CrossRef] [PubMed]
156. Bossut, D.; Maixner, W. Effects of cardiac vagal afferent electrostimulation on the responses of trigerninal and trigeminothalamic neurons to noxious orofacial stimulation. *Pain* **1996**, *65*, 101–109. [CrossRef]
157. Dworkin, B.; Filewich, R.; Miller, N.; Craigmyle, N.; Pickering, T. Baroreceptor activation reduces reactivity to noxious stimulation: Implications for hypertension. *Science* **1979**, *205*, 1299–1301. [CrossRef]
158. Thurston, C.L.; Randich, A. Acute increases in arterial blood pressure produced by occlusion of the abdominal aorta induces antinociception: Peripheral and central substrates. *Brain Res.* **1990**, *519*, 12–22. [CrossRef]
159. Maixner, W.; Touw, K.B.; Brody, M.J.; Gebhart, G.F. Factors influencing the altered pain perception in the spontaneously hypertensive rat. *Brain Res.* **1982**, *237*, 137–145. [CrossRef]
160. D'Antono, B.; Ditto, B.; Sita, A.; Miller, S.B. Cardiopulmonary baroreflex stimulation and blood pressure-related hypoalgesia. *Biol. Psychol.* **2000**, *53*, 217–231. [CrossRef]
161. Dworkin, B.R.; Elbert, T.; Rau, H.; Birbaumer, N.; Pauli, P.; Droste, C.; Brunia, C. Central effects of baroreceptor activation in humans: Attenuation of skeletal reflexes and pain perception. *Proc. Natl. Acad. Sci. USA* **1994**, *91*, 6329–6333. [CrossRef]
162. Rau, H.; Brody, S.; Larbig, W.; Pauli, P.; Vöhringer, M.; Harsch, B.; Kröling, P.; Birbaumer, N. Effects of PRES baroreceptor stimulation on thermal and mechanical pain threshold in borderline hypertensives and normotensives. *Psychophysiology* **1994**, *31*, 480–485. [CrossRef]
163. Angrilli, A.; Mini, A.; Mucha, R.F.; Rau, H. The influence of low blood pressure and baroreceptor activity on pain responses. *Physiol. Behav.* **1997**, *62*, 391–397. [CrossRef]
164. Edwards, L.; McIntyre, D.; Carroll, D.; Ring, C.; France, C.R.; Martin, U. Effects of artificial and natural baroreceptor stimulation on nociceptive responding and pain. *Psychophysiology* **2003**, *40*, 762–769. [CrossRef]
165. Edwards, L.; Ring, C.; McIntyre, D.; Carroll, D. Modulation of the human nociceptive flexion reflex across the cardiac cycle. *Psychophysiology* **2001**, *38*, 712–718. [CrossRef]
166. Bruehl, S.; Chung, O.Y.; Jirjis, J.N.; Biridepalli, S. Prevalence of clinical hypertension in patients with chronic pain compared to nonpain general medical patients. *Clin. J. Pain* **2005**, *21*, 147–153. [CrossRef]
167. Browne, T.J.; Hughes, D.I.; Dayas, C.V.; Callister, R.J.; Graham, B.A. Projection neuron axon collaterals in the dorsal horn: Placing a new player in spinal cord pain processing. *Front. Physiol.* **2020**, *11*, 560802. [CrossRef]
168. Schramm, L.P. Spinal sympathetic interneurons: Their identification and roles after spinal cord injury. *Prog. Brain Res.* **2006**, *152*, 27–37.
169. Clifton, G.L.; Ziegler, M.G.; Grossman, R.G. Circulating catecholamines and sympathetic activity after head injury. *Neurosurgery* **1981**, *8*, 10–14. [CrossRef] [PubMed]
170. Park, E.S.; Park, C.I.; Cho, S.R.; Lee, J.W.; Kim, E.J. Assessment of autonomic nervous system with analysis of heart rate variability in children with spastic cerebral palsy. *Yonsei Med. J.* **2002**, *43*, 65–72. [CrossRef] [PubMed]
171. Kushki, A.; Drumm, E.; Mobarak, M.P.; Tanel, N.; Dupuis, A.; Chau, T.; Anagnostou, E. Investigating the autonomic nervous system response to anxiety in children with autism spectrum disorders. *PLoS ONE* **2013**, *8*, e59730. [CrossRef]
172. Appelhans, B.M.; Luecken, L.J. Heart rate variability as an index of regulated emotional responding. *Rev. Gen. Psychol.* **2006**, *10*, 229. [CrossRef]
173. Webb, S.; Adgey, A.; Pantridge, J. Autonomic disturbance at onset of acute myocardial infarction. *Br. Med. J.* **1972**, *3*, 89–92. [CrossRef]
174. Schwartz, P.J.; Vanoli, E. Cardiac arrhythmias elicited by interaction between acute myocardial ischemia and sympathetic hyperactivity: A new experimental model for the study of antiarrhythmic drugs. *J. Cardiovasc. Pharmacol.* **1981**, *3*, 1251–1259. [CrossRef]

175. Lyon, A.R.; Rees, P.S.; Prasad, S.; Poole-Wilson, P.A.; Harding, S.E. Stress (Takotsubo) cardiomyopathy—A novel pathophysiological hypothesis to explain catecholamine-induced acute myocardial stunning. *Nat. Rev. Cardiol.* **2008**, *5*, 22. [CrossRef]
176. Esler, M.; Lambert, G.; Brunner-La Rocca, H.; Vaddadi, G.; Kaye, D. Sympathetic nerve activity and neurotransmitter release in humans: Translation from pathophysiology into clinical practice. *Acta Physiol. Scand.* **2003**, *177*, 275–284. [CrossRef]
177. Schadt, J.C.; Ludbrook, J. Hemodynamic and neurohumoral responses to acute hypovolemia in conscious mammals. *Am. J. Physiol. -Heart Circ. Physiol.* **1991**, *260*, H305–H318. [CrossRef]
178. Evans, R.G.; Ventura, S.; Dampney, R.A.; Ludbrook, J. John Ludbrook APPS Symposium Neural Mechanisms In The Cardiovascular Responses To Acute Central Hypovolaemia. *Clin. Exp. Pharmacol. Physiol.* **2001**, *28*, 479–487. [CrossRef]
179. Carlson, C.R.; Okeson, J.P.; Falace, D.A.; Nitz, A.J.; Curran, S.L.; Anderson, D. Comparison of psychologic and physiologic functioning between patients with masticatory muscle pain and matched controls. *J. Orofac. Pain* **1993**, *7*.
180. Peters, M.L.; Schmidt, A.J. Psychophysiological responses to repeated acute pain stimulation in chronic low back pain patients. *J. Psychosom. Res.* **1991**, *35*, 59–74. [CrossRef]
181. Collins, G.; Cohen, M.; Naliboff, B.; Schandler, S. Comparative analysis of paraspinal and frontalis EMG, heart rate and skin conductance in chronic low back pain patients and normals to various postures and stress. *Scand. J. Rehabil. Med.* **1982**, *14*, 39–46.
182. Schweinhardt, P.; Bushnell, M.C. Pain imaging in health and disease—How far have we come? *J. Clin. Investig.* **2010**, *120*, 3788–3797. [CrossRef]
183. Baliki, M.N.; Chialvo, D.R.; Geha, P.Y.; Levy, R.M.; Harden, R.N.; Parrish, T.B.; Apkarian, A.V. Chronic pain and the emotional brain: Specific brain activity associated with spontaneous fluctuations of intensity of chronic back pain. *J. Neurosci.* **2006**, *26*, 12165–12173. [CrossRef]
184. Hashmi, J.A.; Baliki, M.N.; Huang, L.; Baria, A.T.; Torbey, S.; Hermann, K.M.; Schnitzer, T.J.; Apkarian, A.V. Shape shifting pain: Chronification of back pain shifts brain representation from nociceptive to emotional circuits. *Brain* **2013**, *136*, 2751–2768. [CrossRef]
185. May, A. Chronic pain may change the structure of the brain. *PAIN®* **2008**, *137*, 7–15. [CrossRef]
186. Hubbard, C.S.; Khan, S.A.; Keaser, M.L.; Mathur, V.A.; Goyal, M.; Seminowicz, D.A. Altered brain structure and function correlate with disease severity and pain catastrophizing in migraine patients. *Eneuro* **2014**, *1*. [CrossRef]
187. Ihara, N.; Wakaizumi, K.; Nishimura, D.; Kato, J.; Yamada, T.; Suzuki, T.; Hashiguchi, S.; Terasawa, Y.; Kosugi, S.; Morisaki, H. Aberrant resting-state functional connectivity of the dorsolateral prefrontal cortex to the anterior insula and its association with fear avoidance belief in chronic neck pain patients. *PLoS ONE* **2019**, *14*, e0221023. [CrossRef]
188. Seminowicz, D.A.; Moayedi, M. The dorsolateral prefrontal cortex in acute and chronic pain. *J. Pain* **2017**, *18*, 1027–1035. [CrossRef]
189. Qi, X.; Cui, K.; Zhang, Y.; Wang, L.; Tong, J.; Sun, W.; Shao, S.; Wang, J.; Wang, C.; Sun, X.; et al. A nociceptive neuronal ensemble in the dorsomedial prefrontal cortex underlies pain chronicity. *Cell Rep.* **2022**, *41*, 111833. [CrossRef]
190. Millan, M.J. Descending control of pain. *Prog. Neurobiol.* **2002**, *66*, 355–474. [CrossRef]
191. Holden, J.E.; Naleway, E. Microinjection of carbachol in the lateral hypothalamus produces opposing actions on nociception mediated by α1-and α2-adrenoceptors. *Brain Res.* **2001**, *911*, 27–36. [CrossRef]
192. Karim, F.; Roerig, S.C. Differential effects of antisense oligodeoxynucleotides directed against Gzα and Goα on antinociception produced by spinal opioid and α2 adrenergic receptor agonists. *Pain* **2000**, *87*, 181–191. [CrossRef]
193. Singewald, N.; Philippu, A. Involvement of biogenic amines and amino acids in the central regulation of cardiovascular homeostasis. *Trends Pharmacol. Sci.* **1996**, *17*, 356–363. [CrossRef]
194. Eisenach, J.C.; De Kock, M.; Klimscha, W. α2-Adrenergic agonists for regional anesthesia: A clinical review of clonidine (1984–1995). *J. Am. Soc. Anesthesiol.* **1996**, *85*, 655–674. [CrossRef]
195. Maze, M.; Tranquilli, W. Alpha-2 adrenoceptor agonists: Defining the role in clinical anesthesia. *J. Am. Soc. Anesthesiol.* **1991**, *74*, 581–605. [CrossRef]
196. Pertovaara, A.; Auppila, T.; Jyväsjärvi, E.; Kalso, E. Involvement of supraspinal and spinal segmental alpha-2-adrenergic mechanisms in the medetomidine-induced antinociception. *Neuroscience* **1991**, *44*, 705–714. [CrossRef] [PubMed]
197. Ghelardini, C.; Galeotti, N.; Bartolini, A. Antinociception induced by amitriptyline and imipramine is mediated by α2A-adrenoceptors. *Jpn. J. Pharmacol.* **2000**, *82*, 130–137. [CrossRef] [PubMed]
198. Green, G.M.; Lyons, L.; Dickenson, A.H. α2-adrenoceptor antagonists enhance responses of dorsal horn neurones to formalin induced inflammation. *Eur. J. Pharmacol.* **1998**, *347*, 201–204. [CrossRef]
199. Xu, M.; Kontinen, V.K.; Kalso, E. Endogenous noradrenergic tone controls symptoms of allodynia in the spinal nerve ligation model of neuropathic pain. *Eur. J. Pharmacol.* **1999**, *366*, 41–45. [CrossRef]
200. Feldman, P.; Felder, R. α-Adrenergic influences on neuronal responses to visceral afferent input in the nucleus tractus solitarius. *Neuropharmacology* **1989**, *28*, 1081–1087. [CrossRef] [PubMed]
201. Brodie, M.S.; Proudfit, H.K. Antinociception induced by local injections of carbachol into the nucleus raphe magnus in rats: Alteration by intrathecal injection of monoaminergic antagonists. *Brain Res.* **1986**, *371*, 70–79. [CrossRef]
202. Budai, D.; Harasawa, I.; Fields, H.L. Midbrain periaqueductal gray (PAG) inhibits nociceptive inputs to sacral dorsal horn nociceptive neurons through α2-adrenergic receptors. *J. Neurophysiol.* **1998**, *80*, 2244–2254. [CrossRef] [PubMed]
203. Cui, M.; Feng, Y.; McAdoo, D.; Willis, W. Periaqueductal gray stimulation-induced inhibition of nociceptive dorsal horn neurons in rats is associated with the release of norepinephrine, serotonin, and amino acids. *J. Pharmacol. Exp. Ther.* **1999**, *289*, 868–876.

204. Adair, J.R.; Hamilton, B.L.; Scappaticci, K.A.; Helke, C.J.; Gillis, R.A. Cardiovascular responses to electrical stimulation of the medullary raphe area of the cat. *Brain Res.* **1977**, *128*, 141–145. [CrossRef] [PubMed]
205. Geerling, J.C.; Shin, J.W.; Chimenti, P.C.; Loewy, A.D. Paraventricular hypothalamic nucleus: Axonal projections to the brainstem. *J. Comp. Neurol.* **2010**, *518*, 1460–1499. [CrossRef] [PubMed]
206. van der Kooy, D.; Koda, L.Y.; McGinty, J.F.; Gerfen, C.R.; Bloom, F.E. The organization of projections from the cortes, amygdala, and hypothalamus to the nucleus of the solitary tract in rat. *J. Comp. Neurol.* **1984**, *224*, 1–24. [CrossRef] [PubMed]
207. Michelini, L.C. The NTS and integration of cardiovascular control during exercise in normotensive and hypertensive individuals. *Curr. Hypertens. Rep.* **2007**, *9*, 214–221. [CrossRef] [PubMed]

International Journal of
Molecular Sciences

MDPI

Article

Aerobic Exercise Prevents Arterial Stiffness and Attenuates Hyperexcitation of Sympathetic Nerves in Perivascular Adipose Tissue of Mice after Transverse Aortic Constriction

Niujin Shi [1,2], Jingbo Xia [1,3], Chaoge Wang [1,4], Jie Zhou [1,2], Junhao Huang [1,5], Min Hu [1,*] and Jingwen Liao [1,5,*]

1 Guangdong Provincial Key Laboratory of Physical Activity and Health Promotion, Guangzhou Sport University, Guangzhou 510500, China
2 Graduate School, Guangzhou Sport University, Guangzhou 510500, China
3 Key Laboratory of Regenerative Medicine, Ministry of Education, Jinan University, Guangzhou 510500, China
4 School of Kinesiology, Shanghai University of Sport, Shanghai 200438, China
5 Scientific Research Center, Guangzhou Sport University, Guangzhou 510500, China
* Correspondence: minhu@gzsport.edu.cn (M.H.); liaojw@gzsport.edu.cn (J.L.)

Abstract: We aimed to investigate the efficacy of exercise on preventing arterial stiffness and the potential role of sympathetic nerves within perivascular adipose tissue (PVAT) in pressure-overload-induced heart failure (HF) mice. Eight-week-old male mice were subjected to sham operation (SHAM), transverse aortic constriction-sedentary (TAC-SE), and transverse aortic constriction-exercise (TAC-EX) groups. Six weeks of aerobic exercise training was performed using a treadmill. Arterial stiffness was determined by measuring the elastic modulus. The elastic and collagen fibers of the aorta and sympathetic nerve distribution in PVAT were observed. Circulating noradrenaline (NE), expressions of β3-adrenergic receptor (β3-AR), and adiponectin in PVAT were quantified. During the recovery of cardiac function by aerobic exercise, thoracic aortic collagen elastic modulus (CEM) and collagen fibers were significantly decreased ($p < 0.05$, TAC-SE vs. TAC-EX), and elastin elastic modulus (EEM) was significantly increased ($p < 0.05$, TAC-SE vs. TAC-EX). Circulating NE and sympathetic nerve distribution in PVAT were significantly decreased ($p < 0.05$, TAC-SE vs. TAC-EX). The expression of β3-AR was significantly reduced ($p < 0.05$, TAC-SE vs. TAC-EX), and adiponectin was significantly increased ($p < 0.05$, TAC-SE vs. TAC-EX) in PVAT. Regular aerobic exercise can effectively prevent arterial stiffness and extracellular matrix (ECM) remodeling in the developmental course of HF, during which sympathetic innervation and adiponectin within PVAT might be strongly implicated.

Keywords: arterial stiffness; sympathetic nerves; PVAT; exercise; heart failure

Citation: Shi, N.; Xia, J.; Wang, C.; Zhou, J.; Huang, J.; Hu, M.; Liao, J. Aerobic Exercise Prevents Arterial Stiffness and Attenuates Hyperexcitation of Sympathetic Nerves in Perivascular Adipose Tissue of Mice after Transverse Aortic Constriction. *Int. J. Mol. Sci.* **2022**, *23*, 11189. https://doi.org/10.3390/ijms231911189

Academic Editors: Yutang Wang and Kate Denton

Received: 7 September 2022
Accepted: 19 September 2022
Published: 23 September 2022

1. Introduction

Arterial stiffness is an independent predictor of cardiovascular events [1], among which heart failure (HF) can be one of the most serious outcomes [2]. The development of arterial stiffness is typically attributed to extracellular matrix (ECM) remodeling within media and adventitia [3], represented by the fragmentation of elastic fibers and deposition of collagen fibers [4]. Recently, perivascular adipose tissue (PVAT), an important cross-linking connective tissue, was highlighted as a mediator of ECM remodeling [5]. PVAT surrounds most of the vasculature and has emerged as an active component of the blood vessel wall [6], regulating vascular homeostasis [7] and affecting the pathogenesis of arterial remodeling [6]. Pathological processes including HF [8] and obesity [9] usually drive hyperexcitation of sympathetic nerves. Sympathetic innervation within PVAT has recently been recognized as a possible regulator of ECM remodeling during the development of arterial stiffness [10]. Previously, the release of adiponectin via activation of β3-adrenergic receptor (β3-AR) in PVAT was reported to be triggered by sympathetic stimulation [11]. Adiponectin, as one of PVAT-derived adipokines, was implicated in arterial stiffness [12] and had beneficial effects in maintaining vascular homeostasis [13].

A meta-analysis of randomized trials supports that exercise-based cardiac rehabilitation should be offered to all HF patients [14], and quantitative evidence from patients with cardiovascular disease also suggests that exercise interventions including aerobic, combined, or isometric trainings are capable of reducing arterial stiffness [15]. Moreover, exercise training would normalize the hyperexcitation of sympathetic nerves in HF [16]. However, the efficacy of exercise on preventing arterial stiffness and the potential role of sympathetic nerves within PVAT are not fully understood.

Therefore, this study was designed to determine the effects of 6 weeks of aerobic exercise training on preventing arterial stiffness in pressure overload-induced HF mice and to explore the potential roles of sympathetic nerves within PVAT. We hypothesized that 6 weeks of aerobic exercise training would be effective in preventing arterial stiffness and ECM remodeling and attenuating the hyperexcitation of sympathetic nerves in PVAT during the developmental course of HF.

2. Results

2.1. The Developmental Course of Heart Failure after TAC Was Delayed by Exercise

Both the ejection fraction (EF) (Figure 1A) and fractional shortening (FS) (Figure 1B) of TAC-SE were significantly reduced ($p < 0.05$) compared with those of SHAM; this reduction was significantly restored ($p < 0.05$) in TAC-EX. The heart weight (HW) normalized to body weight (BW) (Figure 1C) and HW normalized to tibia length (TL) (Figure 1D) of TAC-SE were significantly elevated ($p < 0.05$) compared with those of SHAM, whereas HW/BW and HW/TL of TAC-EX were significantly reduced ($p < 0.05$) compared with those of TAC-SE. The representative wheat germ agglutinin (WGA) staining images of cardiomyocyte (CM) is shown in Figure 1E; the CM size of TAC-SE was significantly increased ($p < 0.05$) compared with that of SHAM, and the CM size of TAC-EX was significantly decreased ($p < 0.05$) compared with that of TAC-SE (Figure 1F).

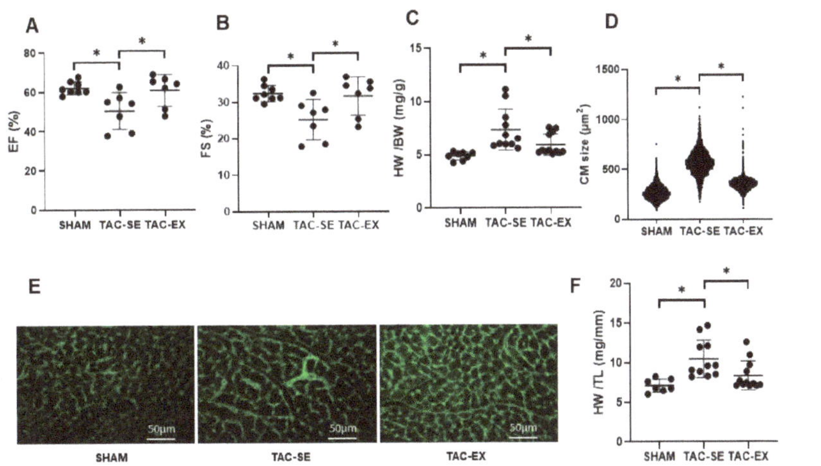

Figure 1. Effects of exercise on cardiac function in mice after TAC. EF (**A**) was calculated by using the formula (EDV - ESV)/EDV × 100%, and FS (**B**) was calculated by using the formula (LVDd − LVDs)/LVDd × 100%. Heart weight per body weight (HW/BW) ratio (**C**) and heart weight per tibia length (HW/TL) ratio (**D**). Original image following WGA staining (**E**), and CM size was measured (**F**). Data were analyzed using one-way ANOVA; values are mean ± SD. * indicates $p < 0.05$. Abbreviations: EF, ejection fraction; EDV, end-diastolic volume; ESV, end-systolic volume; FS, fractional shortening; LVDd, left ventricular end-diastolic dimension; LVDs, left ventricular end-contractile diameter; HW, heart weight; BW, body weight; TL, tibia length; WGA, wheat germ agglutinin; CM, cardiomyocyte; SHAM, sham surgery; TAC-SE, transverse aortic constriction-sedentary; and TAC-EX, transverse aortic constriction-exercise.

2.2. The Remodeling of ECM after TAC Was Prevented by Exercise

EVG staining of the thoracic aorta (Figure 2A) showed that the elastic fibers area of TAC-SE was decreased compared with that of SHAM, and the elastic fibers area of TAC-EX had no statistical difference compared with that of TAC-SE (Figure 2B). Sirius red staining of the thoracic aorta was applied to identify the collagen fibers area (Figure 2C). Compared with SHAM, the collagen fibers area of TAC-SE was significantly larger ($p < 0.05$); compared with TAC-SE, the collagen fibers of TAC-EX were significantly decreased ($p < 0.05$) (Figure 2D). H&E staining of the thoracic aorta (Figure 2E) showed that the aortic wall thickness/diameter of TAC-SE was significantly increased ($p < 0.05$) compared with that of SHAM, and the thickness/diameter of TAC-EX was significantly decreased ($p < 0.05$) compared with that of TAC-SE (Figure 2F).

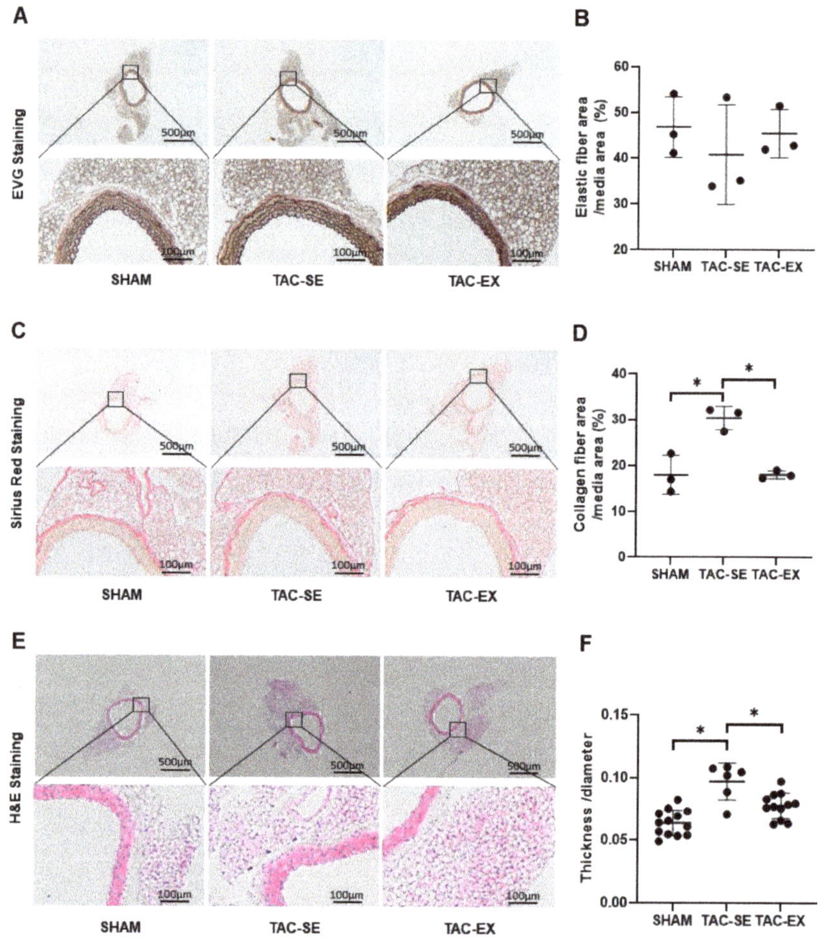

Figure 2. Effects of exercise on aortic elastic/collagen fiber area and aortic morphology in mice after TAC. Elastic fibers of aorta were stained with EVG (**A**,**B**) and collagen fibers with Sirius Red (**C**,**D**) in PVAT. (**E**) Aorta were stained with H&E to compare differences between groups on thickness/diameter (**F**). Data were analyzed using one-way ANOVA; values are mean ± SD. * indicates $p < 0.05$. Abbreviations: EVG, Elastica van Gieson; H&E, hematoxylin and eosin; SHAM, sham surgery; TAC-SE, transverse aortic constriction-sedentary; and TAC-EX, transverse aortic constriction-exercise.

2.3. The Thoracic Arterial Stiffness after TAC Were Prevented by Exercise

A representative image of aortic elastic modulus tests of each group is shown in Figure 3A. The elastin elastic modulus (EEM) of TAC-SE was significantly decreased ($p < 0.05$) compared with that of SHAM, and the EEM of TAC-EX was significantly increased ($p < 0.05$) compared with that of TAC-SE (Figure 3B). The collagen elastic modulus (CEM) of TAC-SE presented as significantly higher than that of SHAM ($p < 0.05$), and the CEM of TAC-EX displayed a significant decrease ($p < 0.05$) compared with that of TAC-SE (Figure 3C).

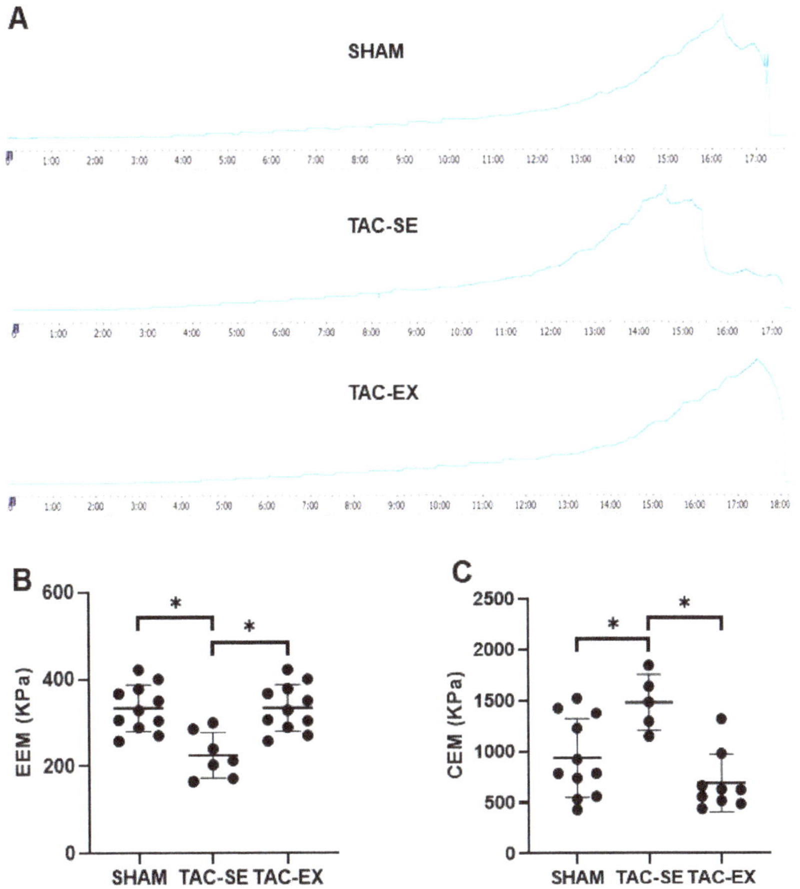

Figure 3. Effects of exercise on aortic stiffness in mice after TAC. Representative images of the force–time curve for each group (**A**); EEM (**B**) and CEM (**C**) were calculated using the slope of stress–strain curves. Data were analyzed using one-way ANOVA; values are mean ± SD. * indicates $p < 0.05$. Abbreviations: EEM, elastin elastic modulus; CEM, collagen elastic modulus; SHAM, sham surgery; TAC-SE, transverse aortic constriction-sedentary; and TAC-EX, transverse aortic constriction-exercise.

2.4. Hyperexcitation of Sympathetic Nerves within PVAT after TAC Was Attenuated by Exercise

The circulating noradrenaline (NE) of TAC-SE was significantly higher ($p < 0.05$) than that of SHAM; the NE of TAC-EX was significantly restored compared with that of TAC-SE ($p < 0.05$) (Figure 4A). The thoracic aortic PVAT was further stained with tyrosine hydroxylase (TH) (a marker of sympathetic neurons) to evaluate the distribution density of sympathetic nerves (Figure 4B,C); the results showed that the sympathetic nerve density of

TAC-SE was significantly higher ($p < 0.05$) than that of SHAM, and the sympathetic nerve density of TAC-EX was significantly lower ($p < 0.05$) compared with that of TAC-SE. The BW and thoracic aortic PVAT of mice were weighed. The PVAT (mg) and PVAT (mg)/BW (g) were significantly increased ($p < 0.05$) after exercise compared with those of the TAC-SE group (Table 1).

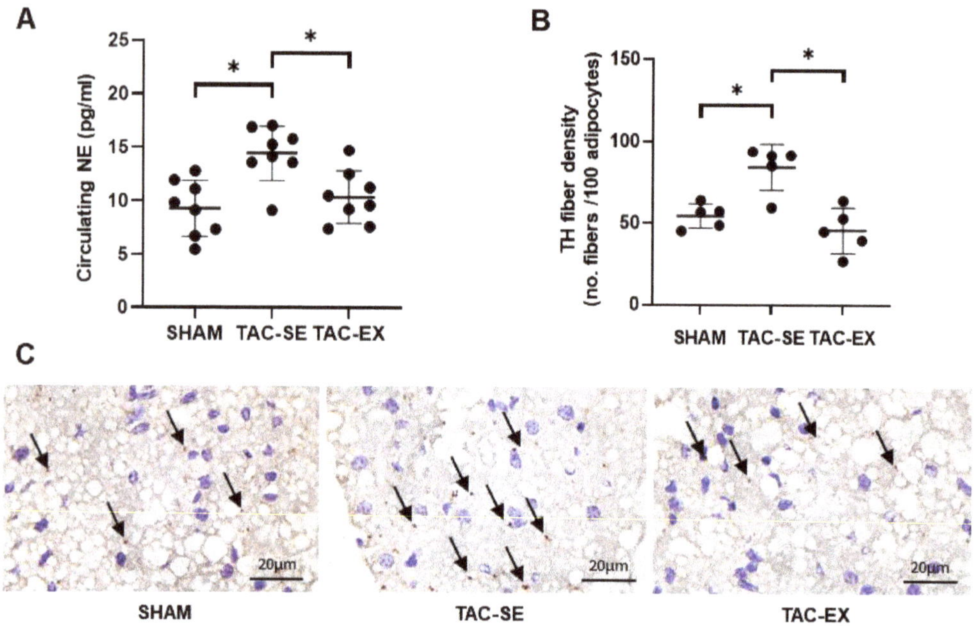

Figure 4. Effects of exercise on hyperexcitation of sympathetic nerves in PVAT of mice after TAC. Circulating levels of NE were determined using ELISA (**A**). Density of TH-immunoreactive parenchymal nerve fibers were calculated by the number of fibers every 100 adipocytes (**B**). Sympathetic nerve fibers were stained with TH antibody in PVAT, as indicated by arrowheads (**C**). Data were analyzed using one-way ANOVA; values are mean ± SD. * indicates $p < 0.05$. Abbreviations: NE, noradrenaline; TH, tyrosine hydroxylase; SHAM, sham surgery; TAC, transverse aortic constriction; TAC-SE, transverse aortic constriction-sedentary; and TAC-EX, transverse aortic constriction-exercise.

Table 1. Effect of exercise on general characteristics of experimental animals after TAC.

	SHAM (n = 15)	TAC-SE (n = 12)	TAC-EX (n = 15)
Body Weight (g)	27.43 ± 2.09	26.65 ± 1.69	26.78 ± 1.90
PVAT (mg)	8.68 ± 1.83	7.31 ± 1.99	10.02 ± 2.29 *
PVAT (mg)/Body Weight (g)	0.318 ± 0.064	0.269 ± 0.076	0.375 ± 0.071 *

Abbreviations: PVAT, perivascular adipose tissue; SHAM, sham surgery; TAC-SE, transverse aortic constriction-sedentary; and TAC-EX, transverse aortic constriction-exercise. Data were analyzed using one-way ANOVA; values are mean ± SD. * indicates $p < 0.05$ vs. TAC-SE.

2.5. β3-Adrenergic Receptor (β3-AR) and Adiponectin within PVAT after TAC Was Influenced by Exercise

The expression of β3-AR in TAC-SE was significantly increased ($p < 0.05$) compared with that of SHAM, and that in TAC-EX was significantly decreased ($p < 0.05$) compared with that of TAC-SE (Figure 5A). There was no significant difference in the expression of adiponectin in PVAT between SHAM and TAC-SE, and a more significant increase ($p < 0.05$) was found in TAC-EX than in TAC-SE (Figure 5B).

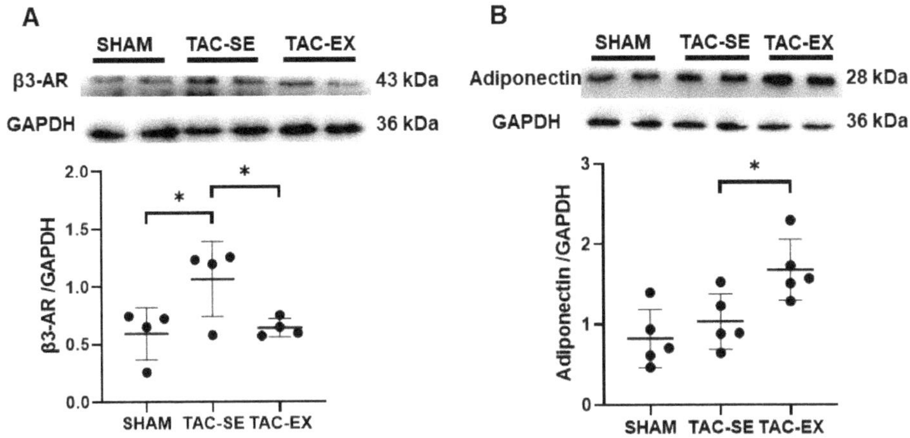

Figure 5. Effects of exercise on the expressions of β3-AR and adiponectin in PVAT of mice after TAC. Representative Western blot assessment of β3-AR (**A**) and adiponectin (**B**) expressions in PVAT normalized to the expressions of GAPDH. Data were analyzed using one-way ANOVA; values are mean ± SD. * indicates $p < 0.05$. Abbreviations: β3-AR, β3-adrenergic receptor; SHAM, sham surgery; TAC-SE, transverse aortic constriction-sedentary; and TAC-EX, transverse aortic constriction-exercise.

3. Discussion

In this study, we examined the efficacy of 6 weeks of aerobic exercise training on preventing arterial stiffness in HF and the contribution of the sympathetic nerve within PVAT to this process. These are the primary findings: (1) aerobic exercise training significantly prevented arterial stiffness represented by ECM remodeling and an elastic modulus shift during HF; (2) sympathetic activation in PVAT may have contributed to the reconstruction of ECM, during which expressions of adiponectin in PVAT significantly changed. These findings provide novel evidence of the effects of aerobic exercise on preventing arterial stiffness during HF and suggest a potential role of sympathetic innervation of PVAT in this process.

Regular exercise prevented arterial stiffening and ECM remodeling. The effects of exercise training on HF was systematically reviewed [17] and the findings supported our results that exercise recovered cardiomyocyte morphology and cardiac ejection fraction. A primary finding of the present study is that aerobic exercise training prevented arterial stiffness during the development of pressure overload-induced HF. Regular aerobic exercise, regardless of intensity or duration, is effective in reducing arterial stiffness [18], which is a critical contributor to HF [19]. The mechanical stiffness of the aorta is dependent on elastin and collagen, which are prominent scaffolding proteins of ECM within the vascular wall [3]. It is generally considered that the elasticity of arteries is facilitated by elastin and stiffness by collagen [3]. Based on the observations of thoracic aortic morphology, our data indicated that the development of HF was accompanied by increased arterial collagen fibers and decreased elastin fibers, whereas aerobic exercise offset those influences. Notably, Ouyang et al. already found that both long-term chronic continuous and interval exercise training could prevent ECM remodeling of the coronary artery in pressure-overload-induced HF mice [20]. We additionally showed that thoracic aortic ECM remodeling and arterial stiffness were prevented by regular aerobic exercise in pressure overload-induced HF.

Sympathetic innervation within PVAT was recently recognized as a possible regulator of ECM remodeling during the development of arterial stiffness [11]; however, this effect has not been confirmed in the setting of exercise intervention. Our results showed that aerobic exercise attenuates the hyperexcitation of sympathetic nerves in PVAT, represented by the reduced expression of circulating NE and the distribution of the sympathetic nerve.

Exercise training has been shown to normalize sympathetic hyperexcitation in patients with HF [16], and to restore PVAT function and prevent vascular complications [21]. Quantitative evidence from HF patients systematically supported this positive influence of exercise [22].

The sympathetic nerve of PVAT could derive NE and activate β3-AR [11]. To the best of our knowledge, the influence of exercise on β3-AR expression was only reported in obesity [21]; β3-AR expression in mesenteric PVAT was downregulated in obesity but improved with exercise [21]. In this study focusing on HF, we observed increased β3-AR expression in aorta PVAT, and aerobic exercise restored its expression. This inconsistence may be explained by difference pathological conditions. A previous study also suggested that β3-AR might undergo desensitization due to sustained sympathetic adrenergic activation in HF [23]. In addition, a recent study also highlighted a role of organic cation transporter-3 on absorbing excessive NE in PVAT, thus preventing β3-AR overactivation by NE [11].

We previously hypothesized that β3-AR mediated adiponectin release in PVAT, because a reduced expression of β3-AR was demonstrated to result in a reduction in adiponectin secretion from mesenteric PVAT [21]. Our results showed that aerobic exercise restored the expression of adiponectin from PVAT, even though the expression of β3-AR decreased. This may be due to the secretion of adiponectin in PVAT being influenced by other factors in addition to β3-AR, including the glucocorticoid receptor in the adipose tissue of obesity [24] and 4-hydroxynonenal (a product of lipid peroxidation) in the PVAT of atherosclerotic patients [25]. The protective role of adiponectin in other conditions were also restored by exercise training: aerobic exercise increased adiponectin expression in the aortic PVAT of type 2 diabetic mice [26]; similar results were also found in the aortic PVAT of Zucker rat under chronic stress [27]. Therefore, our results suggested that aerobic exercise restores adiponectin in PVAT and prevents arterial stiffness during HF development.

The limitation of the present study is that we only quantified the expression of β3-AR in PVAT but did not evaluate the desensitized level of β3-AR or other potential factors involved in regulating adiponectin; therefore, the mechanism of β3-AR-mediated adiponectin release could not be depicted.

In summary, aerobic exercise prevents arterial stiffness during the development of HF, and sympathetic nerve innervation and adiponectin within PVAT are strongly implicated in this process. Our study highlighted PVAT sympathetic innervation as a promising aspect for the prevention of arterial stiffness in HF.

4. Materials and Methods

4.1. Animal Care and Experimental Design

All animal care and experimental procedures were approved by the Animal Experimental Ethics Inspection of Guangzhou Sport University and performed in accordance with the principles of the Declaration of Helsinki. A total of 53 male C57BL/6 mice (aged 8–9 weeks) were obtained from the Animal Center of Guangdong. All animals were housed at 24 °C and 55% humidity under a 12 h alternating light and dark environment and fed with ordinary standard chow throughout the study. Mice were randomly divided into sham operation (SHAM) (n = 15), transverse aortic constriction-sedentary (TAC-SE) (n = 13), and transverse aortic constriction-exercise (TAC-EX) (n = 17) groups.

4.2. Establishment of Transverse Aortic Constriction Model

Transverse aortic constriction (TAC) is a commonly used experimental model for pressure-overload-induced HF [28,29]. Briefly, mice were anesthetized with 2% isoflurane, and a left upper thoracotomy was performed with the pectoralis and intercostal muscles being dissected. Then, a nylon suture was placed around the transverse aorta and loosely tied in a knot. A presterilized 27G blunt needle was then placed into the knot; after the knot was tightened, the needle was withdrawn. Finally, the muscle and skin were sutured. SHAM mice were subjected to the same procedure but without the knot [30].

4.3. Exercise Protocol

Treadmill exercise intervention was performed for two weeks after TAC. Adaptive exercise intervention for 1 week was set at: 8 m/min, 1 h/day, 5 days/week, and slope = 0°; formal exercise intervention for 5 weeks was set at: 12 m/min, 1 h/day, 5 days/week, and slope = 0°. The exercise intervention was performed between 19:30 and 20:30 Animals were allowed to break three times for 3–5 min during each exercise intervention [31,32].

After the exercise intervention, cardiac function was evaluated by echocardiography. Blood samples, heart, thoracic aorta, and PVAT were collected from the mice. The thoracic aortas and hearts were fixed in 4% paraformaldehyde for histological assessments, and thoracic aortic PVAT was stored at −80 °C for western blotting.

4.4. Echocardiography

Echocardiography was performed after exercise using a Vevo2100 system (Visual Sonics Inc., Toronto, ON, Canada) with a 40 MHz MS550D ultrasound transducer. Mice were anesthetized with 2% isoflurane and were depilated with depilatory cream, and then ultrasonic coupling agent was applied. Heart and respiration rates were continuously monitored via stage electrodes. Scanning was initiated with parasternal long- and short-axis views [33,34]. End-diastolic volume (EDV) and end-systolic volume (ESV) were calculated using the Simpson method of disks [35]. The ejection fraction (EF) was calculated: EF (%) = (EDV − ESV)/EDV × 100. The left ventricular end-diastolic dimension (LVDd) and left ventricular end-contractile diameter (LVDs) were measured in one-dimensional mode. Fractional shortening (FS) was calculated: FS (%) = (LVDd − LVDs)/LVDd × 100.

4.5. Histological Assessments

The left ventricle of the heart was collected and fixed with 4% paraformaldehyde, embedded in paraffin, and cut with a thickness of 5 μM. Sections of the heart were stained with WGA kits (Thermo-Fisher, Waltham, MA, USA) to measure myocardial cell size [36]. A cross-section of cardiomyocytes was photographed under a fluorescence microscope, and 1000 cells were counted in each section for analysis.

The thoracic aorta was collected and fixed with 4% paraformaldehyde. The tissue was embedded in paraffin and cut to a thickness of 5 μM. For morphometric analysis, sections of the thoracic aorta were stained with hematoxylin and eosin (H&E) (Bio sharp, Hefei, China) and imaged using an Olympus BX53® (Olympus, Center Valley, PA, USA). The wall thickness and inner diameter of aorta were analyzed using ImageJ software (ImageJ 1.53e, National Institution of Health, Montgomery, MD, USA). For ECM analysis, sections were stained with Sirius Red (Bio sharp, Hefei, China) for collagen fiber distribution and with Elastica-Van Gieson (EVG) staining (Bio sharp, Hefei, China) for elastic fiber distribution [37]. The area of elastic fiber was calculated as the elastic fiber area/cross-sectional area of aorta ×100%, and the expression of collagen fiber was calculated as collagen fiber area/cross-sectional area of aorta ×100%. To observe the sympathetic nerve distribution within PVAT, sections were stained with tyrosine hydroxylase through immunohistochemistry. Sections were then incubated with TH antibody (abs131679, absin, Shanghai, China,) at 4 °C overnight [38,39]. Imaging was captured using a microscope at 1000× magnifications and analyzed using ImageJ software (ImageJ 1.53e, National Institution of Health, Montgomery, MD, USA).

4.6. Mechanical Stiffness Testing

Arterial elasticity and stiffness were assessed by mechanical stiffness testing [40]. Preheated 37 °C Ca^{2+} and Mg^{2+}-free phosphate-buffered saline solution was added into the chamber of a Multi Wire Myograph System (Model 620M, DMT, Aarhus, Denmark), and then the thoracic aorta with no PVAT was cut into a ~2.0 mm arterial ring. The arterial ring was mounted in the chamber and was stretched 1 mm every 3 min until mechanical failure. The elastic modulus was calculated from the stress–strain curve. Strain (λ) = $\Delta d/d(i)$ and stress (t) = $\lambda F/2HD$, where F is the one-dimensional load applied, H is the sample wall thickness, D is the length of the vessel, Δd is the change in diameter, and $d(i)$ is the initial diameter. The thoracic aortic diameter and wall thickness were assessed by H&E staining, and the vessel length was measured with a fine ruler. The elastin region, coinciding with the EEM, was calculated by the slope of the stress–strain curves at the front smooth section linear. The collagen region, coinciding with the CEM was calculated by the slope of the stress–stain curves at the tail steep linear section. The slope of the stress–stain curves was determined using GraphPad Prism (GraphPad Prism 8.02, San Diego, CA, USA).

4.7. Enzyme Linked Immunosorbent Assay (ELISA)

The concentration of NE in plasma was determined by ELISA kits (MM-0876M1, MEIMIAN, Hangzhou, China) according to the manufacturer's protocol [41]. Blood samples were centrifuged to separate plasma. All samples and standards were measured in duplicate, and the optical density of the zero standard was subtracted from each value. Standard curves were fitted using nonlinear regression analysis.

4.8. Western Blotting

The expressions of β3-AR (61033, Sigma, St. Louis, MO, USA), adiponectin (DF7000, Affinity Biosciences, Hangzhou, China), and GAPDH (G9545, Sigma, St. Louis, USA) were quantified by Western blotting. Briefly, the thoracic aortic PVAT was homogenized in 1 mL of lysis buffer by a grinding machine (Servicebio, Wuhan, China), containing 1× RIPA (Beyotime, Shanghai, China), protease inhibitors (Beyotime, Shanghai, China), and phosphatase inhibitors (Cwbio, Beijing, China). The proteins were separated by 10% SDS polyacrylamide gels and transferred to 0.2 μM Immobilon-PSQ PVDF membranes (Millipore-Sigma, Burlington, VT, USA). The membrane was then blocked in 5% BSA for 1.5 h. The primary antibody was incubated at 4°C overnight, and goat anti-rabbit IgG (511203, Zenbio, Chengdu, China) was used as the secondary antibody incubated for one hour at room temperature. Imaging was captured using a TANON-5200Multi imaging system (Ewell, Guangzhou, China) and analyzed using ImageJ software (ImageJ 1.53e, National Institution of Health, Montgomery, MD, USA). Quantitative data were normalized using endogenous GAPDH.

4.9. Statistical Analysis

Data are presented as the mean ± standard deviation (SD). Statistical significance between groups was analyzed using one-way ANOVA, and $p < 0.05$ was considered as statistically significant. All data were analyzed using the GraphPad Prism (GraphPad Prism 8.02, San Diego, CA, USA).

Author Contributions: Conceptualization, J.L. and M.H.; methodology, N.S. and J.X.; software, N.S.; validation, N.S. and J.L.; formal analysis, N.S.; investigation, N.S., J.X., C.W. and J.Z.; resources, J.L. and J.X.; data curation, N.S.; writing—original draft preparation, N.S.; writing—review and editing, N.S., J.L., M.H. and J.H.; visualization, N.S.; supervision, J.L.; project administration, J.L. and M.H.; funding acquisition, J.L., J.X. and M.H. All authors have read and agreed to the published version of the manuscript.

Funding: This research was funded by the National Natural Science Foundation of China (grant numbers: 31971105 and 31801005), the Research Grant of Key Laboratory of Regenerative Medicine, Ministry of Education, Jinan University (No. ZSYXM202206), and the Guangzhou Basic Research Plan, Basic and Applied Basic Research Project (Nos. SL2023A04J00446 and SL2023A04J00488).

Institutional Review Board Statement: The animal study protocol was approved by the Animal Experimental Ethics Inspection of Guangzhou Sport University (2021DWLL-08).

Informed Consent Statement: Not applicable.

Data Availability Statement: The data presented in this study are available on request from the corresponding authors.

Conflicts of Interest: The authors declare no conflict of interest.

References

1. Mitchell, G.F.; Hwang, S.-J.; Vasan, R.S.; Larson, M.G.; Pencina, M.J.; Hamburg, N.M.; Vita, J.A.; Levy, D.; Benjamin, E.J. Arterial Stiffness and Cardiovascular Events. *Circulation* **2010**, *121*, 505–511. [CrossRef] [PubMed]
2. Marti, C.N.; Gheorghiade, M.; Kalogeropoulos, A.P.; Georgiopoulou, V.V.; Quyyumi, A.A.; Butler, J. Endothelial dysfunction, arterial stiffness, and heart failure. *J. Am. Coll. Cardiol.* **2012**, *60*, 1455–1469. [CrossRef] [PubMed]
3. Chirinos, J.A.; Segers, P.; Hughes, T.; Townsend, R. Large.e-Artery Stiffness in Health and Disease: JACC State-of-the-Art Review. *J. Am. Coll. Cardiol.* **2019**, *74*, 1237–1263. [CrossRef] [PubMed]
4. Bonnans, C.; Chou, J.; Werb, Z. Remodelling the extracellular matrix in development and disease. *Nat. Rev. Mol. Cell Biol.* **2014**, *15*, 786–801. [CrossRef] [PubMed]
5. Fleenor, B.S.; Carlini, N.A.; Ouyang, A.; Harber, M.P. Perivascular adipose tissue-mediated arterial stiffening in aging and disease: An emerging translational therapeutic target? *Pharmacol. Res.* **2022**, *178*, 106150. [CrossRef] [PubMed]
6. Qi, X.-Y.; Qu, S.-L.; Xiong, W.-H.; Rom, O.; Chang, L.; Jiang, Z.-S. Perivascular adipose tissue (PVAT) in atherosclerosis: A double-edged sword. *Cardiovasc. Diabetol.* **2018**, *17*, 134. [CrossRef] [PubMed]
7. Szasz, T.; Bomfim, G.F.; Webb, R.C. The influence of perivascular adipose tissue on vascular homeostasis. *Vasc. Health Risk Manag.* **2013**, *9*, 105–116. [CrossRef] [PubMed]
8. Zhang, D.Y.; Anderson, A.S. The sympathetic nervous system and heart failure. *Cardiol. Clin.* **2014**, *32*, 33–45. [CrossRef] [PubMed]
9. Grassi, G.; Biffi, A.; Seravalle, G.; Trevano, F.Q.; Dell'Oro, R.; Corrao, G.; Mancia, G. Sympathetic Neural Overdrive in the Obese and Overweight State. *Hypertension* **2019**, *74*, 349–358. [CrossRef]
10. Nardone, M.; Floras, J.S.; Millar, P.J. Sympathetic neural modulation of arterial stiffness in humans. *Am. J. Physiol. Heart Circ. Physiol.* **2020**, *319*, H1338–H1346. [CrossRef] [PubMed]
11. Saxton, S.N.; Ryding, K.E.; Aldous, R.G.; Withers, S.B.; Ohanian, J.; Heagerty, A.M. Role of Sympathetic Nerves and Adipocyte Catecholamine Uptake in the Vasorelaxant Function of Perivascular Adipose Tissue. *Arter. Thromb Vasc. Biol* **2018**, *38*, 880–891. [CrossRef] [PubMed]
12. Sabbatini, A.R.; Fontana, V.; Laurent, S.; Moreno, H. An update on the role of adipokines in arterial stiffness and hypertension. *J. Hypertens.* **2015**, *33*, 435–444. [CrossRef] [PubMed]
13. Sowka, A.; Dobrzyn, P. Role of Perivascular Adipose Tissue-Derived Adiponectin in Vascular Homeostasis. *Cells* **2021**, *10*, 1485. [CrossRef] [PubMed]
14. Taylor, R.S.; Walker, S.; Smart, N.A.; Piepoli, M.F.; Warren, F.C.; Ciani, O.; Whellan, D.; O'Connor, C.; Keteyian, S.J.; Coats, A.; et al. Impact of Exercise Rehabilitation on Exercise Capacity and Quality-of-Life in Heart Failure. *J. Am. Coll. Cardiol.* **2019**, *73*, 1430–1443. [CrossRef] [PubMed]
15. Lopes, S.; Afreixo, V.; Teixeira, M.; Garcia, C.; Leitão, C.; Gouveia, M.; Figueiredo, D.; Alves, A.J.; Polonia, J.; Oliveira, J.; et al. Exercise training reduces arterial stiffness in adults with hypertension: A systematic review and meta-analysis. *J. Hypertens.* **2021**, *39*, 214–222. [CrossRef] [PubMed]
16. Shen, Y.; Park, J.B.; Lee, S.Y.; Han, S.K.; Ryu, P.D. Exercise training normalizes elevated firing rate of hypothalamic presympathetic neurons in heart failure rats. *Am. J. Physiol. Regul. Integr. Comp. Physiol.* **2019**, *316*, R110–R120. [CrossRef] [PubMed]
17. Leggio, M.; Fusco, A.; Loreti, C.; Limongelli, G.; Bendini, M.G.; Mazza, A.; Coraci, D.; Padua, L. Effects of exercise training in heart failure with preserved ejection fraction: An updated systematic literature review. *Heart Fail. Rev.* **2020**, *25*, 703–711. [CrossRef] [PubMed]
18. Kobayashi, R.; Kasahara, Y.; Ikeo, T.; Asaki, K.; Sato, K.; Matsui, T.; Iwanuma, S.; Ohashi, N.; Hashiguchi, T. Effects of different intensities and durations of aerobic exercise training on arterial stiffness. *J. Phys. Ther. Sci.* **2020**, *32*, 104–109. [CrossRef] [PubMed]
19. Chi, C.; Liu, Y.; Xu, Y.; Xu, D. Association Between Arterial Stiffness and Heart Failure With Preserved Ejection Fraction. *Front. Cardiovasc. Med.* **2021**, *8*, 707162. [CrossRef]
20. Ouyang, A.; Olver, T.D.; Emter, C.A.; Fleenor, B.S. Chronic exercise training prevents coronary artery stiffening in aortic-banded miniswine: Role of perivascular adipose-derived advanced glycation end products. *J. Appl. Physiol.* **2019**, *127*, 816–827. [CrossRef] [PubMed]
21. Saxton, S.N.; Toms, L.K.; Aldous, R.G.; Withers, S.B.; Ohanian, J.; Heagerty, A.M. Restoring Perivascular Adipose Tissue Function in Obesity Using Exercise. *Cardiovasc. Drugs Ther.* **2021**, *35*, 1291–1304. [CrossRef] [PubMed]
22. Saavedra, M.J.; Romero, F.; Roa, J.; Rodriguez-Nunez, I. Exercise training to reduce sympathetic nerve activity in heart failure patients. A systematic review and meta-analysis. *Braz. J. Phys. Ther.* **2018**, *22*, 97–104. [CrossRef] [PubMed]
23. Mahmood, A.; Ahmed, K.; Zhang, Y. β-Adrenergic Receptor Desensitization/Down-Regulation in Heart Failure: A Friend or Foe? *Front. Cardiovasc. Med.* **2022**, *9*, 925692. [CrossRef] [PubMed]
24. Quattrocelli, M.; Wintzinger, M.; Miz, K.; Panta, M.; Prabakaran, A.D.; Barish, G.D.; Chandel, N.S.; McNally, E.M. Intermittent prednisone treatment in mice promotes exercise tolerance in obesity through adiponectin. *J. Exp. Med.* **2022**, 219. [CrossRef] [PubMed]
25. Margaritis, M.; Antonopoulos, A.S.; Digby, J.; Lee, R.; Reilly, S.; Coutinho, P.; Shirodaria, C.; Sayeed, R.; Petrou, M.; De Silva, R.; et al. Interactions between vascular wall and perivascular adipose tissue reveal novel roles for adiponectin in the regulation of endothelial nitric oxide synthase function in human vessels. *Circulation* **2013**, *127*, 2209–2221. [CrossRef] [PubMed]

26. Wang, J.; Polaki, V.; Chen, S.; Bihl, J.C. Exercise Improves Endothelial Function Associated with Alleviated Inflammation and Oxidative Stress of Perivascular Adipose Tissue in Type 2 Diabetic Mice. *Oxidative Med. Cell. Longev.* **2020**, *2020*, 8830537. [CrossRef]
27. DeVallance, E.R.; Branyan, K.W.; Olfert, I.M.; Pistilli, E.E.; Bryner, R.W.; Kelley, E.E.; Frisbee, J.C.; Chantler, P.D. Chronic stress induced perivascular adipose tissue impairment of aortic function and the therapeutic effect of exercise. *Exp. Physiol.* **2021**, *106*, 1343–1358. [CrossRef] [PubMed]
28. Zhang, Y.; Da, Q.; Cao, S.; Yan, K.; Shi, Z.; Miao, Q.; Li, C.; Hu, L.; Sun, S.; Wu, W.; et al. HINT1 (Histidine Triad Nucleotide-Binding Protein 1) Attenuates Cardiac Hypertrophy Via Suppressing HOXA5 (Homeobox A5) Expression. *Circulation* **2021**, *144*, 638–654. [CrossRef] [PubMed]
29. Zhao, D.; Zhong, G.; Li, J.; Pan, J.; Zhao, Y.; Song, H.; Sun, W.; Jin, X.; Li, Y.; Du, R.; et al. Targeting E3 Ubiquitin Ligase WWP1 Prevents Cardiac Hypertrophy Through Destabilizing DVL2 via Inhibition of K27-Linked Ubiquitination. *Circulation* **2021**, *144*, 694–711. [CrossRef]
30. Au-deAlmeida, A.C.; Au-van Oort, R.J.; Au-Wehrens, X.H.T. Transverse Aortic Constriction in Mice. *JoVE* **2010**, e1729. [CrossRef]
31. Fernando, P.; Bonen, A.; Hoffman-Goetz, L. Predicting submaximal oxygen consumption during treadmill running in mice. *Can. J. Physiol. Pharmacol.* **1993**, *71*, 854–857. [CrossRef] [PubMed]
32. Høydal, M.A.; Wisløff, U.; Kemi, O.J.; Ellingsen, O. Running speed and maximal oxygen uptake in rats and mice: Practical implications for exercise training. *Eur. J. Cardiovasc. Prev. Rehabil. Off. J. Eur. Soc. Cardiol. Work. Groups Epidemiol. Prev. Card. Rehabil. Exerc. Physiol.* **2007**, *14*, 753–760. [CrossRef] [PubMed]
33. Kumar, S.; Wang, G.; Zheng, N.; Cheng, W.; Ouyang, K.; Lin, H.; Liao, Y.; Liu, J. HIMF (Hypoxia-Induced Mitogenic Factor)-IL (Interleukin)-6 Signaling Mediates Cardiomyocyte-Fibroblast Crosstalk to Promote Cardiac Hypertrophy and Fibrosis. *Hypertension* **2019**, *73*, 1058–1070. [CrossRef] [PubMed]
34. Malek Mohammadi, M.; Abouissa, A.; Heineke, J. A surgical mouse model of neonatal pressure overload by transverse aortic constriction. *Nat. Protoc.* **2021**, *16*, 775–790. [CrossRef] [PubMed]
35. Clavel, M.A.; Côté, N.; Mathieu, P.; Dumesnil, J.G.; Audet, A.; Pépin, A.; Couture, C.; Fournier, D.; Trahan, S.; Pagé, S.; et al. Paradoxical low-flow, low-gradient aortic stenosis despite preserved left ventricular ejection fraction: New insights from weights of operatively excised aortic valves. *Eur. Heart J.* **2014**, *35*, 2655–2662. [CrossRef]
36. Chang, Z.S.; Xia, J.B.; Wu, H.Y.; Peng, W.T.; Jiang, F.Q.; Li, J.; Liang, C.Q.; Zhao, H.; Park, K.S.; Song, G.H.; et al. Forkhead box O3 protects the heart against paraquat-induced aging-associated phenotypes by upregulating the expression of antioxidant enzymes. *Aging Cell* **2019**, *18*, e12990. [CrossRef] [PubMed]
37. Nishida, N.; Aoki, H.; Ohno-Urabe, S.; Nishihara, M.; Furusho, A.; Hirakata, S.; Hayashi, M.; Ito, S.; Yamada, H.; Hirata, Y.; et al. High Salt Intake Worsens Aortic Dissection in Mice: Involvement of IL (Interleukin)-17A-Dependent ECM (Extracellular Matrix) Metabolism. *Arter. Thromb Vasc. Biol.* **2020**, *40*, 189–205. [CrossRef] [PubMed]
38. Zhang, N.; Feng, B.; Ma, X.; Sun, K.; Xu, G.; Zhou, Y. Dapagliflozin improves left ventricular remodeling and aorta sympathetic tone in a pig model of heart failure with preserved ejection fraction. *Cardiovasc. Diabetol.* **2019**, *18*, 107. [CrossRef] [PubMed]
39. Murano, I.; Barbatelli, G.; Giordano, A.; Cinti, S. Noradrenergic parenchymal nerve fiber branching after cold acclimatisation correlates with brown adipocyte density in mouse adipose organ. *J. Anat.* **2009**, *214*, 171–178. [CrossRef]
40. Ouyang, A.; Garner, T.B.; Fleenor, B.S. Hesperidin reverses perivascular adipose-mediated aortic stiffness with aging. *Exp. Gerontol.* **2017**, *97*, 68–72. [CrossRef] [PubMed]
41. Zhou, D.C.; Su, Y.H.; Jiang, F.Q.; Xia, J.B.; Wu, H.Y.; Chang, Z.S.; Peng, W.T.; Song, G.H.; Park, K.S.; Kim, S.K.; et al. CpG oligodeoxynucleotide preconditioning improves cardiac function after myocardial infarction via modulation of energy metabolism and angiogenesis. *J. Cell. Physiol.* **2018**, *233*, 4245–4257. [CrossRef] [PubMed]

International Journal of
Molecular Sciences

MDPI

Article

Asprosin in the Paraventricular Nucleus Induces Sympathetic Activation and Pressor Responses via cAMP-Dependent ROS Production

Xiao-Li Wang [1], Jing-Xiao Wang [1], Jun-Liu Chen [1], Wen-Yuan Hao [1], Wen-Zhou Xu [2], Zhi-Qin Xu [2], Yu-Tong Jiang [2], Pei-Qi Luo [2], Qi Chen [3], Yue-Hua Li [3], Guo-Qing Zhu [1,*] and Xiu-Zhen Li [2,*]

1 Key Laboratory of Targeted Intervention of Cardiovascular Disease, Collaborative Innovation Center of Translational Medicine for Cardiovascular Disease, Department of Physiology, Nanjing Medical University, Nanjing 211166, China
2 Department of Cardiology and Emergency Department, The Second Affiliated Hospital of Nanjing Medical University, Nanjing 210011, China
3 Department of Pathophysiology, Nanjing Medical University, Nanjing 211166, China
* Correspondence: gqzhucn@njmu.edu.cn (G.-Q.Z.); lixiuzhen0806@njmu.edu.cn (X.-Z.L.)

Abstract: Asprosin is a newly discovered adipokine that is involved in regulating metabolism. Sympathetic overactivity contributes to the pathogenesis of several cardiovascular diseases. The paraventricular nucleus (PVN) of the hypothalamus plays a crucial role in the regulation of sympathetic outflow and blood pressure. This study was designed to determine the roles and underlying mechanisms of asprosin in the PVN in regulating sympathetic outflow and blood pressure. Experiments were carried out in male adult SD rats under anesthesia. Renal sympathetic nerve activity (RSNA), mean arterial pressure (MAP), and heart rate (HR) were recorded, and PVN microinjections were performed bilaterally. Asprosin mRNA and protein expressions were high in the PVN. The high asprosin expression in the PVN was involved in both the parvocellular and magnocellular regions according to immunohistochemical analysis. Microinjection of asprosin into the PVN produced dose-related increases in RSNA, MAP, and HR, which were abolished by superoxide scavenger tempol, antioxidant N-acetylcysteine (NAC), and NADPH oxidase inhibitor apocynin. The asprosin promoted superoxide production and increased NADPH oxidase activity in the PVN. Furthermore, it increased the cAMP level, adenylyl cyclase (AC) activity, and protein kinase A (PKA) activity in the PVN. The roles of asprosin in increasing RSNA, MAP, and HR were prevented by pretreatment with AC inhibitor SQ22536 or PKA inhibitor H89 in the PVN. Microinjection of cAMP analog db-cAMP into the PVN played similar roles with asprosin in increasing the RSNA, MAP, and HR, but failed to further augment the effects of asprosin. Pretreatment with PVN microinjection of SQ22536 or H89 abolished the roles of asprosin in increasing superoxide production and NADPH oxidase activity in the PVN. These results indicated that asprosin in the PVN increased the sympathetic outflow, blood pressure, and heart rate via cAMP–PKA signaling-mediated NADPH oxidase activation and the subsequent superoxide production.

Keywords: asprosin; paraventricular nucleus; reactive oxygen species; sympathetic activity; blood pressure

Citation: Wang, X.-L.; Wang, J.-X.; Chen, J.-L.; Hao, W.-Y.; Xu, W.-Z.; Xu, Z.-Q.; Jiang, Y.-T.; Luo, P.-Q.; Chen, Q.; Li, Y.-H.; et al. Asprosin in the Paraventricular Nucleus Induces Sympathetic Activation and Pressor Responses via cAMP-Dependent ROS Production. *Int. J. Mol. Sci.* **2022**, 23, 12595. https://doi.org/10.3390/ijms232012595

Academic Editors: Yutang Wang and Kate Denton

Received: 24 September 2022
Accepted: 17 October 2022
Published: 20 October 2022

1. Introduction

Sympathetic activity plays a critical role in the regulation of blood pressure. Chronic excessive sympathetic activity contributes to the pathogenesis of hypertension, chronic heart failure, and chronic kidney disease [1–4]. Renal denervation is effective in lowering blood pressure in patients with resistant hypertension [5]. The hypothalamic paraventricular nucleus (PVN) is an important central integrative region in the regulation of cardiovascular activity and a major source of excitatory drive for sympathetic outflow to the spinal cord by both direct and indirect projections [6]. The excessive sympathetic activation in hypertension and chronic heart failure is mainly related to the changes in molecular signaling in the PVN [7,8].

Asprosin, a novel adipokine, was firstly identified in 2016 as a fasting-induced protein hormone that regulates hepatic glucose release [9]. The proprotein of asprosin is 2871 amino acids long and is cleaved by the activated protease furin at its C terminus to produce a 140-amino-acid-long asprosin and mature fibrillin-1 (FBN1) [10]. Asprosin contributes to metabolism and metabolic disorders, including obesity, diabetes, cardiovascular diseases, and polycystic ovary syndrome [11–14]. Asprosin can cross the blood–brain barrier to enhance the activity of orexigenic agouti-related peptide (AgRP) neurons in the hypothalamus to increase appetite [15]. Asprosin is widely distributed in multiple tissues and organs, including the hypothalamus of the brain [16]. Our preliminary study showed that abundant asprosin existed in the PVN. However, it is not known whether asprosin in the PVN is involved in regulating sympathetic activity and blood pressure.

Activation of the cAMP–protein kinase A (PKA) pathway in the PVN increases sympathetic outflow [17–19]. The classical cAMP–PKA signaling contributes to the reactive oxygen species (ROS) production induced by the β-adrenergic receptor agonist isoproterenol in mouse cardiomyocytes [20,21]. NADPH-oxidase-derived superoxide production in the PVN increases sympathetic outflow and mediates the angiotensin II or salusin-β-induced sympathetic activation [22–24]. The increased superoxide production in the PVN contributes to sympathetic overactivity in hypertension and chronic heart failure [7]. It was found that the cAMP–PKA pathway is necessary for asprosin-mediated AgRP neuron activation in the brain [15]. We were interested in whether cAMP and ROS production were involved in the effects of asprosin. Here we investigated the roles and underlined the mechanisms of asprosin in the PVN when regulating sympathetic activity and blood pressure.

2. Results

2.1. Asprosin Expression

The asprosin expressions were examined and compared in several important nuclei of the brain for the regulation of sympathetic outflow and blood pressure. The asprosin mRNA level was higher in the PVN but lower in the caudal ventrolateral medulla (CVLM) compared with that in the rostral ventrolateral medulla (RVLM) (Figure 1A). Western blot analyses showed high asprosin expression in the PVN; moderate expression in the RVLM, nucleus tractus solitaries (NTS), and dorsal motor nucleus of the vagus (DMV); and low expression in the CVLM (Figure 1B). Immunohistochemistry for asprosin at the PVN level of the brain showed that high asprosin expression existed in the PVN, including both the magnocellular and parvocellular neurons in the PVN (Figure 1C). These results suggested a possibility that asprosin may have played crucial roles in the control of the sympathetic outflow and blood pressure.

2.2. Dose Effects and Time Effects of Asprosin in PVN

Bilateral microinjection of asprosin in the PVN caused an immediate increase in the RSNA, MAP, and HR (Figure 2A). Asprosin dose-dependently increased the RSNA, MAP, and HR, and 5 pmol of asprosin almost reached its maximal effects (Figure 2B). The effects of asprosin lasted about 30 min, and the maximal effects occurred approximately 5 min after the PVN microinjection of asprosin. The PVN microinjection of PBS had no significant effects on the RSNA, MAP, and HR (Figure 2C).

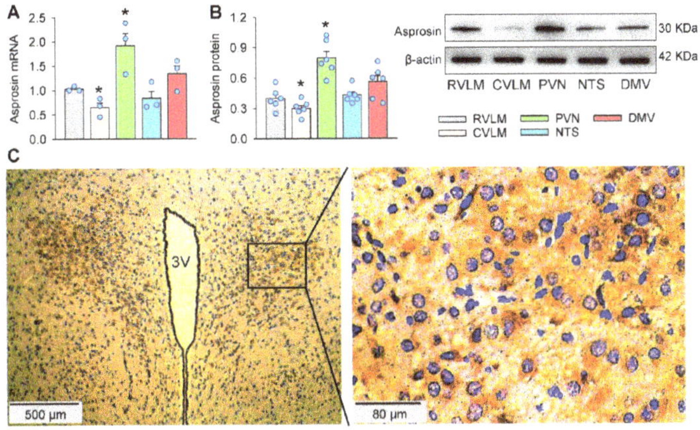

Figure 1. Asprosin expression in the PVN. (**A**) Relative asprosin mRNA levels in some nuclei of the brain. (**B**) Relative asprosin protein levels in some nuclei of the brain. (**C**) Representative images of immunohistochemistry for asprosin (brown color) in the PVN. The nuclei were stained with hematoxylin (blue color). * $p < 0.05$ vs. RVLM. Values are given as mean ± SE. $n = 3$ per group. RVLM, rostral ventrolateral medulla; CVLM, caudal ventrolateral medulla; PVN, paraventricular nucleus of hypothalamus; NTS, nucleus tractus solitaries; DMV, dorsal motor nucleus of the vagus; 3V, the third ventricle.

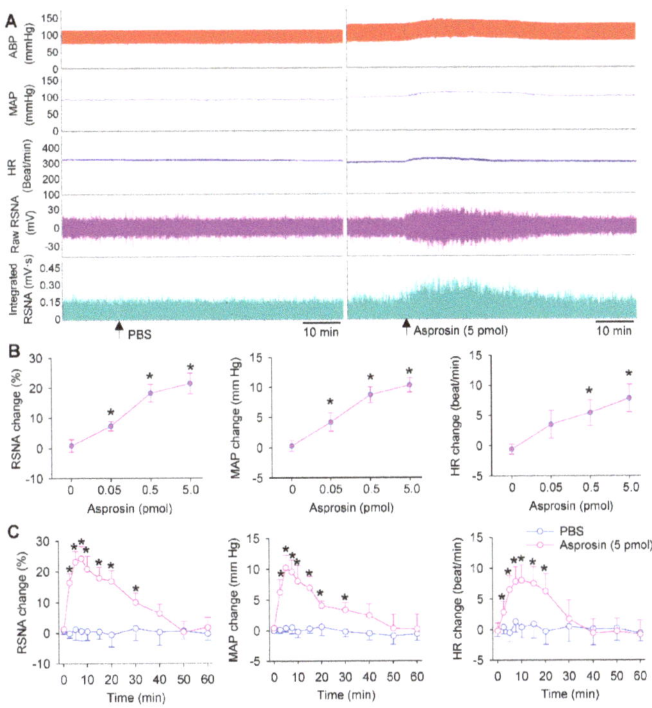

Figure 2. Effects of the microinjection of asprosin in the PVN on the RSNA, MAP and HR. (**A**) Representative recordings showing the effects of microinjection of PBS and asprosin (5 pmol) in the PVN. (**B**) Dose effects of asprosin (0, 0.05, 0.5, 5 pmol) in the PVN. * $p < 0.05$ vs. 0 pmol. (**C**) Time effects of microinjection of PBS and asprosin (5 pmol) in the PVN. * $p < 0.05$ vs. PBS. Values are given as mean ± SE. $n = 6$ per group.

2.3. Effects of Asprosin Antibody in the PVN

Microinjection of specific antibodies in the PVN is a common method that is used to investigate the effects of endogenous active peptides [25–27]. The PVN microinjection of an anti-asprosin antibody to neutralize the endogenous asprosin reduced the RSNA and MAP (Figure 3A) and abolished the asprosin-induced increases in the RSNA, MAP, and HR (Figure 3B). The PVN microinjection of the control antibody had no significant effects. These results suggest that endogenous asprosin in the PVN played a tonic role in enhancing the sympathetic outflow.

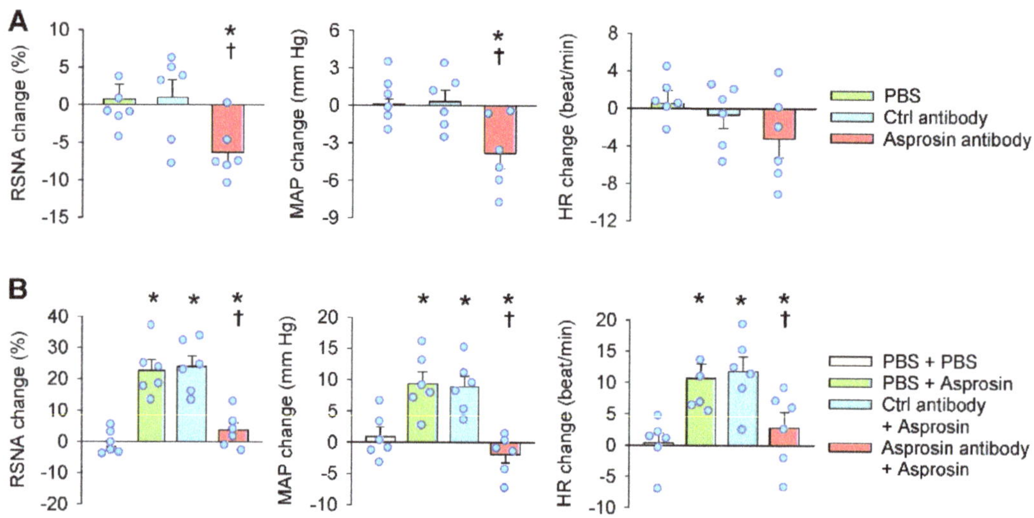

Figure 3. Roles of endogenous asprosin in regulating the RSNA, MAP, and HR. (**A**) Effects of microinjection of anti-asprosin antibody in the PVN on the RSNA, MAP, and HR. * $p < 0.05$ vs. PBS. † $p < 0.05$ vs. control antibody. (B) Effects of PVN pretreatment of anti-asprosin antibody on the roles of asprosin in increasing the RSNA, MAP, and HR. The pretreatment was carried out 10 min before the microinjection of asprosin (5 pmol) in the PVN. * $p < 0.05$ vs. PBS + PBS; † $p < 0.05$ vs. control antibody + asprosin. Values are given as mean ± SE. $n = 6$ per group.

2.4. Roles of Superoxide Production Mediated the Effects of Asprosin

Enhanced superoxide production in the PVN increases the sympathetic outflow [7]. We were interested to know whether superoxide production was involved in the effects of asprosin in the PVN. The microinjection of asprosin in the PVN increased superoxide production and NADPH oxidase activity (Figure 4A). The PVN microinjection of superoxide scavenger tempol, antioxidant N-acetylcysteine (NAC), or NADPH oxidase inhibitor apocynin reduced the RSNA, MAP, and HR (Figure 4B). Pretreatment with tempol, NAC, or apocynin in the PVN almost abolished the effects of asprosin in the PVN (Figure 4C). These results suggested that NADPH-oxidase-dependent superoxide production mediated the effects of asprosin in increasing the RSNA, MAP, and HR.

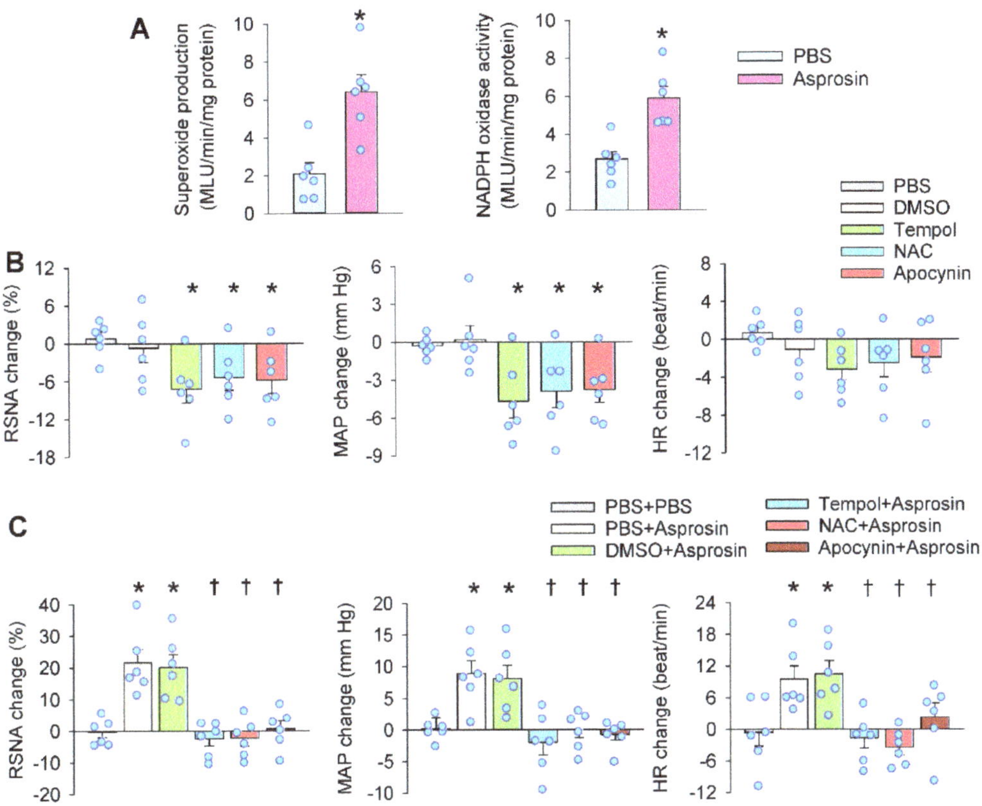

Figure 4. Roles of superoxide production in mediating the effects of asprosin in the PVN on the RSNA, MAP, and HR. (**A**) Effects of PVN microinjection of asprosin on superoxide production and NADPH oxidase activity in the PVN. The measurements were made 15 min after the microinjection. * $p < 0.05$ vs. PBS. (**B**) Effects of superoxide scavenger tempol (20 nmol), antioxidant NAC (40 nmol), and NADPH oxidase inhibitor apocynin (1 nmol) on the RSNA, MAP, and HR. * $p < 0.05$ vs. PBS or DMSO. (**C**) Effects of PVN pretreatment with tempol, NAC, and apocynin on the roles of asprosin in increasing the RSNA, MAP, and HR. The pretreatment was carried out 10 min before the microinjection of asprosin (5 pmol) in the PVN. * $p < 0.05$ vs. PBS + PBS; † $p < 0.05$ vs. PBS + asprosin or DMSO + asprosin. Values are given as mean ± SE. $n = 6$ per group.

2.5. Roles of cAMP–PKA Signaling in Mediating the Effects of Asprosin in the PVN

The activation of cAMP–PKA pathway in the PVN promotes sympathetic activation [17–19]. The microinjection of asprosin into the PVN increased the cAMP level, AC activity, and PKA activity in the PVN (Figure 5A). The PVN microinjection of the cAMP analog dibutyryl-cAMP (db-cAMP) increased the RSNA and MAP, but AC inhibitor SQ22536 or PKA inhibitor H89 reduced the RSNA and MAP (Figure 5B). Pretreatment with SQ22536 or H89 in the PVN abolished the roles of asprosin, but db-cAMP failed to further enhance the roles of asprosin in increasing the RSNA, MAP, and HR (Figure 5C). These results indicated that asprosin-induced sympathetic activation was mediated by the cAMP–PKA signaling pathway in the PVN, which was supported by findings showing that the cAMP–PKA pathway contributes to asprosin-mediated glucose release in the liver [9] and AgRP neuron activation in the brain [15].

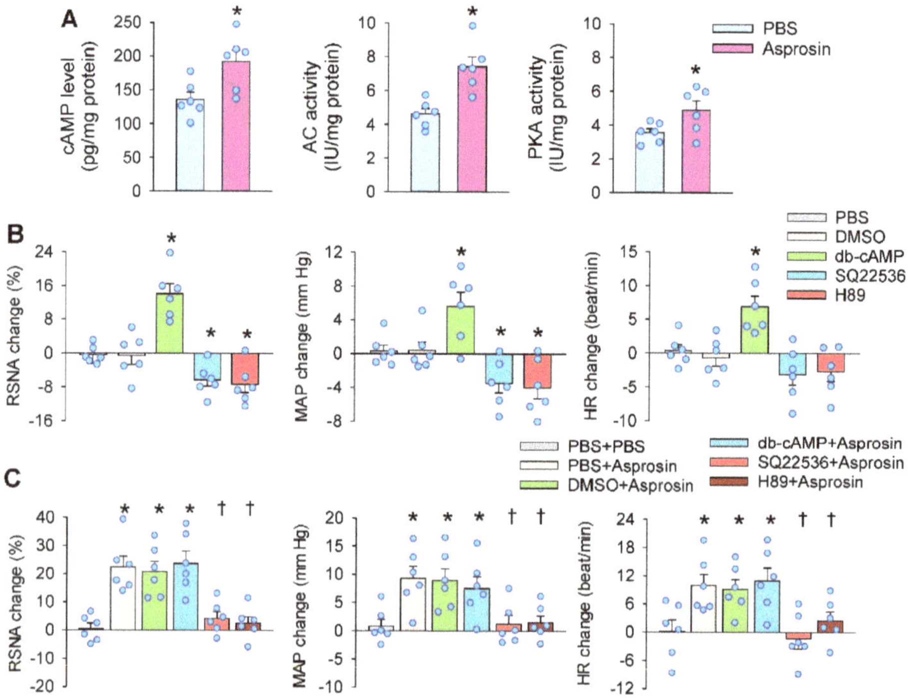

Figure 5. Roles of cAMP signaling in the effects of asprosin in the PVN on the RSNA, MAP, and HR. (**A**) Effects of PVN microinjection of asprosin on the cAMP levels, adenylyl cyclase (AC) activity, and protein kinase A (PKA) activity in the PVN. The measurements were made 15 min after the microinjection. * $p < 0.05$ vs. PBS. (**B**) Effects of the cAMP analog db-cAMP (1 nmol), AC inhibitor SQ22536 (2 nmol), and PKA inhibitor H89 (1 nmol) on the RSNA, MAP, and HR. * $p < 0.05$ vs. PBS or DMSO. (**C**) Effects of PVN pretreatment with db-cAMP, SQ22536, and H89 on the roles of asprosin in increasing the RSNA, MAP, and HR. The pretreatment was carried out 10 min before the microinjection of asprosin (5 pmol) in the PVN. * $p < 0.05$ vs. PBS + PBS; † $p < 0.05$ vs. PBS + asprosin or DMSO + asprosin. Values are given as mean ± SE. $n = 6$ per group.

2.6. cAMP–PKA Signaling Mediated the Effects of Asprosin on Superoxide Production

The cAMP–PKA signaling contributes to ROS production [20,21]. ROS plays a critical role in the cAMP-induced activation of Ras in Leydig cells [28]. We further examined whether cAMP–PKA signaling was involved in the NADPH oxidase activation and superoxide production in the PVN. The microinjection of AC inhibitor SQ22536 or PKA inhibitor H89 attenuated the asprosin-induced superoxide production in the PVN (Figure 6A). The role of SQ22536 or H89 in inhibiting the asprosin-induced superoxide production was involved in both the magnocellular and parvocellular neurons in the PVN (Figure 6B). Similarly, SQ22536 or H89 attenuated the asprosin-induced NADPH oxidase activation in the PVN (Figure 6C). These results indicated that the cAMP–PKA signaling pathway at least partially mediated the asprosin-induced NADPH oxidase activation and superoxide production.

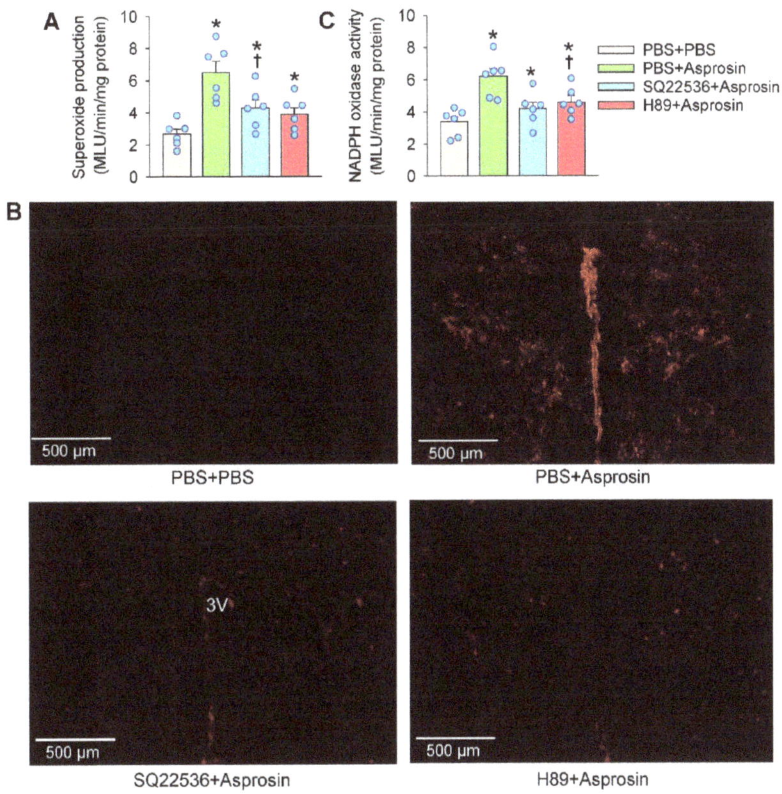

Figure 6. Effects of adenylyl cyclase inhibitor SQ22536 and PKA inhibitor H89 on the asprosin-induced changes in superoxide production and NADPH oxidase activity in the PVN. (**A**) Superoxide production. (**B**) Representative images showing the dihydroethidium (DHE) staining in the PVN. 3V, the third ventricle. (**C**) NADPH oxidase activity. The pretreatment with SQ22536 (2 nmol) or H89 (1 nmol) was carried out 10 min before the microinjection of asprosin (5 pmol) in the PVN. The measurements were made 15 min after the microinjection of asprosin. * $p < 0.05$ vs. PBS + PBS; † $p < 0.05$ vs. PBS + asprosin. Values are given as mean ± SE. $n = 6$ per group.

3. Discussion

Asprosin is an adipokine that is associated with metabolism and metabolic disorder [9]. It serves as a hormone that acts on the hypothalamus to increase appetite by crossing the blood–brain barrier [15]. The primary novel findings in this study were that asprosin in the PVN increased the sympathetic activity, blood pressure, and heart rate via NADPH-oxidase-dependent superoxide production. The cAMP–PKA signaling pathway was involved in the asprosin-induced NADPH oxidase activation and subsequent superoxide production (Figure 7).

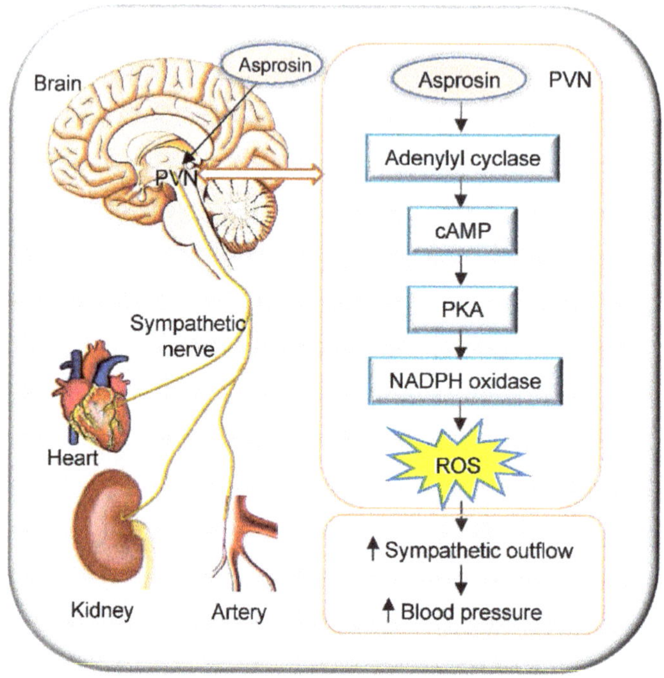

Figure 7. Schematic diagram showing the effects of asprosin in the PVN on the sympathetic outflow and blood pressure in rats and its signaling pathway.

The PVN, which is composed of parvocellular neurons and magnocellular neurons, is an important integrative site in the brain. The parvocellular neurons project to the intermediolateral cell column (IML) of the spinal cord and the RVLM neurons project to the IML and contribute to the regulation of sympathetic activity and blood pressure [29]. Magnocellular neurons express a variety of neuropeptides, including arginine vasopressin and oxytocin, and are involved in the regulation of both hydromineral homeostasis and autonomous functions [30]. The asprosin mRNA and protein expressions existed in several important nuclei in the brain for autonomic regulation, including RVLM, CVLM, PVN, NTS, and DMV, while the highest asprosin expression was found in the PVN. The asprosin-positive neurons included both parvocellular neurons and magnocellular neurons. The PVN microinjection of asprosin increased the RSNA, MAP, and HR, while neutralization of asprosin in the PVN with asprosin antibody reduced the RSNA, MAP, and HR. It was noted that asprosin induced an immediate increase in the RSNA and MAP with similar durations. According to the immediate response, the change range, and the parallel changes of the RSNA and MAP, the asprosin-induced pressor response was primarily caused by sympathetic activation. However, we cannot completely exclude the possibility that vasopressin might play a small role in the asprosin-induced pressor response. Persistent sympathetic activation not only increases blood pressure directly, and the increased norepinephrine from the sympathetic endings promotes extracellular vesicle release from adventitial fibroblasts of arteries [31]. These extracellular vesicles further contribute to vascular remodeling in hypertension [32,33].

Superoxide production in the PVN increased the sympathetic outflow [34]. We found that the microinjection of asprosin in the PVN increased the NADPH oxidase activity and superoxide production. Inhibiting NADPH oxidase or scavenging superoxides in the PVN not only reduced the RSNA, MAP, and HR but also abolished the effects of asprosin in the PVN. These results provided solid evidence that the roles of asprosin in the PVN in increasing the RSNA, MAP, and HR were mediated by NADPH-oxidase-dependent superoxide production. Previous studies showed that the increased superoxide production in the PVN attributes to the excessive sympathetic activation in hypertension [35], chronic heart failure [36], obesity [37], diabetes [38], and metabolic syndrome [23]. It is worthy to study the roles of asprosin in the PVN in the sympathetic over-activation of these diseases.

The cAMP–PKA signaling pathway in the PVN mediates the roles of angiotensin-(1-7) and alamandine in increasing the sympathetic activity [17–19]. Asprosin stimulates glucose release from hepatic cells via activating the cAMP–PKA pathway [39]. We found that the microinjection of asprosin into the PVN increased the cAMP level, AC activity, and PKA activity in the PVN. The inhibition of AC or PKA in the PVN reduced the RSNA and MAP and abolished the roles of asprosin in increasing the RSNA, MAP, and HR. The cAMP analog db-cAMP increased the RSNA and MAP but did not further enhance the roles of asprosin. These results indicated that the cAMP–PKA signaling pathway in the PVN mediated the asprosin-induced sympathetic activation. The inhibition of AC or PKA in the PVN attenuated the asprosin-induced NADPH oxidase activation and superoxide production in the PVN. These results indicated that cAMP–PKA signaling mediated the asprosin-induced superoxide production, which was supported by the findings that superoxides contribute to the cAMP-induced activation of Ras in Leydig cells [28]. It was noted that the inhibition of AC or PKA in the PVN could not completely abolish the effects of asprosin on the NADPH oxidase activity and superoxide production, suggesting that there might be some other signals involved in mediating the asprosin-induced NADPH oxidase activity and superoxide production. It was found that serum asprosin is positively related to angiotensin-converting enzyme inhibitor/angiotensin II receptor blocker therapy in type 2 diabetes patients [40]. However, the interaction of asprosin with angiotensin in the PVN is not known. Our previous studies showed that angiotensin II in the PVN induces sympathetic activation and pressor responses via NADPH-oxidase-derived superoxide production [36], and that Ang-(1-7) contributes to the enhanced sympathetic outflow via the cAMP–PKA pathway in renovascular hypertension [17]. In the present study, we found that asprosin in the PVN increased the sympathetic outflow, blood pressure, and heart rate via cAMP–PKA signaling-mediated NADPH oxidase activation and subsequent superoxide production. These results suggested that the interaction of asprosin with the angiotensin system might exist in their downstream signaling pathway, which needs further investigation.

A previous study showed that asprosin promotes the upregulation of the antioxidant enzyme superoxide dismutase 2 (SOD2) in mesenchymal stromal cells and that asprosin inhibits the hydrogen-peroxide-induced ROS generation and apoptosis via ERK1/2-SOD2 pathway in these cells [41]. This study showed the novel mechanism of asprosin. We found that asprosin induced an immediate sympathetic activation and increased NADPH oxidase activity and superoxide production. The rapid responses must be mediated by the intracellular second messenger rather than SOD2 or other anti-oxidant enzyme expressions; therefore, we did not measure the antioxidant enzyme expressions in the PVN in this study. However, we could not exclude the possibility that long-term asprosin administration may increase the expressions of antioxidant enzymes in the PVN via its direct or secondary mechanism. The different roles of asprosin in modulating the ROS may attribute to the different receptors. So far, three types of asprosin receptors have been reported as follows: (1) asprosin promoted glucose release in the liver through its G-protein-coupled receptor called OR4M1 and its downstream cAMP–PKA pathway [9], (2) the olfactory receptor OLFR734 acted as an asprosin receptor to regulate hepatic glucose production [42] and male fertility [43], and (3) asprosin impairs insulin secretion through toll-like receptor 4 (TLR4) and its downstream JNK-mediated inflammation [44]. Asprosin activated cAMP–PKA pathway to cause sympathetic activation in the present study. We speculate that the effects of asprosin in the PVN may be mediated by OR4M1 or OLFR734 receptors. A limitation of the present study was that the exact receptors of asprosin in the PVN in modulating sympathetic outflow were not identified because specific antagonists of OR4M1 or OLFR734 receptors are not available at present.

4. Materials and Methods

4.1. Animals and General Procedures

Experiments were performed in male SD rats weighing between 300 and 350 g. The rats were kept in an environment under a 12 h cycle of light/dark with controlled temperature and humidity. Standard laboratory chow and tap water were available ad libitum. Rats were anesthetized with an intraperitoneal injection of a mixture of urethane (800 mg/kg) and α-chloralose (40 mg/kg). The animal was kept in a supine position and the trachea and carotid artery were exposed through a vertical incision in the middle of the neck. Endotracheal intubation was made and connected to a small animal ventilator (683, Harvard Apparatus Inc., Holliston, MA, USA) for positive pressure ventilation with room air. A PE50 catheter connected with a pressure transducer was inserted into the right carotid artery for the blood pressure recording. Then, the rats remained in a prone position and fixed on a stereotaxic frame (Stoelting, Chicago, IL, USA) to perform surgery for the PVN microinjection and the renal sympathetic nerve activity (RSNA) recording. Both the blood pressure and RSNA signals were recorded with a data acquisition system (8SP, ADInstruments, Bella Vista, NSW, Australia). The rats were stabilized for at least 30 min before the experiment and finally euthanized via intravenous injection of sodium pentobarbital (100 mg/kg).

4.2. RSNA Recording

A left retroperitoneal incision was made to expose the left renal nerve. The nerve was isolated and cut at its distal end to abolish the renal afferent activity. The central end of the nerve was put on a pair of platinum electrodes and immersed in mineral oil at 37 °C. The nerve signals were amplified using a model DP-304 differential amplifier (Warner Instruments, Hamden, CT, USA). The band-pass filtration was set at 100–3000 Hz. The RSNA was integrated at a time constant of 100 ms with LabChart 8 software (ADInstruments, Bella Vista, NSW, Australia). Background noise was obtained by cutting the central end of the nerve. The percentage change of integrated RSNA from the baseline value was calculated after each intervention [25].

4.3. PVN Microinjection

Stereotaxic coordinates for the PVN microinjection were 0.4 mm lateral to the midline, 1.8 mm caudal from bregma, and 7.9 mm ventral to the dorsal surface according to the Paxinos and Watson rat atlas. Bilateral PVN microinjections were made through glass micropipettes (tip outer diameter 50 μm) with a 0.5 μL microsyringe and completed within 1 min. The microinjection volume for each side of the PVN was 50 nL. In the end, the same volume of Evans Blue was microinjected into the PVN and prepared for the histological identification of the microinjection sites. The rats were excluded if the microinjection site was out of the PVN. A total of 6 rats were excluded from the data analysis in the present study.

4.4. In Situ Detection of the Superoxide Level in the PVN

DHE (Beyotime Biotechnology, Shanghai, China) was used as a specific fluorogenic probe for detecting superoxide levels in the PVN [26]. The rats were euthanized and the brains were rapidly removed, frozen with liquid nitrogen, embedded into the tissue OCT-Freeze Medium, and cryostat-sectioned (30 μm, coronal) onto the chilled microscope slides. The sections were thawed at room temperature, rehydrated with phosphate-buffered saline, and incubated for 5 min in the dark with DHE (1 μmol/L). After washing with phosphate-buffered saline, the DHE fluorescence in sections was detected with a fluorescence microscope (BX51, Olympus, Tokyo, Japan) with an excitation wavelength of 543 nm. The detector and laser settings were kept constant among all samples.

4.5. Measurements of Superoxide Production and NADPH Oxidase Activity

Coronal sections with a thickness of 450 μm at the PVN level were performed with a cryostat microtome (Model CM1900, Leica, Wetzlar, Germany). The PVN tissue was punched out with a 15-gauge needle, homogenized in a lysis buffer, and then centrifuged. The total protein concentration was measured with Bradford assay kit (BCA; Pierce, Santa Cruz, CA, USA). The lucigenin-derived chemiluminescence method was used to measure superoxide production and NADPH oxidase activity in the PVN [27,28]. Photon emission was triggered by adding dark-adapted lucigenin (5 μM) to measure the superoxide production, or both dark-adapted lucigenin (5 μM) and NADPH (100 μM) to measure the NADPH oxidase activity. Light emission was measured with a luminometer (Model 20/20n, Turner, CA, USA) ten times in 10 min. Background chemiluminescence was measured in the buffer containing lucigenin (5 μM). The data were expressed in terms of the mean light unit (MLU)/min/mg protein.

4.6. *Quantitative PCR*

Total RNA in the RVLM, CVLM, PVN, NTS, and DMV of the rat was extracted with the reagent (Life Technologies, Gaithersburg, MD, USA). The purity and concentration of the extracted RNA were determined with a UV spectrophotometer. The RNA samples were subjected to a reverse transcription reaction to synthesize cDNA templates, which were then subjected to PCR amplification. According to the consistency of the standard curve amplification efficiency, the relative expression levels of asprosin mRNA in the samples were analyzed using the ΔΔCt method. GAPDH was used as a normalized control. The primers used are listed in the online Supplementary Materials (Table S1).

4.7. *Western Blot*

Asprosin protein expression in the RVLM, CVLM, PVN, NTS, and DMV was examined with Western blotting. Equal amounts of protein extracts were electrophoresed on 10% SDS-PAGE and transferred onto a PVDF membrane. After blocking with 5% milk, the membranes were incubated with the antibody against asprosin (1:1000) overnight at 4 °C followed by incubation in HRP-conjugated secondary antibody (1:5000) for 1 h at room temperature. The blots were visualized with chemiluminescence. β-actin was employed as a normalized control.

4.8. *Measurements with Commercial Kits*

Measurements were performed following the manufacturer's descriptions of the kits. The cAMP content was measured with a Cyclic Adenosine Monophosphate Assay Kit (Nanjing Jiancheng Bioengineering Institute, Nanjing, China). The AC activity was examined with Adenylate Cyclase Activity Assay Kit (Mlbio Co., Shanghai, China). PKA activity was determined with PKA Activity Kit (Enzo Life Sciences, Ann Arbor, MI, USA). Absorbance was examined at 450 nm using an automatic plate reader (ELx800, Biotek Instruments, Winooski, VT, USA).

4.9. *Immunohistochemistry*

Immunohistochemistry was performed to detect the asprosin expression in the brain sections at the PVN level of rats. The primary anti-asprosin antibody was obtained from FineBiotechCo. (Wuhan, China) and was diluted for use (1:200). Horseradish peroxidase-conjugated goat anti-rabbit antibody was acquired from Santa Cruz Biotechnology Inc. (Santa Cruz, CA, USA). Images were taken with a light microscope (BX-51, Olympus, Tokyo, Japan).

4.10. *Chemicals*

SQ22536, db-cAMP, H89, and asprosin were purchased from Med Chem Express (Monmouth Junction, NJ, USA). Tempol, apocynin, and NAC were obtained from Sigma (St. Louis, MO, USA). Asprosin and tempol were dissolved in PBS, and other chemicals were dissolved in PBS containing 1% DMSO. Vehicles were used as controls.

4.11. Statistics

RSNA, MAP, and HR values were assessed by averaging 1 min data. All data were expressed as mean $\pm$ SE. Student's t-test was used to evaluate the statistical significance of the difference between the two groups. Multiple comparisons were performed using one-way or two-way ANOVA followed by Bonferroni's post hoc analysis. A p-value < 0.05 was considered statistically significant.

5. Conclusions

Asprosin in the PVN increased the sympathetic outflow, blood pressure, and heart rate. The effects of asprosin were mediated by the cAMP–PKA signaling-pathway-mediated NADPH oxidase activation and subsequent superoxide production.

Supplementary Materials: The following supporting information can be downloaded from https://www.mdpi.com/article/10.3390/ijms232012595/s1.

Author Contributions: Conceptualization, X.-L.W., Q.C., Y.-H.L., G.-Q.Z. and X.-Z.L.; formal analysis, X.-L.W., J.-X.W., J.-L.C., W.-Y.H. and G.-Q.Z.; funding acquisition, G.-Q.Z.; investigation, X.-L.W., J.-X.W., J.-L.C., W.-Y.H., W.-Z.X., Z.-Q.X., Y.-T.J. and P.-Q.L.; methodology, X.-L.W., J.-X.W. and G.-Q.Z.; project administration, G.-Q.Z.; resources, G.-Q.Z.; supervision, G.-Q.Z. and X.-Z.L.; validation, G.-Q.Z.; writing—original draft, X.-L.W., G.-Q.Z. and X.-Z.L.; writing—review and editing, Q.C., Y.-H.L., G.-Q.Z. and X.-Z.L. All authors have read and agreed to the published version of the manuscript.

Funding: This research was funded by the National Natural Science Foundation of China (31871148 and 32071106).

Institutional Review Board Statement: The study was conducted in accordance with the NIH guidelines and approved by the Experimental Animal Care and Use Committee of Nanjing Medical University (no. 1811017).

Informed Consent Statement: Not applicable.

Data Availability Statement: Data for this study are available from the corresponding author on reasonable request.

Conflicts of Interest: The authors declare no conflict of interest.

References

1. Grassi, G.; Ram, V.S. Evidence for a critical role of the sympathetic nervous system in hypertension. *J. Am. Soc. Hypertens.* **2016**, *10*, 457–466. [CrossRef] [PubMed]
2. Hering, D.; Schlaich, M. The Role of Central Nervous System Mechanisms in Resistant Hypertension. *Curr. Hypertens. Rep.* **2015**, *17*, 58. [CrossRef] [PubMed]
3. Noh, M.R.; Jang, H.S.; Kim, J.; Padanilam, B.J. Renal Sympathetic Nerve-Derived Signaling in Acute and Chronic kidney Diseases. *Int. J. Mol. Sci.* **2020**, *21*, 1647. [CrossRef]
4. Lymperopoulos, A.; Borges, J.I.; Cora, N.; Sizova, A. Sympatholytic Mechanisms for the Beneficial Cardiovascular Effects of SGLT2 Inhibitors: A Research Hypothesis for Dapagliflozin's Effects in the Adrenal Gland. *Int. J. Mol. Sci.* **2021**, *22*, 7684. [CrossRef]
5. DeLalio, L.J.; Sved, A.F.; Stocker, S.D. Sympathetic Nervous System Contributions to Hypertension: Updates and Therapeutic Relevance. *Can. J. Cardiol.* **2020**, *36*, 712–720. [CrossRef] [PubMed]
6. Dampney, R.A.; Michelini, L.C.; Li, D.P.; Pan, H.L. Regulation of sympathetic vasomotor activity by the hypothalamic paraventricular nucleus in normotensive and hypertensive states. *Am. J. Physiol. Heart Circ. Physiol.* **2018**, *315*, H1200–H1214. [CrossRef]
7. Chen, W.W.; Xiong, X.Q.; Chen, Q.; Li, Y.H.; Kang, Y.M.; Zhu, G.Q. Cardiac sympathetic afferent reflex and its implications for sympathetic activation in chronic heart failure and hypertension. *Acta. Physiol.* **2015**, *213*, 778–794. [CrossRef] [PubMed]
8. Wu, L.L.; Zhang, Y.; Li, X.Z.; Du, X.L.; Gao, Y.; Wang, J.X.; Wang, X.L.; Chen, Q.; Li, Y.H.; Zhu, G.Q.; et al. Impact of Selective Renal Afferent Denervation on Oxidative Stress and Vascular Remodeling in Spontaneously Hypertensive Rats. *Antioxidants* **2022**, *11*, 1003. [CrossRef] [PubMed]
9. Romere, C.; Duerrschmid, C.; Bournat, J.; Constable, P.; Jain, M.; Xia, F.; Saha, P.K.; Del, S.M.; Zhu, B.; York, B.; et al. Asprosin, a fasting-induced glucogenic protein hormone. *Cell* **2016**, *165*, 566–579. [CrossRef] [PubMed]
10. Yuan, M.; Li, W.; Zhu, Y.; Yu, B.; Wu, J. Asprosin: A Novel Player in Metabolic Diseases. *Front. Endocrinol.* **2020**, *11*, 64. [CrossRef]

11. Hekim, M.G.; Kelestemur, M.M.; Bulmus, F.G.; Bilgin, B.; Bulut, F.; Gokdere, E.; Ozdede, M.R.; Kelestimur, H.; Canpolat, S.; Ozcan, M. Asprosin, a novel glucogenic adipokine: A potential therapeutic implication in diabetes mellitus. *Arch. Physiol. Biochem.* **2021**, 1–7, *Online ahead of print*. [CrossRef] [PubMed]
12. Liu, L.J.; Kang, Y.R.; Xiao, Y.F. Increased asprosin is associated with non-alcoholic fatty liver disease in children with obesity. *World J. Pediatr.* **2021**, *17*, 394–399. [CrossRef]
13. Maylem, E.R.S.; Spicer, L.J.; Batalha, I.; Schutz, L.F. Discovery of a possible role of asprosin in ovarian follicular function. *J. Mol. Endocrinol.* **2021**, *66*, 35–44. [CrossRef] [PubMed]
14. Guven, C.; Kafadar, H. Evaluation of Plasma Asprosin Concentration in Patients with Coronary Artery Disease. *Braz. J. Cardiovasc. Surg.* **2022**, *37*, 493–500. [CrossRef]
15. Duerrschmid, C.; He, Y.; Wang, C.; Li, C.; Bournat, J.C.; Romere, C.; Saha, P.K.; Lee, M.E.; Phillips, K.J.; Jain, M.; et al. Asprosin is a centrally acting orexigenic hormone. *Nat. Med.* **2017**, *23*, 1444–1453. [CrossRef]
16. Kocaman, N.; Kuloglu, T. Expression of asprosin in rat hepatic, renal, heart, gastric, testicular and brain tissues and its changes in a streptozotocin-induced diabetes mellitus model. *Tissue Cell* **2020**, *66*, 101397. [CrossRef] [PubMed]
17. Han, Y.; Sun, H.J.; Li, P.; Gao, Q.; Zhou, Y.B.; Zhang, F.; Gao, X.Y.; Zhu, G.Q. Angiotensin-(1-7) in paraventricular nucleus modulates sympathetic activity and cardiac sympathetic afferent reflex in renovascular hypertensive rats. *PLoS ONE* **2012**, *7*, e48966. [CrossRef] [PubMed]
18. Li, P.; Cui, B.P.; Zhang, L.L.; Sun, H.J.; Liu, T.Y.; Zhu, G.Q. Melanocortin 3/4 receptors in paraventricular nucleus modulates sympathetic outflow and blood pressure. *Exp. Physiol.* **2013**, *98*, 435–443. [CrossRef]
19. Shen, Y.H.; Chen, X.R.; Yang, C.X.; Liu, B.X.; Li, P. Alamandine injected into the paraventricular nucleus increases blood pressure and sympathetic activation in spontaneously hypertensive rats. *Peptides* **2018**, *103*, 98–102. [CrossRef] [PubMed]
20. Zhang, J.; Xiao, H.; Shen, J.; Wang, N.; Zhang, Y. Different roles of b-arrestin and the PKA pathway in mitochondrial ROS production induced by acute b-adrenergic receptor stimulation in neonatal mouse cardiomyocytes. *Biochem. Biophys. Res. Commun.* **2017**, *489*, 393–398. [CrossRef]
21. Andersson, D.C.; Fauconnier, J.; Yamada, T.; Lacampagne, A.; Zhang, S.J.; Katz, A.; Westerblad, H. Mitochondrial production of reactive oxygen species contributes to the b-adrenergic stimulation of mouse cardiomycytes. *J. Physiol.* **2011**, *589*, 1791–1801. [CrossRef] [PubMed]
22. Li, H.B.; Yu, X.J.; Bai, J.; Su, Q.; Wang, M.L.; Huo, C.J.; Xia, W.J.; Yi, Q.Y.; Liu, K.L.; Fu, L.Y.; et al. Silencing salusin beta ameliorates heart failure in aged spontaneously hypertensive rats by ROS-relative MAPK/NF-kappaB pathways in the paraventricular nucleus. *Int. J. Cardiol.* **2019**, *280*, 142–151. [CrossRef] [PubMed]
23. Cruz, J.C.; Flor, A.F.; Franca-Silva, M.S.; Balarini, C.M.; Braga, V.A. Reactive oxygen species in the paraventricular nucleus of the hypothalamus alter sympathetic activity during metabolic syndrome. *Front. Physiol.* **2015**, *6*, 384. [CrossRef] [PubMed]
24. Han, Y.; Zhang, Y.; Wang, H.J.; Gao, X.Y.; Wang, W.; Zhu, G.Q. Reactive oxygen species in paraventricular nucleus modulates cardiac sympathetic afferent reflex in rats. *Brain Res.* **2005**, *1058*, 82–90. [CrossRef]
25. Chen, W.W.; Sun, H.J.; Zhang, F.; Zhou, Y.B.; Xiong, X.Q.; Wang, J.J.; Zhu, G.Q. Salusin-b in paraventricular nucleus increases blood pressure and sympathetic outflow via vasopressin in hypertensive rats. *Cardiovasc. Res.* **2013**, *98*, 344–351. [CrossRef]
26. Zheng, F.; Ye, C.; Wan, G.W.; Zhou, B.; Tong, Y.; Lei, J.Z.; Chen, Q.; Li, Y.H.; Kang, Y.M.; Zhu, G.Q. Interleukin-1b in hypothalamic paraventricular nucleus mediates excitatory renal reflex. *Pflugers. Arch.* **2020**, *472*, 1577–1586. [CrossRef]
27. Li, P.; Jie, Y.; YuGen, S.; Yu, W.; Yan, S. High mobility group box-1 in hypothalamic paraventricular nuclei attenuates sympathetic tone in rats at post-myocardial infarction. *Cardiol. J.* **2019**, *26*, 555–563. [CrossRef]
28. Tai, P.; Ascoli, M. Reactive oxygen species (ROS) play a critical role in the cAMP-induced activation of Ras and the phosphorylation of ERK1/2 in Leydig cells. *Mol. Endocrinol.* **2011**, *25*, 885–893. [CrossRef]
29. Badoer, E. Hypothalamic paraventricular nucleus and cardiovascular regulation. *Clin. Exp. Pharmacol. Physiol.* **2001**, *28*, 95–99. [CrossRef]
30. Palasz, A.; Della, V.A.; Saganiak, K.; Worthington, J.J. Neuropeptides of the human magnocellular hypothalamus. *J. Chem. Neuroanat.* **2021**, *117*, 102003. [CrossRef]
31. Ye, C.; Zheng, F.; Xu, T.; Wu, N.; Tong, Y.; Xiong, X.Q.; Zhou, Y.B.; Wang, J.J.; Chen, Q.; Li, Y.H.; et al. Norepinephrine acting on adventitial fibroblasts stimulates vascular smooth muscle cell proliferation via promoting small extracellular vesicle release. *Theranostics* **2022**, *12*, 4718–4733. [CrossRef] [PubMed]
32. Ren, X.S.; Tong, Y.; Qiu, Y.; Ye, C.; Wu, N.; Xiong, X.Q.; Wang, J.J.; Han, Y.; Zhou, Y.B.; Zhang, F.; et al. MiR155-5p in adventitial fibroblasts-derived extracellular vesicles inhibits vascular smooth muscle cell proliferation via suppressing angiotensin-converting enzyme expression. *J. Extracell. Vesicles.* **2020**, *9*, 1698795. [CrossRef] [PubMed]
33. Ye, C.; Zheng, F.; Nan, W.; Zhu, G.Q.; Li, X.Z. Extracellular vesicles in vascular remodeling. *Acta. Pharmacol. Sin.* **2022**, *43*, 2191–2201. [CrossRef] [PubMed]
34. Qiu, Y.; Zheng, F.; Ye, C.; Chen, A.D.; Wang, J.J.; Chen, Q.; Li, Y.H.; Kang, Y.M.; Zhu, G.Q. Angiotensin type 1 receptors and superoxide anion production in hypothalamic paraventricular nucleus contribute to capsaicin-induced excitatory renal reflex and sympathetic activation. *Neurosci. Bull.* **2020**, *36*, 463–474. [CrossRef] [PubMed]
35. Yuan, N.; Zhang, F.; Zhang, L.L.; Gao, J.; Zhou, Y.B.; Han, Y.; Zhu, G.Q. SOD1 gene transfer into paraventricular nucleus attenuates hypertension and sympathetic activity in spontaneously hypertensive rats. *Pflugers. Arch.* **2013**, *465*, 261–270. [CrossRef]

36. Han, Y.; Shi, Z.; Zhang, F.; Yu, Y.; Zhong, M.K.; Gao, X.Y.; Wang, W.; Zhu, G.Q. Reactive oxygen species in the paraventricular nucleus mediate the cardiac sympathetic afferent reflex in chronic heart failure rats. *Eur. J. Heart Fail.* **2007**, *9*, 967–973. [CrossRef]
37. Lu, Q.B.; Sun, J.; Kang, Y.; Sun, H.J.; Wang, H.S.; Wang, Y.; Zhu, G.Q.; Zhou, Y.B. Superoxide Anions and NO in the Paraventricular Nucleus Modulate the Cardiac Sympathetic Afferent Reflex in Obese Rats. *Int. J. Mol. Sci.* **2017**, *19*, 59. [CrossRef]
38. Patel, K.P.; Mayhan, W.G.; Bidasee, K.R.; Zheng, H. Enhanced angiotensin II-mediated central sympathoexcitation in streptozotocin-induced diabetes: Role of superoxide anion. *Am. J. Physiol. Regul. Integr. Comp. Physiol.* **2011**, *300*, R311–R320. [CrossRef]
39. Mazur-Bialy, A.I. Asprosin-a fasting-induced, glucogenic, and orexigenic adipokine as a new promising player. Will it be a new factor in the treatment of obesity, diabetes, or infertility? A review of the literature. *Nutrients* **2021**, *13*, 620. [CrossRef]
40. Wang, R.; Lin, P.; Sun, H.; Hu, W. Increased serum asprosin is correlated with diabetic nephropathy. *Diabetol. Metab. Syndr.* **2021**, *13*, 51. [CrossRef]
41. Zhang, Z.; Tan, Y.; Zhu, L.; Zhang, B.; Feng, P.; Gao, E.; Xu, C.; Wang, X.; Yi, W.; Sun, Y. Asprosin improves the survival of mesenchymal stromal cells in myocardial infarction by inhibiting apoptosis via the activated ERK1/2-SOD2 pathway. *Life Sci.* **2019**, *231*, 116554. [CrossRef]
42. Li, E.; Shan, H.; Chen, L.; Long, A.; Zhang, Y.; Liu, Y.; Jia, L.; Wei, F.; Han, J.; Li, T.; et al. OLFR734 Mediates Glucose Metabolism as a Receptor of Asprosin. *Cell Metab.* **2019**, *30*, 319–328. [CrossRef] [PubMed]
43. Wei, F.; Long, A.; Wang, Y. The Asprosin-OLFR734 hormonal signaling axis modulates male fertility. *Cell Discov.* **2019**, *5*, 55. [CrossRef] [PubMed]
44. Lee, T.; Yun, S.; Jeong, J.H.; Jung, T.W. Asprosin impairs insulin secretion in response to glucose and viability through TLR4/JNK-mediated inflammation. *Mol. Cell Endocrinol.* **2019**, *486*, 96–104. [CrossRef] [PubMed]

International Journal of
Molecular Sciences

MDPI

Article

Relative Contribution of Blood Pressure and Renal Sympathetic Nerve Activity to Proximal Tubular Sodium Reabsorption via NHE3 Activity

Roberto B. Pontes [1], Erika E. Nishi [1], Renato O. Crajoinas [2], Maycon I. O. Milanez [1], Adriana C. C. Girardi [2], Ruy R Campos [1] and Cassia T Bergamaschi [1,]*

[1] Cardiovascular Division, Department of Physiology, Escola Paulista de Medicina, Universidade Federal de São Paulo, São Paulo 04021-001, Brazil
[2] Heart Institute (InCor), University of São Paulo Medical School, São Paulo 01246-903, Brazil
* Correspondence: bergamaschi.cassia@unifesp.br

Abstract: We examined the effects of an acute increase in blood pressure (BP) and renal sympathetic nerve activity (rSNA) induced by bicuculline (Bic) injection in the paraventricular nucleus of hypothalamus (PVN) or the effects of a selective increase in rSNA induced by renal nerve stimulation (RNS) on the renal excretion of sodium and water and its effect on sodium-hydrogen exchanger 3 (NHE3) activity. Uninephrectomized anesthetized male Wistar rats were divided into three groups: (1) Sham; (2) Bic PVN: (3) RNS + Bic injection into the PVN. BP and rSNA were recorded, and urine was collected prior and after the interventions in all groups. RNS decreased sodium (58%) and water excretion (53%) independently of BP changes ($p < 0.05$). However, after Bic injection in the PVN during RNS stimulation, the BP and rSNA increased by 30% and 60% ($p < 0.05$), respectively, diuresis (5-fold) and natriuresis (2.3-fold) were increased ($p < 0.05$), and NHE3 activity was significantly reduced, independently of glomerular filtration rate changes. Thus, an acute increase in the BP overcomes RNS, leading to diuresis, natriuresis, and NHE3 activity inhibition.

Keywords: NHE3; glomerular filtration rate; bicuculline; paraventricular nucleus of the hypothalamus; sympathetic nerve activity

Citation: Pontes, R.B.; Nishi, E.E.; Crajoinas, R.O.; Milanez, M.I.O.; Girardi, A.C.C.; Campos, R.R.; Bergamaschi, C.T. Relative Contribution of Blood Pressure and Renal Sympathetic Nerve Activity to Proximal Tubular Sodium Reabsorption via NHE3 Activity. *Int. J. Mol. Sci.* **2023**, *24*, 349. https://doi.org/10.3390/ijms24010349

Academic Editors: Yutang Wang and Kate Denton

Received: 18 November 2022
Revised: 17 December 2022
Accepted: 22 December 2022
Published: 26 December 2022

1. Introduction

The sympathetic nervous system (SNS) controls the sodium and water balance by the kidneys and blood pressure (BP). The SNS influences renal function by inducing renal vasoconstriction, stimulating renin secretion, and increasing tubular sodium reabsorption [1]. Sympathetic hyperactivity results in excessive vasoconstriction and hypertension [2]. During acute SNS activation, both arterial hypertension and sympathetic actions in the kidneys control sodium and water excretion. However, the relative contribution to salt and water balance by the kidneys of each of these mechanisms is not completely understood.

The brain controls SNS activity through specific nuclei, such as the paraventricular nucleus of the hypothalamus (PVN). Experimentally inhibiting PVN decreases the BP and sympathetic vasomotor tone [3], whereas activation of PVN has the opposite effects [4]. Cardiovascular actions can be evaluated by targeting this area with the GABAa receptor antagonist bicuculline (Bic), which blocks the inhibitory synaptic influence on this crucial cardiovascular control area [5].

Renal nerve stimulation (RNS) is a technique used to examine the specific influence of SNS on renal function without altering the BP. For instance, low-frequency (LF) RNS causes sodium and water reabsorption at the tubular level without affecting renal blood flow and the glomerular filtration rate [6]. Additionally, Healey et al. showed that LF-RNS results in sodium retention, primarily by increasing proximal tubule sodium reabsorption, which is mediated by sodium hydrogen exchanger isoform 3 (NHE3) [7]. Because RNS can be applied at different frequencies, this technique enabled us to demonstrate that SNS activation in the kidney can impact renal tubular function independently of hemodynamic changes [8]. We found that ability of NHE3 to increase LF-RNS in the renal proximal tubule was accompanied by activation of intrarenal but not circulating angiotensin II (Ang II); this effect was completely abolished when the animals were pretreated with the Ang II type 1 receptor antagonist losartan [8] and occurred independently of BP changes. Therefore, NHE3 hyperactivation in response to RNS depends on Ang II-induced Ang II type 1 receptor activation in the kidneys [8,9].

NHE3 is the renal apical transporter responsible for most sodium and water reabsorption at the proximal tubular level [10,11]. Thus, NHE3 plays an essential role in maintaining the extracellular water balance, and natriuresis and diuresis are physiologically required under increased BP to eliminate extracellular fluid. The primary role of NHE3 in pressure natriuresis was recently demonstrated by Li et al. using proximal tubule-specific NHE3 knockout mice (PT-Nhe3$^{-/-}$). The authors demonstrated that Ang II-induced hypertension was attenuated in PT-Nhe3$^{-/-}$ and that the mice displayed more prominent pressure natriuresis in response to increased renal perfusion pressure [12].

Increased renal sympathetic nerve activity (rSNA) activates NHE3-mediated sodium reabsorption [7,8], whereas an increased BP reduces this effect [11,13]. However, how proximal tubule NHE3-mediated sodium reabsorption is regulated in the presence of both rSNA stimulation and increased BP remains unclear. Importantly, sympathetic vasomotor activation and arterial hypertension occur in different experimental hypertension models [14–17] and hypertensive patients [18–20]. The PVN plays a major role in central cardiovascular and volume control [2]. Therefore, we target the PVN and examined the effect of acute increases in BP and rSNA induced by Bic injection or the effects of a selective increase in rSNA by RNS on the renal excretion of sodium and water and its potential effect on the regulation of NHE3 activity in uninephrectomized Wistar rats.

2. Results

2.1. First and Second Series of Experiments

2.1.1. Bic Injection into the PVN Increases BP and rSNA

We evaluated the effect of Bic injection into the PVN on the BP and rSNA. The BP and rSNA were recorded in the sham group, in this group there was no injection of bicuculline in the PVN. The Bic PVN group was injected with Bic into the PVN at baseline, which led to an increase in both rSNA and BP.

Figure 1A shows representative tracers of the pulsatile BP and rSNA in the sham group (throughout the experiment) and Bic PVN group (at baseline before Bic infusion). Figure 1B shows the tracers after Bic infusion. The sham and Bic PVN groups before Bic infusion showed similar values for the mean arterial pressure (MAP) and rSNA. After Bic PVN infusion, the MAP and rSNA increased by ~30% (Figure 1C) and ~60% (Figure 1D), respectively (MAP: sham 121 ± 6 mmHg, pre-Bic PVN 123 ± 5 mmHg, post-Bic PVN 160 ± 5 mmHg; and rSNA: sham 100 ± 3, pre-Bic PVN 95 ± 3, post-Bic PVN 159 ± 7%).

2.1.2. Bic Injection in the PVN Increases Urinary Flow Rate and Urinary Sodium Rate but Not Creatinine Clearance

The effect of acute increases in the BP and rSNA on renal excretory function was evaluated as creatinine clearance (CrCl, to estimate the glomerular filtration rate), urinary flow rate (UFR), and urinary sodium rate (UNa). Soon after the BP and rSNA increased, both UFR and UNa increased significantly, with no change in the CrCl.

The CrCl values of rats in the sham and Bic PVN groups before and after Bic infusion were similar (0.0018 ± 0.0002, 0.0018 ± 0.0002, and 0.0018 ± 0.0001 mL/min/g, respectively; Figure 2A). UFR and UNa showed similar values in the sham and Bic PVN groups before Bic infusion. After Bic infusion, the UFR was 5-fold higher than that at baseline (0.0349 ± 0.0114, 0.0300 ± 0.0037, and 0.1500 ± 0.0182 mL/min/g in the sham, pre-Bic PVN, and post-Bic PVN groups, respectively; Figure 2B), and UNa was increased by ~2.3-fold

(1.72 ± 0.06, 1.71 ± 0.07, and 3.95 ± 0.16 μEq/min/g in the sham, pre-Bic PVN, and post-Bic PVN groups, respectively), Figure 2C.

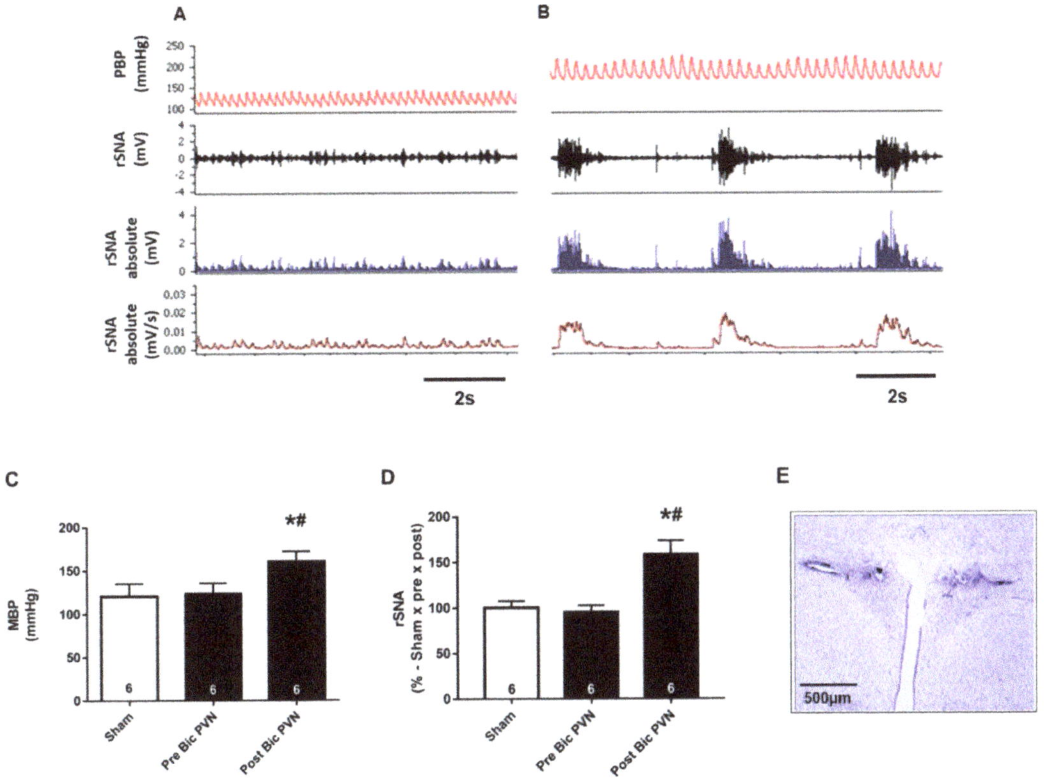

Figure 1. Effect of bicuculline injection into the paraventricular nucleus of the hypothalamus (Bic PVN) on mean arterial pressure (MAP) and renal sympathetic nerve activity (rSNA). (**A**) Typical trace containing pulsatile blood pressure (PBP) and rSNA of sham (whole experiment) and Bic PVN (at baseline before bicuculline infusion) groups. (**B**) PBP of the Bic PVN group after bicuculline infusion. Filtered signal from rSNA (black trace, second from top to bottom in (**A**,**B**) was first converted to positive absolute values (rSNA absolute, blue traces third from top) and then to mV/s (rSNA absolute mV/s, red lines bottom traces); see methods for more details. White bar represents the MAP of the sham group during with the experiment, and black bars represent the MAP of the Bic PVN group before and after Bic injection (**C**). White bar represents the MAP of the sham group during the experiment; black bars represent the rSNA of the Bic PVN group before and after Bic infusion (**D**). Representative histological image: arrow indicates the site of microinjection into the PVN. Scale bar = 500 μm. 3 V, third ventricle (**E**). The number of rats is indicated inside the bars. * $p < 0.001$ compared to sham. # $p < 0.001$ compared to pre-Bic PVN. One-way analysis of variance for multiple comparisons followed by Tukey's test.

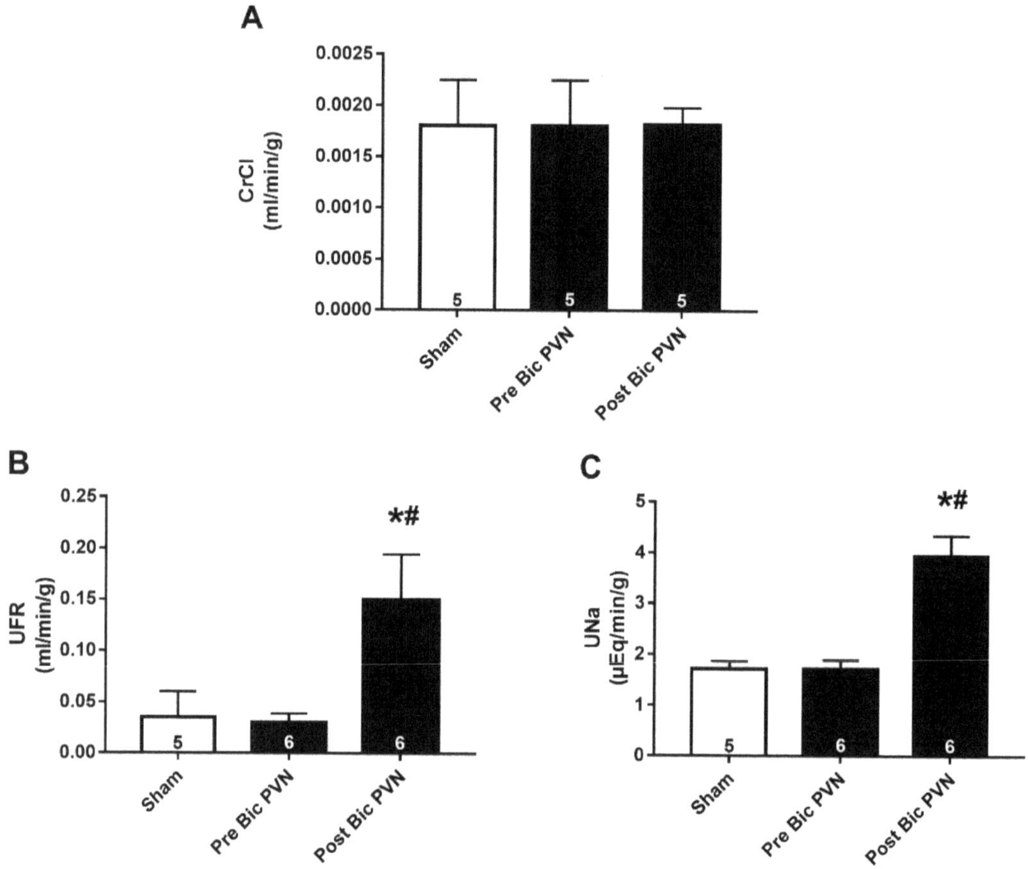

Figure 2. Effect of bicuculline on the paraventricular nucleus of the hypothalamus (Bic PVN) on renal function. (**A**) Creatinine clearance (CrCl) of rats in the sham, pre-Bic PVN, and post-Bic PVN groups. (**B**) White bar represents the urinary flow rate (UFR) of rats in the sham group during the experiment, and black bars represent the UFR of rats in the Bic PVN group before and after Bic injection. (**C**) White bar represents the urinary sodium rate (UNa) of rats in the sham group during the experiment, and black bars represent the UFR of rats in the Bic PVN group before and after Bic injection. The number of rats is shown inside the bars. * $p < 0.001$ compared to sham. # $p < 0.001$ compared to pre-Bic PVN. One-way analysis of variance for multiple comparisons followed by Tukey's test.

2.1.3. Bic Injection in the PVN Increases NHE3 Phosphorylation Levels at Serine 552

To investigate the mechanism of diuretic and natriuretic pressure in more detail, we evaluated whether these effects were associated with NHE3 inhibition. We assessed the NHE3 phosphorylation levels at serine 552, a surrogate for NHE3 inactivation [21–24]. Figure 3A shows representative immunoblotting and semi-quantitative results for the NHE3 total abundance, NHE3 phosphorylated at serine 552 (NHE3-PS552), and actin in the renal cortex of sham and Bic PVN rats. Total NHE3 expression did not differ between groups (sham: 100 ± 10% and Bic PVN: 122 ± 15%, Figure 3B), whereas the ratio of NHE3-PS552 to total NHE3 was 49% higher in the Bic PVN group than in the sham group (Figure 3C).

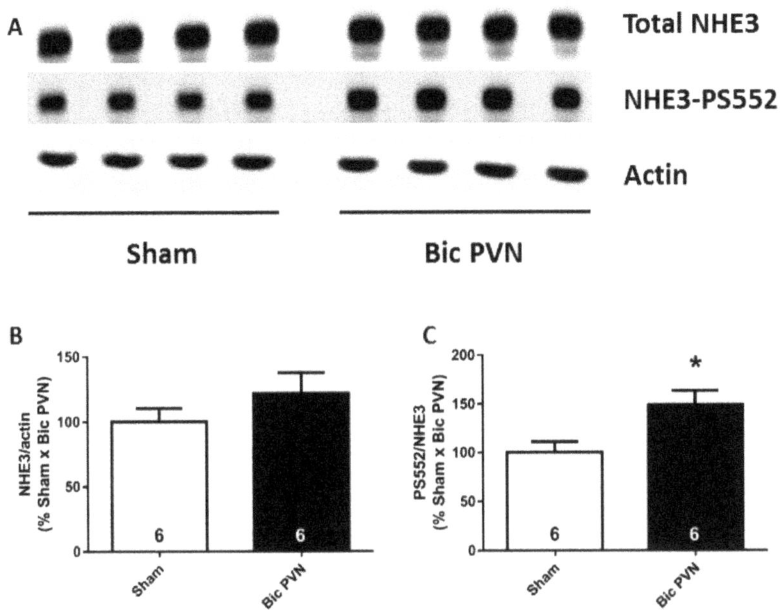

Figure 3. Effect of bicuculline on the paraventricular nucleus of the hypothalamus (Bic PVN) on sodium-hydrogen exchanger isoform 3 (NHE3). (**A**) Blots from the renal cortex of total NHE3, NHE3 phosphorylated at serine 552 (NHE3-PS552), and actin. The total amount of NHE3 corrected by actin did not distinguish between the sham and Bic PVN groups (**B**). Ratio of PS552/total NHE3 was significantly higher in the Bic PVN group (**C**). The number of rats is shown inside the bars. * $p < 0.05$ compared to sham. Student's *t*-test.

2.2. Third Series of Experiments

2.2.1. Effect of RNS and RNS + Bic PVN on BP

To evaluate the diuretic and natriuretic pressure under conditions known to decrease urinary excretion, we used a separate group of anesthetized animals. After baseline BP and rSNA recordings, the animals were first subjected to RNS, which did not affect the BP. Under RNS, animals were injected with Bic into the PVN, following which the BP increased.

The bars in Figure 4 represent the MAP values obtained from the same rats and from the experiment at three different time points. The MAP values were similar at baseline and under RNS. Under RNS, Bic PVN increased the MAP by approximately 33% (122 ± 3 mmHg vs. 125 ± 3 mmHg vs. 163 ± 5 mmHg, respectively).

2.2.2. Effect of RNS and RNS + Bic PVN on Renal Function: CrCl, UFR, and UNa

To investigate the role of RNS alone and with a concomitant increase in BP caused by Bic, we evaluated the CrCl, UFR, and Una at three different periodstime points (baseline, RNS and RNS + Bic). RNS decreased both the UFR and UNa. Under RNS after Bic PVN, the increase in BP was followed by diuresis and natriuresis. The CrCl remained unchanged throughout the experiment.

The bars in Figure 5A represent the CrCl at three different time points in the experiment (baseline: 0.001943 ± 0.0002 mL/min/g, RNS: 0.002029 ± 0.0001 mL/min/g, and RNS + Bic PVN: 0.002072 ± 0.0001 mL/min/g). Compared to the value at baseline, the UFR was significantly decreased by RNS and increased by Bic PVN (0.0358 ± 0.0012 vs. 0.0168 ± 0.0008 vs. 0.0594 ± 0.0020 mL/min/g, respectively) (Figure 5B). Compared to the value at baseline, the UNa was significantly decreased by RNS and increased by Bic PVN (1.54 ± 0.05 vs. 0.64 ± 0.05 vs. 4.06 ± 0.11 μEq/min/g, respectively) (Figure 5C).

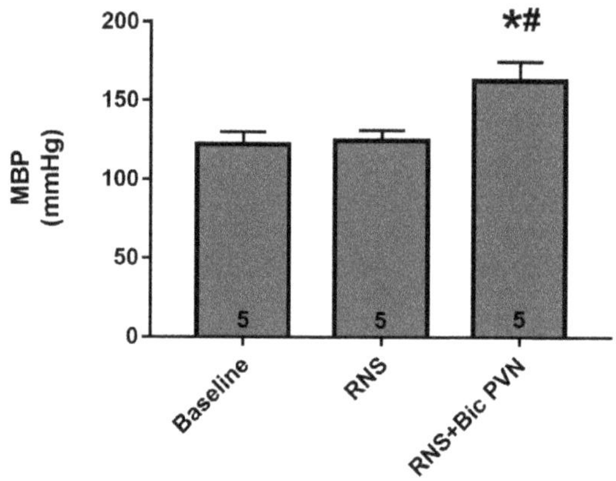

Figure 4. Mean arterial blood pressure (MBP) during the experiment at three different time points. Bars represent the mean arterial pressure (MAP) during the baseline period, under renal nerve stimulation (RNS), and under RNS plus bicuculline into the PVN (Bic PVN). The number of rats is shown inside the bars. * $p < 0.001$ compared to baseline. # $p < 0.001$ compared to RNS. One-way analysis of variance for multiple comparisons followed by Tukey's test.

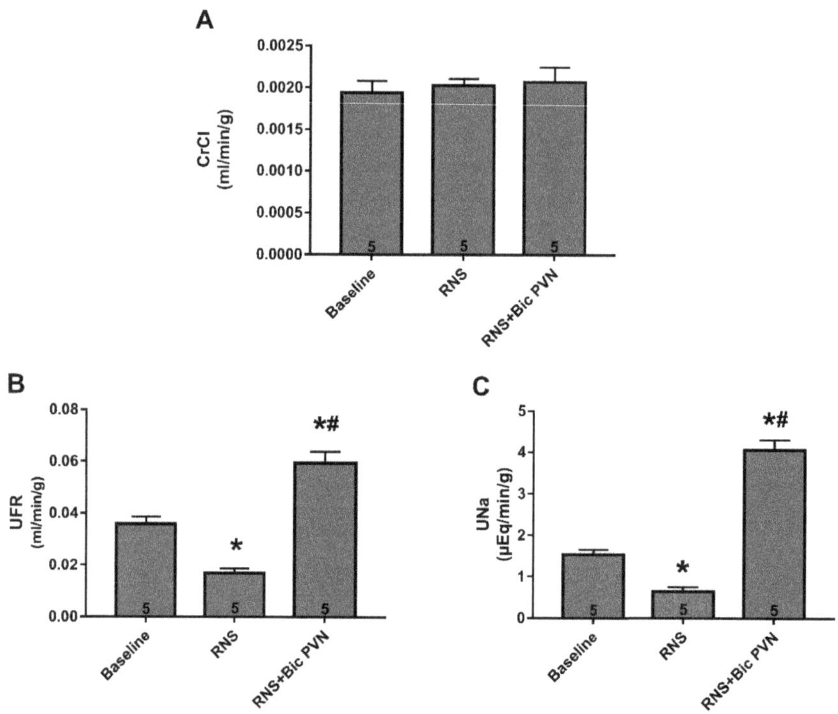

Figure 5. Creatinine clearance (CrCl), urinary flow rate (UFR), and urinary sodium rate (UNa) at three time points. (**A**) CrCl at baseline, under renal nerve stimulation (RNS), and under RNS plus bicuculline injection into the paraventricular nucleus of the hypothalamus (Bic PVN). (**B**) UFR at baseline, under RNS, and under RNS + Bic PVN. (**C**) UNa at baseline, under RNS, and under RNS + Bic PVN. The number of rats samples is shown inside the bars. * $p < 0.001$ compared to baseline. # $p < 0.001$ compared to RNS. One-way analysis of variance for multiple comparisons followed by Tukey's test.

3. Discussion

We found that bicuculline injection in the PVN increased the rSNA and MAP; the acute increase in BP was followed by diuresis and natriuresis, whereas CrCl remained unchanged. In rats administered Bic in the PVN, NHE3 showed increased phosphorylation at serine 552. Additionally, RNS decreased the UFR and UNa independently of changes in the BP. Under these conditions, an acute increase in the BP (by Bic into the PVN) reverted the effects of RNS and resulted in diuresis and natriuresis.

Disinhibition of the PVN, a vital nucleus containing pre-motor sympathetic neurons, results in a pressor response. The acute increase in BP resulted in substantial pressure diuresis and natriuresis, accompanied by increased NHE3 phosphorylation at serine 552. We showed that in a model of acute increase in centrally generated BP, NHE3 phosphorylation is necessary to counteract the sympathetic antidiuretic and anti-natriuretic effects. We injected Bic into the PVN to increase the BP rather than using vasoactive drugs such as phenylephrine to avoid a reflex decrease in rSNA via the baroreceptor reflex [25,26]. As mentioned previously, PVN inhibition decreases the BP [3], whereas PVN activation increases the BP [4], which is sympathetically mediated. There are two different groups of PVN neurons: magnocellular and parvocellular. Porter and Brody stimulated one of two different groups of PVN neurons and observed that magnocellular stimulation led to vasodilation, whereas parvocellular stimulation resulted in vasoconstriction [27]. In this study, we targeted parvocellular neurons because they project into the medulla and spinal cord, which are also involved in autonomic BP regulation [2,28]. Based on our histological images combined with autonomic and pressure responses, we predominantly activated parvocellular neurons.

We found that the acute centrally generated increased BP was rapidly followed by diuresis and natriuresis associated with higher renal cortical NHE3-PS552 levels. This effect was achieved despite the increase in rSNA, a well-known stimulus of sodium and water retention, without a change in BP [1,6]. Importantly, a correlation between NHE3-PS552 and proximal tubule NHE3 activity has been demonstrated in several experimental models, including in hypertensive animals [8,22–24,29–31]. Indeed, adult hypertensive SHRs exhibit higher NHE3 activity, which correlates with higher renal cortical NHE3-PS552 [22,23]. Another study by the same group showed that a vitamin D-free diet for 30 days led to hypertension in Wistar rats. Molecular experiments revealed higher NHE3-PS552 levels in the renal cortex and medulla of vitamin D-deficient rats than in control rats, which may be necessary to counteract the increased sodium reabsorption mediated by other renal sodium apical transporters [32]. McDonough's laboratory also focused on determining the role of the proximal renal tubule in response to pressure natriuresis. In Ang II-dependent hypertensive rats, increased diuresis and natriuresis were accompanied by higher NHE3-PS552 [13]. In another study, female rats excreted sodium more rapidly than did male rats in response to saline challenge, and female rats displayed higher levels of NHE3-PS552 compared to males [24]. Our findings complement those of earlier studies and support that NHE3 phosphorylation at serine 552 is an essential mechanism underlying the pressure natriuresis response.

We assessed the role of RSN alone and its association with increased BP in renal excretion. RNS is a widely used method that mimics increased rSNA and can be used to determine the role of the SNS in renal function [1,6]. Many studies have demonstrated that RNS increases sodium and water tubular reabsorption independently of glomerular or hemodynamic changes [1,33]. We previously showed that RNS results in antidiuresis and anti-natriuresis and is associated with a reduction in NHE3-PS552. Based on these previous findings and the observation that Bic PVN produced pressor diuresis and natriuresis associated with higher renal cortical NHE3-PS552 levels, we assessed a separate group of rats under three different conditions to distinguish between BP and sRNA regulation of sodium and water reabsorption. First, the BP of the rats was recorded for 1 h, following which the renal nerve was stimulated using a frequency that causes tubular sodium and water retention [8,9]. Finally, under RNS, Bic was injected into the PVN to cause an immediate increase in BP. Urine samples were collected at this time point. RNS increased sodium and water reabsorption compared to that at the baseline period, as previously demonstrated [8]. Furthermore, Bic in the PVN produced renal sodium and fluid excretion values greater than those of the two previous periods in the same rats. Although RNS is a potent stimulus of renal sodium and water retention, the increase in BP is a more powerful stimulus for controlling extracellular fluid. However, we did not examine the role of the renal nerve opposing natriuresis and diuresis under conditions such as chronic hypertension, in which both rSNA and BP are chronically increased. A protocol using renal denervation may clarify this mechanism.

3.1. Limitations of the Present Study

The experiments were carried out in anesthetized rats, and it is not possible to state that the same findings will be obtained in awake rats. Assessing the PVN and recording sympathetic activity simultaneously in non-anesthetized rats is a difficult task and only new experiments will be able to clarify this question. However, to overcome this limitation, the rats used in the three independent series of experiments were submitted to the same experimental conditions, including the anesthesia protocol.

3.2. Conclusions

The major new finding of the present study is that an acute increase in the BP induced by Bic in the PVN overcomes RNS, leading to diuresis, natriuresis, and increase in the NHE3 phosphorylation levels at serine 552, Figure 6 schematically shows this hypothesis. NHE3-mediated sodium reabsorption by renal sympathetic activation may be a new therapeutic target to treat hypertension.

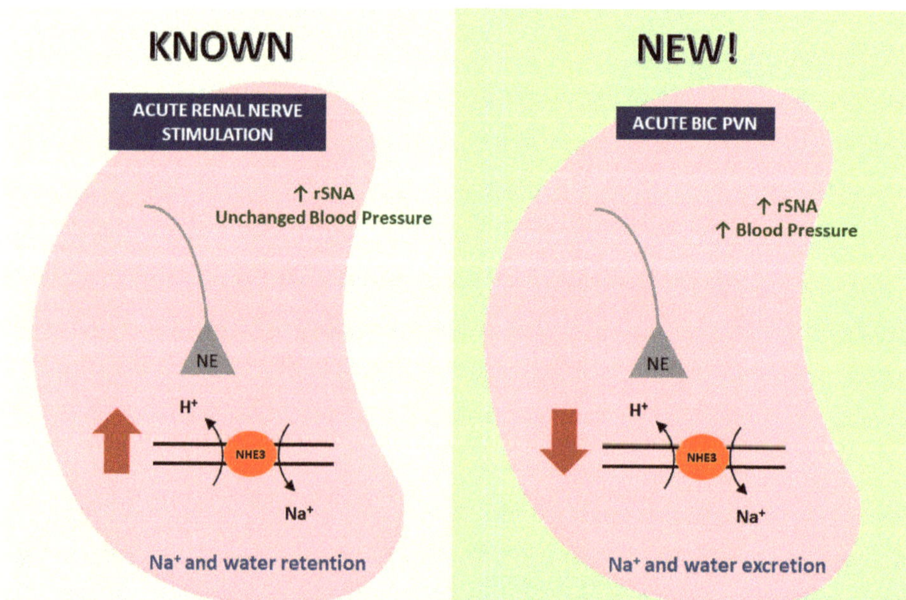

Figure 6. Schematic representation on the left of the figure shows increased NHE3 activity during selective renal nerve stimulation leading to increased sodium and water retention as previously reported [8]. On the right of the figure, shows that the renal nerve stimulation accompanied by increased blood pressure during PVN activation leads to reduction of NHE3 activity and consequently, increased salt and water excretion.

4. Materials and Methods

4.1. Materials

Ketamine and xylazine were obtained from Syntec (Cotia, Brazil). Sodium thiopental was obtained from Cristália (Itapira, Brazil). The monoclonal antibody to NHE3 was a gift from Dr. Peter Aronson (Yale University, New Haven, CT, USA). A phosphor-specific monoclonal antibody that recognizes NHE3 only when it is phosphorylated at serine 552 [21] was purchased from Santa Cruz Biotechnology (Dallas, TX, USA), and a monoclonal antibody to actin (clone JLA20) was purchased from Merck (Darmstadt, Germany). Horseradish peroxidase-conjugated goat anti-mouse and goat anti-rabbit secondary antibodies were purchased from Life Technologies (Carlsbad, CA, USA). All other reagents were purchased from Sigma Chemical (St. Louis, MO, USA) unless otherwise stated.

4.2. Animals

All experimental procedures were performed according to the guidelines established by the Institutional Animal Care Committee of the Ethics in Research Committee of the Universidade Federal de São Paulo (protocol no. 0361/11). Male Wistar rats (n = 19, 250–300 g, 8–9 weeks old) were obtained from the animal care facility at the Universidade Federal de São Paulo. All animals were housed in group cages, provided free access to rat chow and water, and maintained in a temperature-controlled environment (23 °C) on a 12:12 h light-dark cycle. The animals were subjected to right nephrectomy, as described previously [8], to avoid any influence from the contralateral kidney. The animals were administered meloxicam (0.1 mg/kg) on the day of surgery and on the two following days.

4.3. Study Design

The rats were divided into three groups for the experiments: (1) sham, in which the BP and rSNA were recorded, and urine was collected over 2 h; (2) Bic PVN, in which the BP and rSNA of the rats were recorded, and urine was collected for 1 h at baseline conditions and during 1 h after the GABA-A antagonist Bic was infused into the PVN; and (3) RNS + bicuculine injection into the PVN, after which the BP and rSNA were recorded, and urine was collected for 1 h after RNS and 1 h after Bic injection into the PVN.

In three series of experiments 7–8 days after uninephrectomy (right kidney) to avoid possible effects from the contralateral kidney by the reno-renal reflex [1], the rats were anesthetized (sodium thiopental; 60 mg/kg intraperitoneally) and tracheotomized. The femoral vein was catheterized with a PE-10 polyethylene tube connected to a PE-50 for administration of additional anesthetic (10 mg (kg·h) intravenously) and isotonic saline at 10 mL/(kg·h) in the three independent series of experiments. The femoral artery was cannulated to directly measure the arterial pressure. The temperature of the rats was maintained at 37 °C using a rectal probe connected to a servo-controlled electric blanket (Letica®, Woonsocket, RI, USA). The femoral artery catheter was connected to a pressure transducer, and the arterial pressure was recorded online (PowerLab, ADInstruments, Sydney, Australia). The ureter was catheterized with polyethylene tubing (PE-10) to collect the urine.

4.4. rSNA Signal Acquisition

A retroperitoneal incision was made to expose the left kidney and record rSNA. Using a dissecting microscope and fine forceps, the renal nerve was localized, freed from the connective tissue, and positioned on a bipolar silver recording electrode. The renal nerve and electrode were covered with paraffin oil. The signal of the renal nerve was displayed on an oscilloscope, and nerve activity was amplified (gain 20,000, Neurolog, Digitimer, Hertfordshire, UK), filtered with a band-pass filter (100–1000 Hz), and collected for display and analysis using a PowerLab data-acquisition system (ADInstruments).

In the third series of experiments (RNS + BicPVN), as previously described [8,9], after BP was recorded for 1 h (baseline), we applied RNS for 1 h using parameters known to cause sodium and water retention without altering the renal blood flow and glomerular filtration rate [6]. Furthermore, for an additional 1 h, the RNS was maintained, and Bic was injected into the PVN to increase the BP (RNS + Bic PVN). Urine samples were collected separately during each of the three periods. At the end of the experiment, plasma samples were stored as described above and are shown in Supplementary Figures S1 and S2. The animals were euthanized by injecting 5% KCl intravenously into the bolus.

4.5. RNS

To stimulate the renal nerve, we first localized and dissected this nerve, as described above. The renal nerve was placed on the bipolar electrode and then cut at the proximal part, so that the stimulation was only efferent. The electrode was connected to an electrical stimulator system (Grass S88,RI, USA), and the distal portion of the renal nerve was stimulated for 15, 1.5, and 0.5 ms.

4.6. Bic PVN Injection

Bic PVN microinjections were performed as described previously [34]. The rats were placed in a stereotaxic frame, and the PVN was located 1.8 mm caudal to the bregma, 0.5 mm lateral, and 7.8 mm deep (bite bar: −3.4 mm). Unilateral microinjections into the PVN were performed using glass micropipettes with tip diameters of 10–20 μm connected to a nitrogen pressure injector (MicroData Instruments, South Plainfield, NJ, USA). Microinjections were comprised of Bic methiodide (4 mM) in a volume of 100 nL. All drugs were dissolved in sterile saline.

4.7. Analysis of Baseline rSNA

Renal nerve activity was analyzed offline using a software (PowerLab, ADInstruments). The filtered nerve signal was rectified using only positive absolute values and then smoothed into a single line to obtain the mV/s. We assumed that the baseline signal represented 100%. We then compared the period in which the rSNA changed after Bic PVN. The background level of rSNA was determined by intravenous hexamethonium bromide administration (30 mg/Kg) as previously reported [14].

4.8. Plasma and Urine Analyses

At the end of the experiment, blood and urine samples and the kidneys were collected for subsequent analysis. A blood sample (approximately 1 mL) was collected from the femoral artery, and the left kidney was extracted. Blood samples were centrifuged at 3000× *g* for 20 min at 4 °C, and the plasma was stored at −80 °C. The kidneys were kept for up to 5 h in ice-cold PBS (150 mM NaCl, 2.8 mM sodium phosphate monobasic, 7.2 mM sodium phosphate dibasic at pH 7.4) containing protease (0.7 mg/mL pepstatin, 0.5 mg/mL leupeptin, and 40 mg/mL PMSF) and phosphatase (15 mM sodium fluoride and 50 mM sodium pyrophosphate) inhibitors.

The UFR was determined by measuring the urinary volumes normalized to the body weight obtained over a period. UNa concentrations were measured with an electrolyte analyzer (AVL Medical Instruments, Saint Quen l'Aumone, France) and then multiplied by the UFR to obtain the UNa. The CrCl was used to estimate glomerular filtration rates. Plasma and urinary creatinine concentrations were measured using a kinetic method (Labtest, Minas Gerais, Brazil) with a Thermo Plate Analyzer Plus (Thermo Fisher Scientific, Waltham, MA, USA).

4.9. Renal Cortical Membrane Protein Isolation

The kidneys were homogenized in ice-cold PBS. The kidney cortices were isolated at 4 °C and homogenized in PBS. The homogenate was centrifuged at 4000× *g* for 10 min at 4 °C, and the supernatant was removed and subjected to an additional 90 min of centrifugation at 28,000× *g* at 4 °C. The supernatant was discarded, and the pellets were resuspended in fresh PBS containing protease and phosphatase inhibitors. Protein concentrations were measured using the Lowry method [35].

4.10. Sodium Dodecyl Sulfate-Polyacrylamide Gel Electrophoresis and Immunoblotting

Renal cortical membrane proteins were solubilized in sample buffer (2% sodium dodecyl sulfate, 20% glycerol, 100 mM 2-mercaptoethanol, 50 mM Tris, pH 6.8, 0.05% bromophenol blue) and separated using sodium dodecyl sulfate-polyacrylamide gel electrophoresis on a 7.5% polyacrylamide gel. Proteins were then transferred from the polyacrylamide gel to a polyvinylidene difluoride microporous membrane (Immobilon, Millipore, Billerica, MA, USA). The membranes were blocked for 1 h with blocking solution (5% non-fat dry milk and 0.1% Tween 20 in PBS, pH 7.4) and incubated overnight with 1:1000 anti-NHE3, 1:1000 anti-PS552-NHE3 (phosphor specific anti-NHE3 monoclonal antibody directed against phospho-serine 552), or 1:5000 anti-actin. The membranes were washed five times with blocking solution and incubated for 1 h with the appropriate horseradish peroxidase-conjugated immunoglobulin secondary antibody (1:2.000). After five washes in blocking solution, the membranes were rinsed twice in PBS and incubated with enhanced chemiluminescence reagent for 1 min. The membranes were digitized on an ImageScanner LAS 4000 mini (GE Healthcare, Little Chalfont, UK) and quantified using ImageJ software (National Institutes of Health, Bethesda, MD, USA).

4.11. Statistics Analysis

All results are reported as the mean ± SE. Comparisons among groups were performed using one-way analysis of variance for multiple comparisons followed by Tukey's test or Student's *t*-test. A p-value < 0.05 was considered to indicate statistically significant results.

5. Conclusions

Our results suggest that an acute increase in BP has a more powerful influence on renal sodium and water absorption than the effect of acute stimulation of rSNA. Under both conditions, NHE3 phosphorylation regulation at serine 552 appears to play a critical role in sodium reabsorption.

Supplementary Materials: The following supporting information can be downloaded at: https://www.mdpi.com/article/10.3390/ijms24010349/s1.

Author Contributions: Conceptualization, C.T.B., A.C.C.G. and R.R.C.; methodology, E.E.N. and R.B.P.; software, M.I.O.M.; validation, C.T.B. and R.B.P.; formal analysis, R.B.P., M.I.O.M. and R.O.C.; writing—review and editing, R.B.P., C.T.B. and R.R.C.; supervision, C.T.B.; project administration, C.T.B. and A.C.C.G.; funding acquisition, C.T.B. and R.R.C. All authors have read and agreed to the published version of the manuscript.

Funding: This research was funded by FAPESP (Fundação de Amparo à Pesquisa do Estado de São Paulo—Brazil—Grant number: 18/02671-3, 19/25295-0) and Capes (Coordenação de Aperfeiçoamento de Pessoal de Nível Superior—Financial Code 001). CTB, ACCG, and RRC are recipients of CNPq (Conselho Nacional de Ensino Superior) fellowships.

Institutional Review Board Statement: The animal study protocol was approved by Ethics Committee from the Federal University of São Paulo—Brazil.

Informed Consent Statement: Not applicable.

Data Availability Statement: Not applicable.

Acknowledgments: FAPESP: CAPES and CNPq.

Conflicts of Interest: The authors declare no conflict of interest. The funders had no role in the design of the study; in the collection, analyses, or interpretation of data; in the writing of the manuscript; or in the decision to publish the results.

References

1. DiBona, G.F.; Kopp, U.C. Neural control of renal function. *Physiol. Rev.* **1997**, *77*, 75–197. [CrossRef] [PubMed]
2. Guyenet, P.G. The sympathetic control of blood pressure. *Nat. Rev. Neurosci.* **2006**, *7*, 335–346. [CrossRef] [PubMed]

3. Akine, A.; Montanaro, M.; Allen, A.M. Hypothalamic paraventricular nucleus inhibition decreases renal sympathetic nerve activity in hypertensive and normotensive rats. *Auton Neurosci.* **2003**, *108*, 17–21. [CrossRef] [PubMed]
4. Dampney, R.A.; Michelini, L.C.; Li, D.P.; Pan, H.L. Regulation of sympathetic vasomotor activity by the hypothalamic paraventricular nucleus in normotensive and hypertensive states. *Am. J. Physiol. Heart Circ. Physiol.* **2018**, *315*, H1200–H1214. [CrossRef]
5. Chen, Q.H.; Haywood, J.R.; Toney, G.M. Sympathoexcitation by PVN-injected bicuculline requires activation of excitatory amino acid receptors. *Hypertension* **2003**, *42*, 725–731. [CrossRef]
6. Bell-Reuss, E.; Trevino, D.L.; Gottschalk, C.W. Effect of renal sympathetic nerve stimulation on proximal water and sodium reabsorption. *J. Clin. Investig.* **1976**, *57*, 1104–1107. [CrossRef]
7. Healy, V.; Thompson, C.; Johns, E.J. The adrenergic regulation of proximal tubular Na^+/H^+ exchanger 3 in the rat. *Acta Physiol.* **2014**, *210*, 678–689. [CrossRef]
8. Pontes, R.B.; Crajoinas, R.O.; Nishi, E.E.; Oliveira-Sales, E.B.; Girardi, A.C.; Campos, R.R.; Bergamaschi, C.T. Renal nerve stimulation leads to the activation of the Na+/H+ exchanger isoform 3 via angiotensin II type I receptor. *Am. J. Physiol. Renal. Physiol.* **2015**, *308*, F848–F856. [CrossRef]
9. Pontes, R.B.; Girardi, A.C.; Nishi, E.E.; Campos, R.R.; Bergamaschi, C.T. Crosstalk between the renal sympathetic nerve and intrarenal angiotensin II modulates proximal tubular sodium reabsorption. *Exp. Physiol.* **2015**, *100*, 502–506. [CrossRef]
10. Aronson, P.S. Ion exchangers mediating NaCl transport in the renal proximal tubule. *Cell Biochem. Biophys.* **2002**, *36*, 147–153. [CrossRef]
11. McDonough, A.A. Mechanisms of proximal tubule sodium transport regulation that link extracellular fluid volume and blood pressure. *Am. J. Physiol. Regul. Integr. Comp. Physiol.* **2010**, *298*, R851–R861. [CrossRef] [PubMed]
12. Li, X.C.; Zhu, D.; Chen, X.; Zheng, X.; Zhao, C.; Zhang, J.; Soleimani, M.; Rubera, I.; Tauc, M.; Zhou, X.; et al. Proximal Tubule-Specific Deletion of the NHE3 (Na+/H+ Exchanger 3) in the Kidney Attenuates Ang II (Angiotensin II)-Induced Hypertension in Mice. *Hypertension* **2019**, *74*, 526–535. [CrossRef] [PubMed]
13. Nguyen, M.T.; Lee, D.H.; Delpire, E.; McDonough, A.A. Differential regulation of Na+ transporters along nephron during ANG II-dependent hypertension: Distal stimulation counteracted by proximal inhibition. *Am. J. Physiol. Renal. Physiol.* **2013**, *305*, F510–F519. [CrossRef] [PubMed]
14. Nishi, E.E.; Almeida, V.R.; Amaral, F.G.; Simon, K.A.; Futuro-Neto, H.A.; Pontes, R.B.; Cespedes, J.G.; Campos, R.R.; Bergamaschi, C.T. Melatonin attenuates renal sympathetic overactivity and reactive oxygen species in the brain in neurogenic hypertension. *Hypertens. Res.* **2019**, *42*, 1683–1691. [CrossRef] [PubMed]
15. Oliveira-Sales, E.B.; Nishi, E.E.; Carillo, B.A.; Boim, M.A.; Dolnikoff, M.S.; Bergamaschi, C.T.; Campos, R.R. Oxidative stress in the sympathetic premotor neurons contributes to sympathetic activation in renovascular hypertension. *Am. J. Hypertens.* **2009**, *22*, 484–492. [CrossRef] [PubMed]
16. Oliveira-Sales, E.B.; Toward, M.A.; Campos, R.R.; Paton, J.F. Revealing the role of the autonomic nervous system in the development and maintenance of Goldblatt hypertension in rats. *Auton. Neurosci.* **2014**, *183*, 23–29. [CrossRef] [PubMed]
17. Zambrano, L.I.; Pontes, R.B.; Garcia, M.L.; Nishi, E.E.; Nogueira, F.N.; Higa, E.M.S.; Cespedes, J.G.; Bergamaschi, C.T.; Campos, R.R. Pattern of sympathetic vasomotor activity in a model of hypertension induced by nitric oxide synthase blockade. *Physiol. Rep.* **2019**, *7*, e14183. [CrossRef] [PubMed]
18. Esler, M.D.; Krum, H.; Sobotka, P.A.; Schlaich, M.P.; Schmieder, R.E.; Böhm, M.; Symplicity HTN-2 Investigators. Renal sympathetic denervation in patients with treatment-resistant hypertension (The Symplicity HTN-2 Trial): A randomised controlled trial. *Lancet* **2010**, *376*, 1903–1909.
19. Krum, H.; Schlaich, M.; Whitbourn, R.; Sobotka, P.A.; Sadowski, J.; Bartus, K.; Kapelak, B.; Walton, A.; Sievert, H.; Thambar, S.; et al. Catheter-based renal sympathetic denervation for resistant hypertension: A multicentre safety and proof-of-principle cohort study. *Lancet* **2009**, *373*, 1275–1281. [CrossRef]
20. Schlaich, M.P.; Sobotka, P.A.; Krum, H.; Lambert, E.; Esler, M.D. Renal sympathetic-nerve ablation for uncontrolled hypertension. *N. Engl. J. Med.* **2009**, *361*, 932–934. [CrossRef]
21. Kocinsky, H.S.; Girardi, A.C.; Biemesderfer, D.; Nguyen, T.; Mentone, S.; Orlowski, J.; Aronson, P.S. Use of phospho-specific antibodies to determine the phosphorylation of endogenous Na+/H+ exchanger NHE3 at PKA consensus sites. *Am. J. Physiol. Renal. Physiol.* **2005**, *289*, F249–F258. [CrossRef] [PubMed]
22. Crajoinas, R.O.; Lessa, L.M.; Carraro-Lacroix, L.R.; Davel, A.P.; Pacheco, B.P.; Rossoni, L.V.; Malnic, G.; Girardi, A.C. Posttranslational mechanisms associated with reduced NHE3 activity in adult vs. young prehypertensive SHR. *Am. J. Physiol. Renal. Physiol.* **2010**, *299*, F872–F881. [CrossRef] [PubMed]
23. Crajoinas, R.O.; Pessoa, T.D.; Rodrigues, M.V.; Malnic, G.; Girardi, A.C. Changes in the activity and expression of protein phosphatase-1 accompany the differential regulation of NHE3 before and after the onset of hypertension in spontaneously hypertensive rats. *Acta Physiol.* **2014**, *211*, 395–408. [CrossRef] [PubMed]
24. Veiras, L.C.; Girardi, A.C.C.; Curry, J.; Pei, L.; Ralph, D.L.; Tran, A.; Castelo-Branco, R.C.; Pastor-Soler, N.; Arranz, C.T.; Yu, A.S.L.; et al. Sexual Dimorphic Pattern of Renal Transporters and Electrolyte Homeostasis. *J. Am. Soc. Nephrol.* **2017**, *28*, 3504–3517. [CrossRef]

25. Garcia, M.L.; Pontes, R.B.; Nishi, E.E.; Ibuki, F.K.; Oliveira, V.; Sawaya, A.C.; Carvalho, P.O.; Nogueira, F.N.; Franco, M.D.; Campos, R.R.; et al. The antioxidant effects of green tea reduces blood pressure and sympathoexcitation in an experimental model of hypertension. *J. Hypertens.* **2017**, *35*, 348–354. [CrossRef] [PubMed]
26. Fardin, N.M.; Oyama, L.M.; Campos, R.R. Changes in baroreflex control of renal sympathetic nerve activity in high-fat-fed rats as a predictor of hypertension. *Obesity* **2012**, *20*, 1591–1597. [CrossRef] [PubMed]
27. Porter, J.P.; Brody, M.J. A comparison of the hemodynamic effects produced by electrical stimulation of subnuclei of the paraventricular nucleus. *Brain Res.* **1986**, *375*, 20–29. [CrossRef]
28. Dampney, R.A. Functional organization of central pathways regulating the cardiovascular system. *Physiol. Rev.* **1994**, *74*, 323–364. [CrossRef] [PubMed]
29. Borges-Júnior, F.A.; Silva Dos Santos, D.; Benetti, A.; Polidoro, J.Z.; Wisnivesky, A.C.T.; Crajoinas, R.O.; Antônio, E.L.; Jensen, L.; Caramelli, B.; Malnic, G.; et al. Empagliflozin Inhibits Proximal Tubule NHE3 Activity, Preserves GFR, and Restores Euvolemia in Nondiabetic Rats with Induced Heart Failure. *J. Am. Soc. Nephrol.* **2021**, *32*, 1616–1629. [CrossRef]
30. Crajoinas, R.O.; Polidoro, J.Z.; Carneiro de Morais, C.P.; Castelo-Branco, R.C.; Girardi, A.C. Angiotensin II counteracts the effects of cAMP/PKA on NHE3 activity and phosphorylation in proximal tubule cells. *Am. J. Physiol. Cell Physiol.* **2016**, *311*, C768–C776. [CrossRef]
31. Martins, F.L.; Bailey, M.A.; Girardi, A.C.C. Endogenous Activation of Glucagon-Like Peptide-1 Receptor Contributes to Blood Pressure Control: Role of Proximal Tubule Na+/H+ Exchanger Isoform 3, Renal Angiotensin II, and Insulin Sensitivity. *Hypertension* **2020**, *76*, 839–848. [CrossRef] [PubMed]
32. Luchi, W.M.; Crajoinas, R.O.; Martins, F.L.; Castro, P.C.; Venturini, G.; Seguro, A.C.; Girardi, A.C.C. High blood pressure induced by vitamin D deficiency is associated with renal overexpression and hyperphosphorylation of Na+-K+-2Cl- cotransporter type 2. *J. Hypertens.* **2021**, *39*, 880–891. [CrossRef] [PubMed]
33. Bello-Reuss, E.; Pastoriza-Muñoz, E.; Colindres, R.E. Acute unilateral renal denervation in rats with extracellular volume expansion. *Am. J. Physiol.* **1977**, *232*, F26–F32. [CrossRef] [PubMed]
34. Carillo, B.A.; Oliveira-Sales, E.B.; Andersen, M.; Tufik, S.; Hipolide, D.; Santos, A.A.; Tucci, P.J.; Bergamaschi, C.T.; Campos, R.R. Changes in GABAergic inputs in the paraventricular nucleus maintain sympathetic vasomotor tone in chronic heart failure. *Auton. Neurosci.* **2012**, *171*, 41–48. [CrossRef] [PubMed]
35. Lowry, O.H.; Rosebrough, N.J.; Farr, A.L.; Randall, R.J. Protein measurement with the Folin phenol reagent. *J. Biol. Chem.* **1951**, *193*, 265–275. [CrossRef] [PubMed]

International Journal of
Molecular Sciences

MDPI

Article

Acute Severe Heart Failure Reduces Heart Rate Variability: An Experimental Study in a Porcine Model

Jan Naar [1,*], Mikulas Mlcek [2], Andreas Kruger [1], Dagmar Vondrakova [1], Marek Janotka [1], Michaela Popkova [2], Otomar Kittnar [2], Petr Neuzil [1] and Petr Ostadal [1]

1 Department of Cardiology, Na Homolce Hospital, 150 30 Prague, Czech Republic
2 Department of Physiology, First Faculty of Medicine, Charles University, 128 00 Prague, Czech Republic
* Correspondence: jan.naar@seznam.cz; Tel.: +420-257-272-208; Fax: +420-257-272-342

Abstract: There are substantial differences in autonomic nervous system activation among heart (cardiac) failure (CF) patients. The effect of acute CF on autonomic function has not been well explored. The aim of our study was to assess the effect of experimental acute CF on heart rate variability (HRV). Twenty-four female pigs with a mean body weight of 45 kg were used. Acute severe CF was induced by global myocardial hypoxia. In each subject, two 5-min electrocardiogram segments were analyzed and compared: before the induction of myocardial hypoxia and >60 min after the development of severe CF. HRV was assessed by time-domain, frequency-domain and nonlinear analytic methods. The induction of acute CF led to a significant decrease in cardiac output, left ventricular ejection fraction and an increase in heart rate. The development of acute CF was associated with a significant reduction in the standard deviation of intervals between normal beats (50.8 [20.5–88.1] ms versus 5.9 [2.4–11.7] ms, $p < 0.001$). Uniform HRV reduction was also observed in other time-domain and major nonlinear analytic methods. Similarly, frequency-domain HRV parameters were significantly changed. Acute severe CF induced by global myocardial hypoxia is associated with a significant reduction in HRV.

Keywords: acute heart failure; experimental model; heart rate variability; pig

Citation: Naar, J.; Mlcek, M.; Kruger, A.; Vondrakova, D.; Janotka, M.; Popkova, M.; Kittnar, O.; Neuzil, P.; Ostadal, P. Acute Severe Heart Failure Reduces Heart Rate Variability: An Experimental Study in a Porcine Model. *Int. J. Mol. Sci.* **2023**, *24*, 493. https://doi.org/10.3390/ijms24010493

Academic Editors: Yutang Wang and Kate Denton

Received: 1 November 2022
Revised: 22 December 2022
Accepted: 25 December 2022
Published: 28 December 2022

1. Introduction

Neurohumoral activation, including sympathetic overactivity and vagal tone withdrawal, plays an important role in the pathogenesis and progression of heart failure with reduced ejection fraction (HFrEF) [1,2]. Pharmacotherapy that aims to reduce renin-angiotensin-aldosterone axis activation and autonomic nervous system imbalance continues to be the cornerstone of therapy [3–5]. Nevertheless, the prognosis of HFrEF remains poor [6]. Recently, several neuromodulation strategies have been proposed in HFrEF patients; they directly target the autonomic nervous system to improve residual autonomic imbalance. Despite promising preclinical results, neuromodulation has failed to prove beneficial in randomized clinical trials [7–9].

Heart rate variability (HRV) is a parameter that reflects autonomic nervous system activation. It has been shown that therapeutic approaches that decrease mortality in HFrEF patients (physical activity, pharmacotherapy, cardiac resynchronization therapy) also increase HRV [10–13]. Thus, HRV improvement may be a desirable therapeutic aim of neuromodulation in the heart (cardiac) failure (CF) population. There are, however, substantial differences in HRV among CF patients. Some CF patients, who are characterized by a lower HRV, reflecting higher cardiac sympathetic nervous activity and more severe CF, may benefit from neuromodulation therapy [14]. The failure of neuromodulation therapy in clinical settings due to the enrollment of subjects with insufficiently advanced disease is debatable. Clinical data related to HRV in acute CF are scarce [15], and the effect of acute severe CF on HRV has not been experimentally explored in detail before.

The main objective of this study was to test the hypothesis that acute severe experimental CF significantly decreases HRV. We further aimed to describe individual differences in reaction to acute CF development using a broad range of HRV metrics.

2. Results

Electrocardiogram (ECG) data from 23 subjects were eligible for HRV analysis. One subject with continuous arrhythmia was excluded. The development of severe acute CF resulted in a significant decrease in systemic arterial blood pressure, cardiac output, and left ventricular ejection fraction, while heart rate significantly increased. Hemodynamic details related to the development of acute CF are outlined in Table 1.

Table 1. Main hemodynamic and vital signs variables at baseline and after the induction of acute heart failure.

	Baseline	Acute Heart Failure	*p*-Value
Heart rate (bpm)	88.9 ± 26.0	107.5 ± 26.4	0.01
Mean aortic pressure (mmHg)	94.9 ± 9.3	67.9 ± 23.1	<0.001
Systolic aortic pressure (mmHg)	115.4 ± 12.3	84.1 ± 22.4	<0.001
Diastolic aortic pressure (mmHg)	79.4 ± 8.5	59.9 ± 21.3	0.005
Central venous pressure (mmHg)	3.6 ± 1.5	5.1 ± 2.8	0.03
Mean pulmonary arterial pressure (mmHg)	18.7 ± 4.7	22.8 ± 4.6	0.005
Diastolic pulmonary arterial pressure (mmHg)	13.1 ± 4.6	19.1 ± 4.7	<0.001
Pulse oximetry (tail; %)	98.8 ± 1.6	97 ± 2.4	0.017
Cerebral tissue oximetry (NIRS; %)	62.3 ± 6.7	53.2 ± 14.5	0.046
Hind limb tissue oximetry (NIRS; %)	56.5 ± 7.3	54.3 ± 7.2	0.051
Left ventricular dP/dt_{max} (mmHg/s)	1740 ± 454	949 ± 316	<0.001
Left ventricular end-diastolic volume (mL)	126.9 ± 11.1	135.8 ± 15.9	0.058
Stroke volume (mL)	74.8 ± 9.8	23.8 ± 7.4	<0.001
Cardiac output (L/min)	7.2 ± 1.8	2.6 ± 0.7	<0.001
Left ventricular ejection fraction (%)	57.9 ± 4.9	17.2 ± 5.0	<0.001

Data are expressed as mean and standard deviation. dP/dt_{max} = maximal rate of pressure rise during ventricular contraction; NIRS = near-infrared spectroscopy.

2.1. Effect of Experimental Acute Heart Failure on Time-Domain Parameters of Heart Rate Variability

Acute CF led to a prominent and uniform decrease in the standard deviation of intervals between normal beats (SDNN; 50.8 [20.5–88.1] ms versus 5.9 [2.4–11.7] ms, $p < 0.001$), as well as the square root of the mean of the sum of the squares of differences between adjacent NN intervals (RMSSD; 46.3 [10.1–87.2] ms versus 2.0 [21.8–6.2] ms, $p < 0.001$) and a number of pairs of adjacent NN intervals differing by more than 50 ms divided by the total number of all NN intervals (pNN50; 4.0 [0.0–15.2]% versus 0.0 [0.0–0.0]%, $p < 0.001$; Figure 1). A representative example of time-domain HRV analysis is displayed in Figure 2.

2.2. Effect of Experimental Acute Heart Failure on Frequency-Domain Parameters of Heart Rate Variability

Spectral frequency analysis of HRV also revealed significant changes following acute CF induction compared with the baseline. Significant increase in low frequency (LF) expressed in normalized units (n.u.; 5.3 [2.7–15.4] n.u. versus 9.5 [4.1–14.2] n.u., $p = 0.01$) and a low frequency over high-frequency ratio (LF/HF; 0.06 [0.03–0.09] versus 0.10 [0.04–0.17], $p = 0.008$) was observed, whereas the high frequency (HF) significantly decreased (94.7 [91.7–97.3] n.u. versus 90.5 [85.8–95.9] n.u., $p = 0.01$; Figure 3).

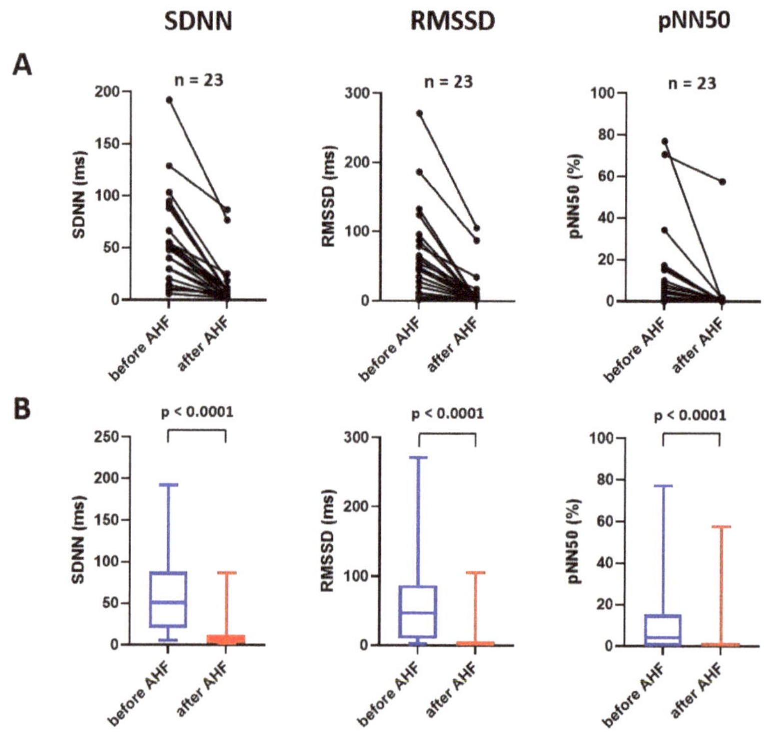

Figure 1. Effect of acute heart failure on time-domain parameters of heart rate variability displayed as an individual graph (**A**) and as median with interquartile range, minimum and maximum value (**B**). AHF = acute heart failure; pNN50 = number of pairs of adjacent NN intervals differing by more than 50 ms divided by the total number of all NN intervals; RMSSD = square root of the mean of the sum of the squares of differences between adjacent NN intervals; SDNN = standard deviation of intervals between normal beats.

Figure 2. Representative example of HRV analysis comparing time-domain HRV parameters before (**A**) and after (**B**) acute CF induction in one study subject. BPM = beats per minute; pNN50 = number of pairs of adjacent NN intervals differing by more than 50 ms divided by the total number of all NN

intervals; RMSSD = square root of the mean of the sum of the squares of differences between adjacent NN intervals; RR = interval between adjacent R waves; SDNN = standard deviation of intervals between normal beats.

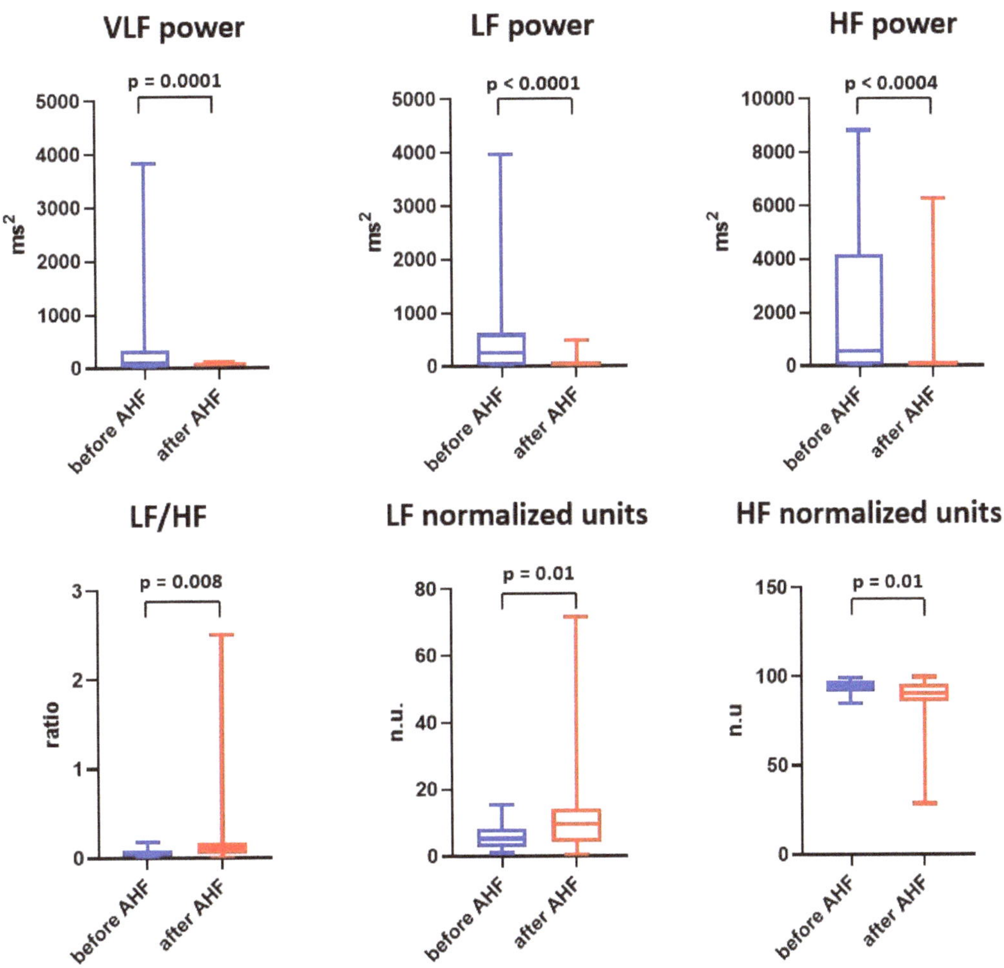

Figure 3. Effect of acute heart failure on frequency-domain parameters of heart rate variability. Data are expressed as median with interquartile range, minimum and maximum value. AHF = acute heart failure; HF = high frequency; LF = low frequency; n.u. = normalized units; VLF = very low frequency.

2.3. *Effect of Experimental Acute Heart Failure on Nonlinear Analysis of Heart Rate Variability*

The basic parameters from the nonlinear analysis of HRV (SD1, SD2) also prominently and uniformly decreased with acute CF (SD1, 37.8 [7.2–61.7] ms versus 1.4 [1.3–4.4] ms, $p < 0.001$; SD2, 50.8 [21.1–93.8] ms versus 8.3 [3.0–15.0] ms, $p < 0.001$; Figure 4).

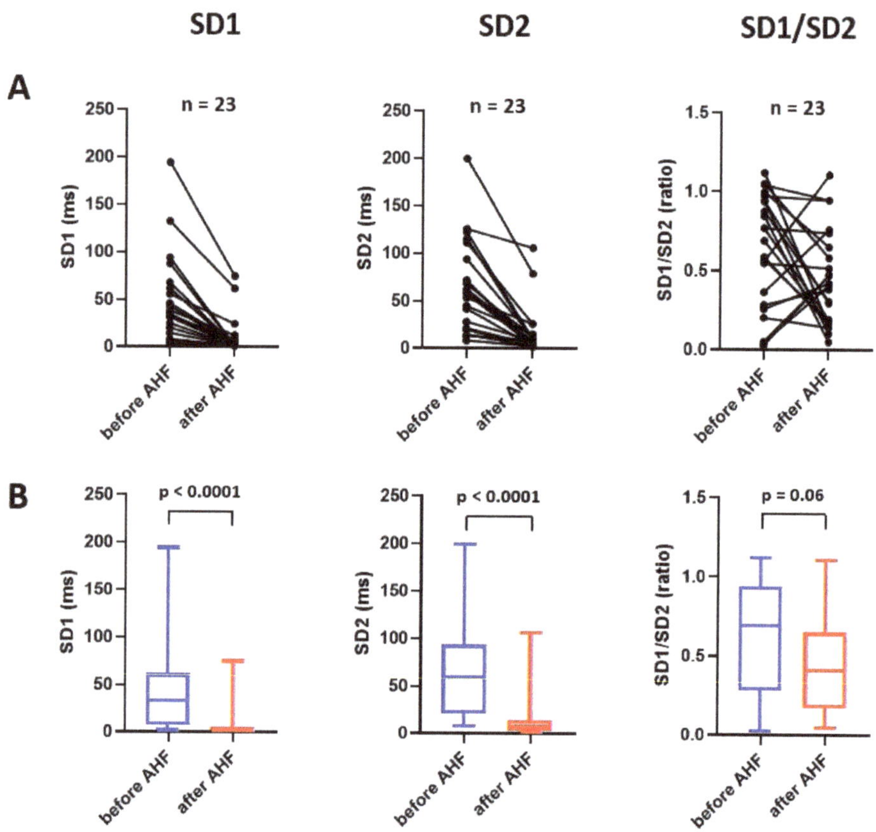

Figure 4. Effect of acute heart failure on nonlinear analysis of heart rate variability displayed as individual graph (**A**) and as median with interquartile range, minimum and maximum value (**B**). AHF = acute heart failure; SD1 = standard deviation of the points along short semi-axis of an ellipse fitted to Poincaré plot; SD2 = standard deviation of the points along long semi-axis of an ellipse fitted to Poincaré plot.

3. Discussion

The major observation of this study is that acute severe CF significantly influences HRV. To the best of our knowledge, this is the first demonstration that the development of acute CF is associated with a reduction in HRV. Changes in the time-domain and main nonlinear analytic parameters were particularly uniform and pronounced, but changes in frequency-domain parameters were also significant.

It is well known that HFrEF patients have a lower HRV compared with healthy individuals [16]. The significant prognostic value of HRV in the CF population has also been extensively explored [17,18]. Likewise, decreased HRV has been documented in an experimental model of chronic (progressive) CF, primarily induced by rapid right ventricular pacing [19,20]. Surprisingly, the effect of acute CF on HRV has not been described in detail before.

The results of the present study are in agreement with previous experiments that documented the relationship between an acute pathological situation, such as hemorrhagic shock or sepsis, and a decrease in HRV [21–23]. In sepsis, the decrease in HRV preceded hemodynamic changes [21,22]. In hemorrhagic shock, hemodynamic and metabolic variables were better discriminants in revealing survivors than HRV metrics [23]. Thus, HRV may not always serve as a marker of the severity of a pathological situation. Nevertheless, in subjects suffering from chronic CF, clinical and veterinary data suggest that HRV negatively correlates with disease severity [24,25]. Our findings support these data and also imply a close relationship between acute severe CF and pronounced autonomic imbalance.

Moreover, according to time-domain and nonlinear HRV parameters, severe acute CF reduces HRV in every single subject. It supports our hypothesis that the patients with high HRV enrolled in neuromodulation studies had insufficiently impaired autonomic balance, signaling a less severe disease. HRV or other autonomic balance estimating tools should be involved in the recruitment process when testing the therapy which affects the autonomic nervous system in CF patients.

Evaluating the effect of our acute CF model on the parasympathetic versus sympathetic nervous system is challenging, as the relationship between both branches of the autonomic nervous system is complex and not simply competitive [26]. Moreover, the interpretation of individual HRV parameters may be problematic, as none of the HRV parameters reflects solely sympathetic nervous system activity change [27]. Nevertheless, a pronounced decrease in time-domain methods, SD1 and HF power signals that a reduction of parasympathetic nervous system activity was the predominant change in the autonomic balance after the development of acute CF in our model [28]. Baroreceptor reflex activation due to loss of myocardial contractility with subsequent systemic hypotension was probably the main vagal activity-modulating factor [29]. Reduction of parasympathetic nervous system activity also probably contributed to a significant decrease in very low frequency (VLF) power [30].

Study Limitations

Our study has several limitations. Extrapolation of our findings to the clinical setting in HFrEF may be problematic, as we used a model of acute CF, specifically severe acute CF, with features of cardiogenic shock, as opposed to chronic CF. It is, therefore, not clear that changes in HRV would also be as pronounced and uniform in less severe CF.

The hypoxic model of acute CF may be associated with transient hypoxemia in the carotid arteries, beginning after the initiation of hypoxia and lasting until the loss of left ventricular contractility, which is followed by prevalent perfusion of the carotid region by oxygenated blood from the ECMO. During the experiment, decreased cerebral tissue saturation < 40% was detected by near-infrared spectroscopy oximetry in the majority of animals. Despite the fact that we did not observe any severe clinical signs of hypoxic brain damage (seizures or impaired brain stem reflexes) and subcortical structures involved in autonomic nervous system circuits are less susceptible to hypoxia than the cortex, we cannot entirely exclude the influence of short-term hypoxia on the autonomic nervous system. Using standard HRV measures, it is not well possible to differentiate the effect of cerebral hypoxia on HRV from the compensatory HRV reaction due to the loss of myocardial contractility. HRV response can be non-specific regarding the etiology of the insult, and HRV changes are also dependent on the insult severity [31]. Nevertheless, the pattern of HRV change following the development of acute CF was identical in all study subjects, although some experienced minimal, whereas others had longer or deeper cortical hypoxia. Therefore, we believe that changes in HRV predominantly reflect autonomic imbalance caused by acute CF. Moreover, apart from possible transient cerebral hypoxic hypoxia linked to our acute CF model, severe acute CF or cardiogenic shock is frequently accompanied by circulatory cerebral hypoxia. From this aspect, brain hypoxia is an integrative and inseparable part of severe acute CF syndrome and reflects a common clinical scenario.

HRV frequency-domain parameters may also be impaired by changes in ventilation parameters during the experiment (lower tidal volume and respiratory frequency) [32]. However, time-domain analyses of HRV should not be affected by ventilation [33].

Only two 5-min ECG segments were chosen for HRV analysis in each subject (before and after the development of acute CF), with an emphasis on the quality of the traces, as this experimental model is associated with frequent rhythmic disturbances. We did not endeavor to map HRV timeline changes following acute CF induction but aimed to describe the extent of the response.

We expressed HRV according to customary practice in non-indexed values. Nevertheless, there is a negative exponential relationship between heart rate and HRV [34]. As heart rate was significantly higher after the induction of acute CF, when using HRV metrics corrected for heart rate, the change in HRV would be slightly smaller.

Finally, we cannot entirely exclude the influence of sex on HRV, as only female pigs were enrolled due to the easier cannulation of the urinary tract in comparison with boars. In healthy human subjects, significant differences in HRV have been previously described between males and females [35,36]. We may hypothesize that baseline HRV behaves similarly in a porcine model and thus, we cannot exclude that the extent of the changes in HRV would be different in males.

4. Materials and Methods

4.1. Animal Model

Twenty-four female domestic pigs (*Sus scrofa domestica*, Large White × Landrace crossbreed), 4–5 months of age, with a mean body weight of 45 kg, were enrolled. The animal model used in the present study has been described in detail previously [37,38]. Briefly, after 24 h of fasting, general anesthesia was induced by intramuscular administration of midazolam (0.3 mg/kg) and ketamine hydrochloride (15–20 mg/kg). Following an intravenous bolus of propofol and morphine, endotracheal intubation was performed. General anesthesia was maintained by continuous intravenous administration of propofol, midazolam and morphine. Drop rates of infusions with normal saline and unfractionated heparin were adjusted to maintain a mean central venous pressure of 5–7mmHg and an activated clotting time of 200–250 s, respectively.

4.2. Hypoxic Model of Acute Heart Failure

The hypoxic model of severe acute CF used in the present study has been described previously [38,39]. Global myocardial hypoxia was induced by perfusing the anterior body with deoxygenated blood. This was achieved by reducing ventilation support after the initiation of veno-arterial extracorporeal membrane oxygenation (VA-ECMO). In detail, tidal volume was reduced to 100 mL, respiratory frequency to 5 breaths per minute and fraction of inspired oxygen to 0.21. Positive end-expiratory pressure was maintained at 6 cm H_2O. Ventilation was partially increased to prevent circulatory collapse if needed. Global myocardial hypoxia led to a rapid decrease in ventricular contractility, resulting in reduced ejection fraction, cardiac output and systemic arterial blood pressure. The Hamilton G5 (Hamilton Medical AG, Bonaduz, Switzerland) mechanical ventilator and VA-ECMO in femoro-femoral configuration with a 21 Fr venous cannula, 15–18 Fr arterial cannula and i-cor (Xenios AG, Heilbronn, Germany) or Cardiohelp (Getinge, Rastatt, Germany) systems were used.

4.3. Vital Signs and Hemodynamic Monitoring

Vital signs and invasive hemodynamics were monitored throughout the experiment, including continuous ECG, pulse oximetry measured at the tail, central venous pressure, arterial pressure obtained invasively from the aortic arch using a pigtail catheter and standard fluid-field pressure transducer, pulmonary arterial pressure measured by Swan-Ganz catheter, and cerebral and peripheral tissue oximetry using near-infrared spectroscopy (INVOS, Medtronic, Minneapolis, MN, USA). The Life Scope TR (Nihon Kohden, Tokyo, Japan) and Vigilance II (Edwards Lifescience, Irvine, CA, USA) hemodynamic monitors were used for continuous hemodynamic data recording. Left ventricular performance was assessed by pressure-volume 7 Fr pigtail conductance catheter Scisense (Transonic, Ithaca, NY, USA) inserted into the left ventricle through the aortic valve via the left carotid artery.

4.4. Study Protocol and Heart Rate Variability Assessment

ECG was continuously recorded during the experiment. In each subject, two 5-min ECG segments were selected and compared: (i) 5–30 min after insertion of VA-ECMO before induction of myocardial hypoxia and (ii) >60 min after the development of severe CF [40]. Beat-to-beat control of the selected ECG segments was performed to confirm sinus rhythm and avoid irregularities (premature complexes, arrhythmias, artifacts). HRV analysis was automated using LabChart software version 8.1.19 (ADInstruments, Dunedin, New Zealand) in the animal mode for pigs. Time-domain and frequency-domain parameters, as well as nonlinear analytic methods, were used for HRV assessment. In detail, SDNN, RMSSD and pNN50 were chosen as representatives of time-domain measures. HF power, LF power, VLF power expressed in ms^2, as well as LF/HF and HF and LF bands expressed in normalized units (relative power of particular frequency band related to total power without VLF component), were used to describe the spectral frequency distribution of HRV. In the present porcine model, the HF band was defined as 0.09–2 Hz, the LF band as 0.01–0.09 Hz and the VLF band as <0.01 Hz. Nonlinear analysis was derived from the Poincaré plot that displays the dependence of the RR interval on the preceding RR interval. The plot was fitted by an ellipse, where SD1 represents the standard deviation of the points along the short semi-axis and SD2 the standard deviation of the points along the main semi-axis [22].

4.5. Statistical Analyses

Hemodynamic parameters were expressed as mean and standard deviation and analyzed using a two-tailed paired *t*-test. The normality of HRV data was evaluated with the Shapiro-Wilk and D'Agostino-Pearson tests. As all HRV data had non-normal distribution, the Wilcoxon two-tailed matched pairs test was applied and are presented as median and interquartile range. All statistical analyses were performed using GraphPad Prism version 8.3.0 (GraphPad Software, La Jolla, CA, USA). p-values < 0.05 were considered statistically significant.

5. Conclusions

Acute severe heart failure induced by global myocardial hypoxia is associated with a significant reduction in heart rate variability, which is probably driven predominantly by the reduction of parasympathetic nervous system activity via baroreflex activation.

Author Contributions: J.N. performed data analysis, data interpretation and drafting of the manuscript. M.M., M.P. and P.O. participated in the conception and design of the study, performed the experimental procedures, participated in data interpretation and critically revised the manuscript. O.K. and P.N. participated in the conception and design of the study and critically revised the manuscript. A.K., D.V. and M.J. participated in the study procedures and data analysis and critically revised the manuscript. All authors have read and agreed to the published version of the manuscript.

Funding: This work was supported by the Ministry of Health, Czech Republic, conceptual development of research organization (NHH, 00023884): IG180502, IG200505.

Institutional Review Board Statement: This study was approved by the Charles University, First Faculty of Medicine Animal Care and Use Committee, and was performed at the Animal Laboratory, Department of Physiology, First Faculty of Medicine, Charles University in Prague and Na Homolce Hospital, Prague, Czech Republic, in accordance with Act No. 246/1992 Coll. for the protection of animals against cruelty. The investigation and protocol conformed to the Guide for the Care and Use of Laboratory Animals published by the United States National Institutes of Health (8th edition, 2011).

Data Availability Statement: The data regarding the presented study are available from the corresponding author upon reasonable request.

Conflicts of Interest: The authors J.N., A.K. and D.V. received a speaker honorarium from Edwards Lifescience. P.O. received a speaker honorarium from Getinge, Edwards Lifescience and Xenios AG. The authors M.M., M.J., M.P., O.K. and P.N. declare that they have no conflict of interest.

References

1. Azevedo, E.R.; Parker, J.D. Parasympathetic control of cardiac sympathetic activity: Normal ventricular function versus congestive heart failure. *Circulation* **1999**, *100*, 274–279. [CrossRef] [PubMed]
2. Packer, M. The neurohormonal hypothesis: A theory to explain the mechanism of disease progression in heart failure. *J. Am. Coll. Cardiol.* **1992**, *20*, 248–254. [CrossRef] [PubMed]
3. Swedberg, K.; Kjekshus, J. CONSENSUS Trial Study Group. Effects of enalapril on mortality in severe congestive heart failure. Results of the Cooperative North Scandinavian Enalapril Survival Study (CONSENSUS). *N. Engl. J. Med.* **1987**, *316*, 1429–1435.
4. Packer, M.; Bristow, M.R.; Cohn, J.N.; Colucci, W.S.; Fowler, M.B.; Gilbert, E.M.; Shusterman, N.H.; U.S. Carvedilol Heart Failure Study Group. The effect of carvedilol on morbidity and mortality in patients with chronic heart failure. *N. Engl. J. Med.* **1996**, *334*, 1349–1355. [CrossRef] [PubMed]
5. Pitt, B.; Zannad, F.; Remme, W.J.; Cody, R.; Castaigne, A.; Perez, A.; Palensky, J.; Wittes, J.; Randomized Aldactone Evaluation Study Investigators. The effect of spironolactone on morbidity and mortality in patients with severe heart failure. *N. Engl. J. Med.* **1999**, *341*, 709–717. [CrossRef]
6. Spitaleri, G.; Lupón, J.; Domingo, M.; Santiago-Vacas, E.; Codina, P.; Zamora, E.; Cediel, G.; Santesmases, J.; Diez-Quevedo, C.; Troya, M.I.; et al. Mortality trends in an ambulatory multidisciplinary heart failure unit from 2001 to 2018. *Sci. Rep.* **2021**, *11*, 732. [CrossRef]
7. Zipes, D.P.; Neuzil, P.; Theres, H.; Caraway, D.; Mann, D.L.; Mannheimer, C.; Van Buren, P.; Linde, C.; Linderoth, B.; Kueffer, F.; et al. Determining the Feasibility of Spinal Cord Neuromodulation for the Treatment of Chronic Systolic Heart Failure: The DEFEAT-HF Study. *JACC Heart Fail.* **2016**, *4*, 129–136. [CrossRef]
8. Zannad, F.; De Ferrari, G.M.; Tuinenburg, A.E.; Wright, D.; Brugada, J.; Butter, C.; Klein, H.; Stolen, C.; Meyer, S.; Stein, K.M.; et al. Chronic vagal stimulation for the treatment of low ejection fraction heart failure: Results of the NEural Cardiac TherApy foR Heart Failure (NECTAR-HF) randomized controlled trial. *Eur. Heart J.* **2015**, *36*, 425–433. [CrossRef]
9. Gold, M.R.; Van Veldhuisen, D.J.; Hauptman, P.J.; Borggrefe, M.; Kubo, S.H.; Lieberman, R.A.; Milasinovic, G.; Berman, B.J.; Djordjevic, S.; Neelagaru, S.; et al. Vagus Nerve Stimulation for the Treatment of Heart Failure: The INOVATE-HF Trial. *J. Am. Coll. Cardiol.* **2016**, *68*, 149–158. [CrossRef]
10. Pearson, M.J.; Smart, N.A. Exercise therapy and autonomic function in heart failure patients: A systematic review and meta-analysis. *Heart Fail. Rev.* **2018**, *23*, 91–108. [CrossRef]
11. Ozdemir, M.; Arslan, U.; Türkoğlu, S.; Balcioğlu, S.; Cengel, A. Losartan improves heart rate variability and heart rate turbulence in heart failure due to ischemic cardiomyopathy. *J. Card. Fail.* **2007**, *13*, 812–817. [CrossRef] [PubMed]
12. Pousset, F.; Copie, X.; Lechat, P.; Jaillon, P.; Boissel, J.P.; Hetzel, M.; Fillette, F.; Remme, W.; Guize, L.; Le Heuzey, J.Y. Effects of bisoprolol on heart rate variability in heart failure. *Am. J. Cardiol.* **1996**, *77*, 612–617. [CrossRef] [PubMed]
13. Fantoni, C.; Raffa, S.; Regoli, F.; Giraldi, F.; La Rovere, M.T.; Prentice, J.; Pastori, F.; Fratini, S.; Salerno-Uriarte, J.A.; Klein, H.U.; et al. Cardiac resynchronization therapy improves heart rate profile and heart rate variability of patients with moderate to severe heart failure. *J. Am. Coll. Cardiol.* **2005**, *46*, 1875–1882. [CrossRef] [PubMed]
14. Naar, J.; Jaye, D.; Neuzil, P.; Doskar, P.; Malek, F.; Linderoth, B.; Lind, G.; Stahlberg, M. Acute effect of spinal cord stimulation on autonomic nervous system function in patients with heart failure. *J. Appl. Biomed.* **2021**, *19*, 133–141. [CrossRef]
15. Ozdemir, O.; Alyan, O.; Kacmaz, F.; Kaptan, Z.; Ozbakir, C.; Geyik, B.; Cagirci, G.; Soylu, M.; Demir, A.D. Evaluation of effects of intra aortic balloon counterpulsation on autonomic nervous system functions by heart rate variability analysis. *Ann. Noninvasive Electrocardiol.* **2007**, *12*, 38–43. [CrossRef]
16. Casolo, G.; Balli, E.; Taddei, T.; Amuhasi, J.; Gori, C. Decreased spontaneous heart rate variability in congestive heart failure. *Am. J. Cardiol.* **1989**, *64*, 1162–1167. [CrossRef]
17. Hadase, M.; Azuma, A.; Zen, K.; Asada, S.; Kawasaki, T.; Kamitani, T.; Kawasaki, S.; Sugihara, H.; Matsubara, H. Very low frequency power of heart rate variability is a powerful predictor of clinical prognosis in patients with congestive heart failure. *Circ. J.* **2004**, *68*, 343–347. [CrossRef]
18. Ponikowski, P.; Anker, S.D.; Chua, T.P.; Szelemej, R.; Piepoli, M.; Adamopoulos, S.; Webb-Peploe, K.; Harrington, D.; Banasiak, W.; Wrabec, K.; et al. Depressed heart rate variability as an independent predictor of death in chronic congestive heart failure secondary to ischemic or idiopathic dilated cardiomyopathy. *Am. J. Cardiol.* **1997**, *79*, 1645–1650. [CrossRef]
19. Motte, S.; Mathieu, M.; Brimioulle, S.; Pensis, A.; Ray, L.; Ketelslegers, J.M.; Montano, N.; Naeije, R.; van de Borne, P.; Entee, K.M. Respiratory-related heart rate variability in progressive experimental heart failure. *Am. J. Physiol. Heart Circ. Physiol.* **2005**, *289*, H1729–H1735. [CrossRef]
20. Zhou, S.X.; Lei, J.; Fang, C.; Zhang, Y.L.; Wang, J.F. Ventricular electrophysiology in congestive heart failure and its correlation with heart rate variability and baroreflex sensitivity: A canine model study. *Europace* **2009**, *11*, 245–251. [CrossRef]
21. Jarkovska, D.; Valesova, L.; Chvojka, J.; Benes, J.; Danihel, V.; Sviglerova, J.; Nalos, L.; Matejovic, M.; Stengl, M. Heart-rate variability depression in porcine peritonitis-induced sepsis without organ failure. *Exp. Biol. Med.* **2017**, *242*, 1005–1012. [CrossRef] [PubMed]
22. Jarkovska, D.; Valesova, L.; Chvojka, J.; Benes, J.; Sviglerova, J.; Florova, B.; Nalos, L.; Matejovic, M.; Stengl, M. Heart Rate Variability in Porcine Progressive Peritonitis-Induced Sepsis. *Front. Physiol.* **2016**, *6*, 412. [CrossRef] [PubMed]

23. Salomão, E., Jr.; Otsuki, D.A.; Correa, A.L.; Fantoni, D.T.; dos Santos, F.; Irigoyen, M.C.; Auler, J.O., Jr. Heart Rate Variability Analysis in an Experimental Model of Hemorrhagic Shock and Resuscitation in Pigs. *PLoS ONE* **2015**, *10*, e0134387. [CrossRef] [PubMed]
24. Casolo, G.C.; Stroder, P.; Sulla, A.; Chelucci, A.; Freni, A.; Zerauschek, M. Heart rate variability and functional severity of congestive heart failure secondary to coronary artery disease. *Eur. Heart J.* **1995**, *16*, 360–367. [CrossRef]
25. Calvert, C.A.; Wall, M. Effect of severity of myocardial failure on heart rate variability in Doberman pinschers with and without echocardiographic evidence of dilated cardiomyopathy. *J. Am. Vet. Med. Assoc.* **2001**, *219*, 1084–1088. [CrossRef]
26. Billman, G.E. The LF/HF ratio does not accurately measure cardiac sympatho-vagal balance. *Front. Physiol.* **2013**, *4*, 26. [CrossRef]
27. Hayano, J.; Yuda, E. Pitfalls of assessment of autonomic function by heart rate variability. *J. Physiol. Anthropol.* **2019**, *38*, 3. [CrossRef]
28. Shaffer, F.; Ginsberg, J.P. An Overview of Heart Rate Variability Metrics and Norms. *Front. Public Health* **2017**, *5*, 258. [CrossRef]
29. Vaschillo, E.; Lehrer, P.; Rishe, N.; Konstantinov, M. Heart rate variability biofeedback as a method for assessing baroreflex function: A preliminary study of resonance in the cardiovascular system. *Appl. Psychophysiol. Biofeedback* **2002**, *27*, 1–27. [CrossRef]
30. Taylor, J.A.; Carr, D.L.; Myers, C.W.; Eckberg, D.L. Mechanisms underlying very-low-frequency RR-interval oscillations in humans. *Circulation* **1998**, *98*, 547–555. [CrossRef]
31. Botek, M.; Krejčí, J.; De Smet, S.; Gába, A.; McKune, A.J. Heart rate variability and arterial oxygen saturation response during extreme normobaric hypoxia. *Auton. Neurosci.* **2015**, *190*, 40–45. [CrossRef] [PubMed]
32. Frazier, S.K.; Moser, D.K.; Stone, K.S. Heart rate variability and hemodynamic alterations in canines with normal cardiac function during exposure to pressure support, continuous positive airway pressure, and a combination of pressure support and continuous positive airway pressure. *Biol. Res. Nurs.* **2001**, *2*, 167–174. [CrossRef] [PubMed]
33. Schipke, J.D.; Arnold, G.; Pelzer, M. Effect of respiration rate on short-term heart rate variability. *J. Clin. Basic Cardiol.* **1999**, *2*, 92–95.
34. de Geus, E.J.C.; Gianaros, P.J.; Brindle, R.C.; Jennings, J.R.; Berntson, G.G. Should heart rate variability be "corrected" for heart rate? Biological, quantitative, and interpretive considerations. *Psychophysiology* **2019**, *56*, e13287. [CrossRef] [PubMed]
35. Koenig, J.; Thayer, J.F. Sex differences in healthy human heart rate variability: A meta-analysis. *Neurosci. Biobehav. Rev.* **2016**, *64*, 288–310. [CrossRef]
36. Sammito, S.; Böckelmann, I. Reference values for time- and frequency-domain heart rate variability measures. *Heart Rhythm.* **2016**, *13*, 1309–1316. [CrossRef]
37. Ostadal, P.; Mlcek, M.; Kruger, A.; Hala, P.; Lacko, S.; Mates, M.; Vondrakova, D.; Svoboda, T.; Hrachovina, M.; Janotka, M.; et al. Increasing venoarterial extracorporeal membrane oxygenation flow negatively affects left ventricular performance in a porcine model of cardiogenic shock. *J. Transl. Med.* **2015**, *13*, 266. [CrossRef]
38. Ostadal, P.; Mlcek, M.; Strunina, S.; Hrachovina, M.; Kruger, A.; Vondrakova, D.; Janotka, M.; Hala, P.; Kittnar, O.; Neuzil, P. Novel porcine model of acute severe cardiogenic shock developed by upper-body hypoxia. *Physiol. Res.* **2016**, *65*, 711–715. [CrossRef]
39. Lacko, S.; Mlcek, M.; Hala, P.; Popkova, M.; Janak, D.; Hrachovina, M.; Kudlicka, J.; Hrachovina, V.; Ostadal, P.; Kittnar, O. Severe acute heart failure—experimental model with very low mortality. *Physiol. Res.* **2018**, *67*, 555–562. [CrossRef]
40. Heart rate variability—Standards of measurement, physiological interpretation, and clinical use—Task Force of The European Society of Cardiology and The North American Society of Pacing and Electrophysiology. *Eur. Heart J.* **1996**, *17*, 354–381. [CrossRef]

International Journal of
Molecular Sciences

MDPI

Article

The Effect of Renal Denervation on T Cells in Patients with Resistant Hypertension

Marta Kantauskaite [1], Oliver Vonend [1,2], Mina Yakoub [1], Philipp Heilmann [1], Andras Maifeld [3,4], Peter Minko [5], Lars Schimmöller [5], Gerald Antoch [5], Dominik N. Müller [3,4], Claudia Schmidt [1], Blanka Duvnjak [1], Ulf Zierhut [1], Sebastian A. Potthoff [1], Lars C. Rump [1], Johannes C. Fischer [6,†] and Johannes Stegbauer [1,*,†]

1 Department of Nephrology, Medical Faculty, University Hospital Düsseldorf, Heinrich-Heine-University Düsseldorf, 40225 Düsseldorf, Germany
2 Nierenzentrum Wiesbaden, 65191 Wiesbaden, Germany
3 Charité-Universitätsmedizin Berlin, Corporate Member of Freie Universität Berlin and Humboldt-Universität zu Berlin, 10117 Berlin, Germany
4 Experimental and Clinical Research Center, a Cooperation of Charité-Universitätsmedizin Berlin and Max Delbruck Center for Molecular Medicine, 13125 Berlin, Germany
5 Department for Diagnostic and Interventional Radiology, Medical Faculty, University Hospital Düsseldorf, Heinrich-Heine-University Düsseldorf, 40225 Düsseldorf, Germany
6 Institute of Transplantation Diagnostics and Cell Therapeutics, Medical Faculty, University Hospital Düsseldorf, Heinrich-Heine-University Düsseldorf, 40225 Düsseldorf, Germany
* Correspondence: johannes.stegbauer@med.uni-duesseldorf.de; Tel.: +49-211-81-17726
† These authors share the last authorship.

Abstract: (1) Background: Sympathetic overactivity is a major contributor to resistant hypertension (RH). According to animal studies, sympathetic overactivity increases immune responses, thereby aggravating hypertension and cardiovascular outcomes. Renal denervation (RDN) reduces sympathetic nerve activity in RH. Here, we investigate the effect of RDN on T-cell signatures in RH. (2) Methods: Systemic inflammation and T-cell subsets were analyzed in 17 healthy individuals and 30 patients with RH at baseline and 6 months after RDN. (3) Results: The patients with RH demonstrated higher levels of pro-inflammatory cytokines and higher frequencies of CD4+ effector memory (T_{EM}), CD4+ effector memory residential (T_{EMRA}) and CD8+ central memory (T_{CM}) cells than the controls. After RDN, systolic automated office blood pressure (BP) decreased by -17.6 ± 18.9 mmHg. Greater BP reductions were associated with higher CD4+ T_{EM} (r -0.421, $p = 0.02$) and CD8+ T_{CM} (r -0.424, $p = 0.02$) frequencies at baseline. The RDN responders, that is, the patients with $\geq$10mmHg systolic BP reduction, showed reduced pro-inflammatory cytokine levels, whereas the non-responders had unchanged inflammatory activity and higher CD8+ T_{EMRA} frequencies with increased cellular cytokine production. (4) Conclusions: The pro-inflammatory state of patients with RH is characterized by altered T-cell signatures, especially in non-responders. A detailed analysis of T cells might be useful in selecting patients for RDN.

Keywords: sympathetic activity; renal denervation; resistant hypertension; immune response; T cells; lymphocytes; pro-inflammatory cytokines; inflammation

Citation: Kantauskaite, M.; Vonend, O.; Yakoub, M.; Heilmann, P.; Maifeld, A.; Minko, P.; Schimmöller, L.; Antoch, G.; Müller, D.N.; Schmidt, C.; et al. The Effect of Renal Denervation on T Cells in Patients with Resistant Hypertension. *Int. J. Mol. Sci.* **2023**, *24*, 2493. https://doi.org/10.3390/ijms24032493

Academic Editors: Yutang Wang and Kate Denton

Received: 24 December 2022
Revised: 21 January 2023
Accepted: 23 January 2023
Published: 27 January 2023

1. Introduction

By affecting more than 1.5 billion people worldwide, hypertension is one of the most important global health concerns [1,2]. This is mostly because its complications, including stroke, heart failure and kidney disease, are the leading causes of cardiovascular morbidity and mortality [3,4]. Large randomized studies have shown the effectiveness of antihypertensive treatments [5]. However, despite the increasing awareness and the vast number of effective antihypertensive agents, hypertension control rates remain unsatisfactory. Typically, less than 50% of treated patients achieve their blood pressure targets [6,7]. A growing number of these cases are being called resistant hypertension (RH), which is defined as a blood pressure above 140/90mmHg despite the concurrent use of three antihypertensive agents, including a diuretic [8]. In this regard, patients with RH have the highest risk of cardiovascular events [9,10].

One of the cornerstones for the development and persistence of hypertension in RH is an increased activation of the sympathetic nervous system (SNS) [11,12]. In addition to SNS activation at a systemic level, renal sympathetic nerve activation seems to play an outstanding role in the pathogenesis of RH, as it leads to sodium and water retention, the activation of the renin–angiotensin system and vasoconstriction [13,14]. Therefore, the reduction of SNS activity at the renal level using catheter-based renal denervation (RDN) has become an interventional treatment approach for RH. By reducing sympathetic activity, a significant and longstanding blood pressure reduction in 60–70 % of cases can be achieved [15–17]. However, about one-third of patients do not respond to the procedure with respect to a significant blood pressure reduction. The reasons for non-response however are not well understood. More importantly, there are no reliable clinical or biochemical factors that can be applied to predict the success of the procedure [15,18].

For several years, there has been evidence supported mainly by experimental animal studies that SNS interacts with the immune system and modulates blood pressure, as well as the course of cardiovascular disease. It is known that primary and secondary lymphatic organs are innervated by SNS fibers [19]. Increased SNS activity, as in the case of RH, has been shown to drive immune cell functions and signatures into pro-inflammatory phenotypes and foster immune cell infiltration into target organs, such as the kidney or heart [20,21]. In this regard, norepinephrine released from sympathetic neurons alters the phenotypes and functions of monocytes, macrophages and T cells [19,22–24]. Specifically, CD8+ cells produce more pro-inflammatory cytokines and display enhanced survival upon increased SNS activity [22,24,25]. In fact, CD8+ T cells play a major role in the development of hypertension, as a lack of these immune cells is associated with protection against experimental hypertension in mice [26]. Unfortunately, despite a large body of experimental evidence, there are no human studies to date demonstrating that sympathetic overactivity in the presence of RH affects the T-cell response or that RDN can effectively modulate it.

In the present study, we aimed to describe the immune cell signatures of patients with RH and to compare them to those of healthy controls. Next, we aimed to determine whether these changes were modulated by RDN. Lastly, to improve patient selection for RDN, we sought to characterize the differences in immune cell phenotypes between the patients who responded to the RDN procedure and those who did not.

2. Results

2.1. Baseline Characteristics

A total of 30 patients with resistant hypertension (RH) with an average age of 61.1 ± 10.9 years were included in this study, and there was a slightly higher number of males (18 males and 12 females). According to 24-h blood pressure measurements, they had an average systolic blood pressure of 157 ± 15 mmHg and a diastolic blood pressure of 90 ± 11 mmHg before the renal denervation procedure (RDN). The antihypertensive regime consisted of 6 ± 1 drugs, and this regime was stable for a minimum of 4 weeks (Table 1). After inclusion in the study, all of the patients underwent RDN. In addition, we included 17 healthy controls, who presented with normal blood pressure and were 42.2 ± 11.3 years old.

2.2. Patients with Resistant Hypertension Display High Levels of Inflammatory Markers

The patients with RH displayed significantly higher concentrations of high-sensitivity CRP than the healthy controls (hsCRP: 3691 (1126–7945) ng/mL vs. 513 (116–2993) ng/mL, $p < 0.05$). In addition, the patients with RH had higher levels of pro-inflammatory cytokines (TNF-α: 2.1 (1.5–2.8) pg/mL vs. 0.8 (0.7–0.9) pg/mL; IFN-γ: 2.7 (1.5–3.8) pg/mL vs. 0.7 (0.5–1.5) pg/mL, $p < 0.001$) than the healthy controls, as demonstrated in Figure 1A. Moreover, higher IL-6 levels were associated with higher diastolic blood pressure values among the patients with RH (r 0.404, $p = 0.03$). No relationship between cytokines and the systolic blood pressure values was observed.

Table 1. Characteristics of responders and non-responders before and after the renal denervation procedure.

Parameter	RH		RH Treatment	
	Responder	Non-Responder	Responder	Non-Responder
Age, y	66.0 ± 5.8	51.2 ± 12.2 ***	66.5 ± 5.8	51.7 ± 12.2 ***
Sex (M:F)	12:8	5:5	12:8	5:5
AOBPM, mmHg	169 ± 15/92±18	167 ± 18/92 ± 19	139 ± 14/78 ± 10 ###	167 ± 16/89 ± 10 **
ABPM, mmHg	159 ± 17/90 ±12	150 ± 6/90 ± 8	139 ± 14/78 ± 10 ###	155 ± 8/85 ± 10 **
Antihypertensive, N	5.0 ± 1.0	6.0 ± 1.0	5.0 ± 2.0	6.0 ± 1.0
Antihypertensive class				
— ACEi/ARB	17 (85%)	10 (100%)	18 (90%)	10 (100%)
— MRA	4 (20%)	2 (20%)	7 (35%)	3 (30%)
— CCB	17 (85%)	8 (80%)	16 (80%)	8 (80%)
— Diuretic	19 (95%)	10 (100%)	15 (75%)	10 (100%)
— β-Blocker	12 (60%)	8 (80%)	14 (70%)	8 (80%)
— α-Blocker	7 (35%)	4 (40%)	6 (30%)	2 (20%)
— Other	17 (85%)	8 (80%)	16 (80%)	7 (70%)
BMI	30.2 ± 5.8	30.8 ± 5.8	30.2 ± 5.8	30.8 ± 5.8
OSAS, N (%)	14 (70%)	4 (40%) *	14 (70%)	4 (40%) *
Diabetes, N (%)	5 (25%)	3 (30%)	5 (25%)	3 (30%)
Hb1AC, %	6.0 ± 0.9	5.9 ± 0.7	6.0 ± 0.9	5.8 ± 0.7
eGFR, mL/min	72 ± 27	75 ± 28	72 ± 27	75 ± 28
hsCRP, ng/mL	4041 (1261–8023)	3063 (1043–8823)	1909 (499–5110) #	2745 (1331–10,099)
TNFα, pg/mL	2.1 (1.5–2.9)	2.6 (1.4–3.7)	2.3 (1.8–2.8)	2.7 (1.7–5.4)
IL-6, pg/mL	2.8 (1.5–3.8)	2.5 (1.7–3.6)	1.9 (1.5–2.5) #	3.2 (2.0–3.7) *
CD4, %	65 ± 17	61 ± 14	65 ±15	63 ± 14
CD4 T_{CM}, %	39 ± 17	38 ± 15	36 ± 15	39 ± 16
CD4 T_{EM}, %	39 ± 18	29 ± 12 *	38 ± 19	25 ± 8 *
CD4 T_{EMRA}, %	4 (2–8)	7 (2–12)	5 (3–9)	4 (2–10)
CD4 naïve, %	19 ± 13	27 ± 7 **	20 ± 15	26 ± 7
CD8, %	35 ± 17	39 ± 14	35 ± 15	37 ± 14
CD8 T_{CM}, %	17 ± 8	15 ± 9	14 ± 7 #	17 ± 8
CD8 T_{EM}, %	26 ± 13	24 ± 9	25 ± 12	21 ± 8
CD8 T_{EMRA}, %	11 (7–20)	24 (10–39) *	11 (7–17)	25 (9–44) **
CD8 naïve, %	38 ± 13	37 ± 17	38 ± 15	33 ± 14
CD4/CD8	2.3 (1.1–4.4)	1.4 (0.9–2.9)	2.2 (1.0–3.5)	1.5 (1.0–3.3)

RH—resistant hypertension, AOBPM—automated office blood pressure measurement, ABPM—ambulatory blood pressure measurement, ACEi/ARB—angiotensin-converting enzyme inhibitor and angiotensin receptor blocker, MRA—mineralocorticoid receptor antagonist, CCB—calcium channel blocker, BMI—body mass index, OSAS—obstructive sleep apnea syndrome, Hb1AC—hemoglobin A1c, eGFR—estimated glomerular filtration rate, hsCRP—high-sensitivity C-reactive protein, TNF-α—tumor necrosis factor alfa, IL-6—interleukin 6, CD4+—helper T cells, CD8+—cytotoxic T cells, T_{CM}—central memory cells, T_{EM}—effector memory cells, T_{EMRA}—effector memory residential cells. * indicates differences between responders and non-responders; # indicates differences between before and after treatment within the responder or non-responder group. * $p < 0.05$, ** $p < 0.01$, *** $p < 0.001$, or # $p < 0.05$, ### $p < 0.001$ using unpaired or paired t-test, Mann–Whitney or Wilcoxon test where applicable.

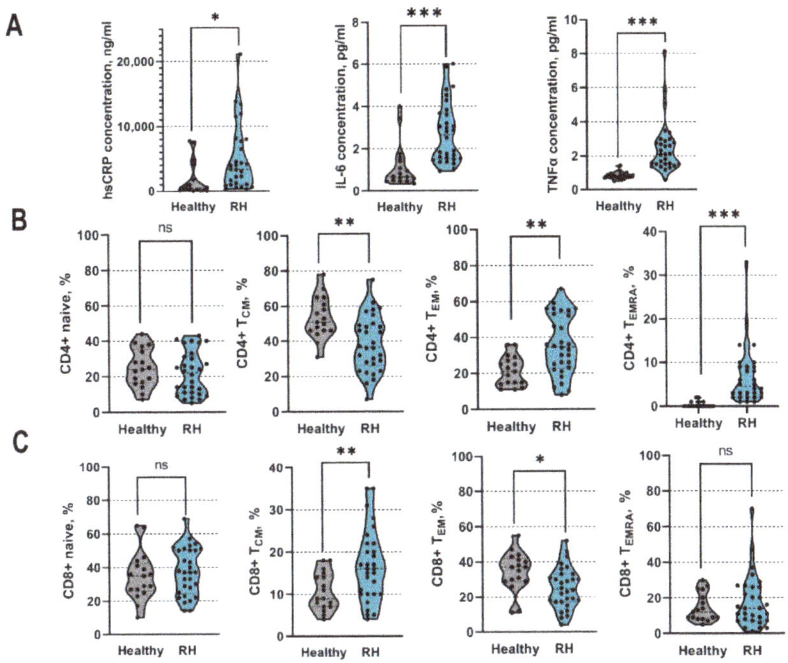

Figure 1. Comparison of inflammatory activity between patients with resistant hypertension and healthy controls. (**A**)—Concentrations of various pro-inflammatory cytokines and proteins among study groups at entry. (**B**,**C**)—The frequencies of different T-cell subsets among healthy individuals and patients with resistant hypertension at study entry. RH—resistant hypertension, hsCRP—high-sensitivity C-reactive protein, TNF-α—tumor necrosis factor alfa, IL-6—interleukin 6, CD4+—T helper cells, CD8+—T cytotoxic cells, T_{CM}—central memory cells, T_{EM}—effector memory cells, T_{EMRA}—effector memory residential cells. *** represents significant difference between the groups with $p < 0.001$, ** $p < 0.01$, * $p < 0.05$, ns – not significant using unpaired *t*-test or Mann–Whitney test.

2.3. Differences in T-Cell Subsets between Patients with Hypertension and Healthy Controls

Next, we analyzed the T-cell subsets and compared them between the healthy subjects and the patients with RH. The patients with RH presented higher frequencies of CD4+ effector memory T cells (T_{EM}) (36 ± 16% vs. 21 ± 8.3%, $p < 0.01$) and CD4+ effector memory residential cells (T_{EMRA}) than the control group (4.5 (2–9)% vs. 0.2 (0.1–0.6)%, $p < 0.001$, Figure 1B). In addition, the patients with RH had lower frequencies of CD4+ T_{CM} cells than the healthy controls (Figure 1B). As for CD8+ T cells, the patients with RH had a higher frequency of CD8+ T_{CM} cells (16 ± 8% vs. 11 ± 4%, $p < 0.01$) and a reduced fraction of CD8+ T_{EM} cells (25 ± 12% vs. 34 ± 11%, $p < 0.05$) compared to the healthy controls. There were no differences in other CD4+ or CD8+ subsets or in the frequencies of total CD4+ and CD8+ cells (Figure 1B,C).

2.4. Effect of RDN on Blood Pressure and Low-Grade Inflammation

Six months after RDN, there was an average decrease in systolic automated office blood pressure of −17.6 ± 18.9 mmHg, whereas diastolic ambulatory blood pressure decreased by −8.7 ± 11.9 mmHg. Twenty patients (67%) achieved at least a 10 mmHg reduction in their systolic office blood pressure and were assigned to the responder group (Figure 2A). The systolic ambulatory BP reduction was −24.5 ± 12.8 mmHg in the responders and 3.9 ± 8.3 mmHg in the non-responders ($p < 0.05$). The responders and non-responders were similar with respect to their clinical parameters, except for age. The responders were significantly older patients, with an average age of 66.0 years (Table 1). Next, we focused on the changes in the inflammatory marker levels before and after RDN. Of note, the responders demonstrated significant reductions in hsCRP (4041 (1261–8023) ng/mL and 1909 (499–5110) ng/mL, $p < 0.05$, before and after RDN, respectively) and IL-6 (2.8 (1.5–3.8) pg/mL vs. 1.9 (1.5–2.5) pg/mL, $p < 0.05$, before and after RDN, respectively) concentrations following the treatment (Figure 2B). TNF-α levels were not affected by the treatment.

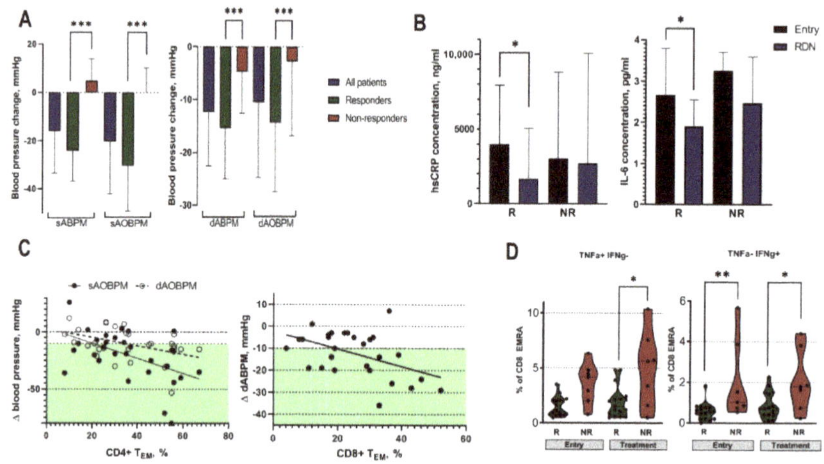

Figure 2. The effect of renal denervation procedure on blood pressure and inflammatory activity. (**A**)—Changes in ambulatory and automated office blood pressure among patients with resistant hypertension. Responders were defined by a systolic office blood pressure reduction of at least 10 mmHg following the renal denervation procedure. (**B**)—High-sensitivity CRP and IL-6 levels at baseline and 6 months following RDN among responders and non-responders. (**C**)—A negative correlation between CD4+ effector memory cell frequencies and automated office blood pressure was observed (r −0.421, p = 0.02, and r −0.413, p = 0.02, for systolic and diastolic blood pressure, respectively), as well as a negative correlation between CD8+ effector memory and diastolic ambulatory blood pressure (r −0.424, p = 0.02). Green colored area represents a blood pressure reduction ≥10mmHg. (**D**)—CD8+ T_{EMRA} cells positive for pro-inflammatory cytokines before and after the treatment among responders and non-responders. On the left, CD8+ T_{EMRA} cells positive for TNF-α and, on the right, the ones positive for IFN-γ. sABPM—systolic ambulatory blood pressure measurement, sAOBPM—systolic automated office blood pressure measurement, dABPM—diastolic ambulatory blood pressure measurement, dAOBPM—diastolic automated office blood pressure measurement, hsCRP—high-sensitivity C-reactive protein, IL-6—interleukin 6, RDN—renal denervation, R—responders, NR—non-responders, Δ—change, T_{EM}—effector memory cells, T_{EMRA}—effector memory residential cells, TNF-α—tumor necrosis factor alfa, IFN-γ—interferon gamma. *** represents significant difference between the groups with $p < 0.001$, ** $p < 0.01$, * $p < 0.05$ using unpaired t-test or Mann–Whitney test.

2.5. Effect of RDN on T-Cell Signatures

To analyze the effect of RDN on T-cell signatures, we first investigated the differences between responders and non-responders at baseline. Before RDN, the responders were characterized by significantly higher CD4+ T_{EM} (39 ± 18% vs. 29 ± 12%, $p < 0.05$), lower CD4+ naïve (19 ± 13% vs. 27 ± 7%, $p < 0.05$), and lower CD8+ T_{EMRA} cell frequencies than the non-responders (11 (7–20)% vs. 24 (10–39)%, $p < 0.05$). Furthermore, we analyzed the capacity of CD8+ T_{EMRA} cells to produce pro-inflammatory cytokines upon stimulation in the responders and non-responders. Not only did the non-responders have higher levels of CD8+ T_{EMRA} cells, but they also produced more IFN-γ before RDN (Figure 2D), suggesting that non-responders have a higher capacity of pro-inflammatory T cells, which promotes the chronic low-grade inflammation observed in these patients.

Further, we characterized the T-cell subsets 6 months after the RDN procedure. The responders demonstrated a significant decrease in CD8+ T_{CM} cell frequencies (Table 1), which was more pronounced among the male responders (Suppl. Figure 1). Moreover, the capacity of the CD8+ T_{EMRA} cells to produce the pro-inflammatory cytokines IFN-γ and TNF-α was significantly higher in the non-responders than in the responders (Figure 2D), suggesting that RDN did not affect T-cell function in the non-responders.

As the T-cell response showed differences between the responders and non-responders, we further evaluated the potential predictors of a successful outcome after RDN. Higher CD4+ T_{EM} cell frequencies at study entry were associated with higher systolic and diastolic ambulatory blood pressure reductions (r −0.421, $p = 0.02$ and r −0.413, $p = 0.02$, for systolic and diastolic blood pressure, respectively, Figure 2C). Higher pre-procedural CD8+ T_{EM} cell frequencies were related to a higher ambulatory diastolic blood pressure reduction (r −0.424, $p = 0.02$, Figure 2C), suggesting that higher levels of T effector memory cells are predictors of a successful outcome after RDN.

3. Discussion

In the present study, we showed that patients with resistant hypertension (RH) have a more pronounced pro-inflammatory immune response than healthy controls. Furthermore, we demonstrated that non-responders to RDN are characterized by a pro-inflammatory T-cell signature that is unlike that of responders.

Several studies have demonstrated that patients with hypertension show a more pronounced chronic low-grade inflammation than healthy controls [27,28]. Here, we confirmed these results and showed that patients with RH are characterized by higher levels of pro-inflammatory cytokines, such as hsCRP, IL-6 and TNF-α, than healthy controls. In addition to this observation, we found that patients with RH exhibit a pro-inflammatory T-cell signature compared to control subjects, suggesting that, at least in part, a pro-inflammatory T-cell response drives this low-grade inflammation. In detail, the patients with RH had higher frequencies of CD4+ T_{EM} and T_{EMRA} cells, as well as CD8+ T_{CM} cells, than controls. Higher circulating levels of CD4+ memory cells have been described to be associated with senescence [29,30], atherosclerosis and cardiovascular risks [31–34]. Moreover, higher levels of CD8+ T_{CM} cells have been observed among patients with high cardiovascular risks [35]. These observations may explain why patients with RH have the highest cardiovascular risk [10]. However, since the patients with RH included in this study were older than the healthy controls, we cannot exclude the possibility that these differences in T-cell signatures are, in part, due to older age.

Next, we investigated the effect of RDN on the immune response in the patients with RH. In our cohort, 67% of the patients with RH showed a significant blood pressure reduction 6 months after RDN. These results are in accordance with those of previously published works [15–17]. Interestingly, older patients tended to have a more successful response to RDN. Previously published works do not present consistent results with regard to age as a predicting factor for RDN [15,36–38]. In general, our study cohort was older than that in other studies. It is difficult to determine to what extent the immunological effects of aging (immunosenescence) or other factors, such as differences in vascular function or the renin–angiotensin system, affect this association, as we did not measure these factors. Among the RDN responders, decreases in hsCRP and IL-6 concentrations following the procedure were observed. Several studies have previously demonstrated reduced inflammatory activity in RDN responders 6 to 12 months after the procedure [29,36,39,40]. These associations suggest that reducing sympathetic nerve activity via RDN might modulate the inflammatory state observed among patients with hypertension. However, it is not clear whether this effect is related to a direct interaction between the SNS and the immune system or to the blood pressure reduction itself, since antihypertensive drug treatment also reduces chronic inflammation in essential hypertension [41,42].

In order to analyze whether the reduced inflammation among the RDN responders was related to a modulated immune cell phenotype and/or function, we further compared the

T cytotoxic and helper lymphocyte subsets among the responders and the non-responders. The characterization of immune signatures within the responder group showed that the responders initially had higher CD4+ T_{EM} cell frequencies. In addition to this, higher CD4+ T_{EM} frequencies were associated with a more pronounced decrease in systolic and diastolic blood pressure following RDN. Initially, the RDN responders had lower CD4+ naïve cell frequencies. It should be noted that the significantly higher level of CD4+ naïve T cells in the non-responders in our cohort was also higher than the level among the healthy controls (27±7% vs. 25±11%). This interesting observation suggests that this cell population with a naïve phenotype consists mainly of atypical effector memory cells [43]. These effector memory cells can produce IL-2, TNF-α and IFN-γ as cytokine-producing naive T cells, or memory stem cells. They are likely to rapidly migrate into inflammatory foci and then secrete type 1 cytokines. Unfortunately, we did not further characterize these cells.

Animal studies have suggested an important role of cytotoxic CD8+ T cells in the development of hypertension. Moreover, these T cells seem to be particularly well-activated by SNS overexcitation. [24,26,44,45]. Here, we observed several differences within the CD8+ T-cell subsets among the RDN responders and non-responders before and after RDN, highlighting the interaction between cytotoxic CD8+ T cells, blood pressure control and renal sympathetic nerve activation. First, higher frequencies of CD8+ T_{CM} cells at study entry were associated with higher blood pressure reductions. Second, following RDN, a slight but significant reduction in CD8+ T_{CM} cells among the responders was observed. This reduction might be mainly related to oligoclonal CD8+ T cells participating in hypertension development and persistence [21,26,46]. To determine whether this is the case, additional in-depth CD8+ T-cell phenotyping would need to be performed. Third, the non-responders had significantly higher frequencies of CD8+ T_{EMRA} cells. Interestingly, the CD8+ T_{EMRA} cells of non-responders produced more pro-inflammatory cytokines than those of the responders before and especially after RDN. This very interesting finding could be explained by the fact that norepinephrine binds to different adrenoreceptors expressed on CD8+ T_{EMRA} cells and activates pro-inflammatory cytokine release [24]. Thus, increased norepinephrine release induced by persisting SNS overactivity among the non-responders would lead to an increased release of pro-inflammatory cytokines, as observed in our study [47–49]. However, based on this observation and the fact that non-responders are characterized by higher inflammatory markers, it also seems feasible that non-responders to RDN have an overactivated immune response that partially contributes to the difficulty in treating patients with RH [43,50]. Furthermore, one can speculate that an overactivated immune response attenuates the response to RDN, and in this case, the reduction in sympathetic activity would not have a substantial impact on blood pressure. It is noteworthy that we only measured the T-cell subsets in the peripheral blood and not in the target organs. Further studies are necessary to investigate the differences between the abundance and function of activated T cells in the blood and target organs [25]. In any case, it seems feasible that an analysis of immune cell signatures before RDN may provide additional information on the success rates of RDN.

The present study has several strengths and limitations. Although we present results on lymphocyte subsets from patients with known high sympathetic activity, our study population is small, and we used an observational design. Due to ongoing interfering antihypertensive medication, we did not measure aldosterone, renin or norepinephrine during the study. The study lacks an age-matched healthy control group, which would decrease the possibility of biased associations. Furthermore, although the observed changes in the T-cell subsets show a direct association between RDN and T-cell signatures, we cannot exclude that the changes in T-cell subsets are a result of RDN-induced blood pressure reductions. However, the strength of the study is that, for the first time, we demonstrate changes in T-cell subsets following the RDN procedure. Secondly, all patients received 24 h ambulatory blood pressure measurements before and after the treatment, allowing for an objective evaluation of the RDN effect. Moreover, the study was performed in one center, ensuring a strict work-up and follow-up schedule, as well as consistent and uniform pre-analytical sample processing.

To conclude, we were able to confirm the hypothesis derived from animal studies indicating that there is an interplay between sympathetic overactivity and immune response. A specific T-cell signature leads to higher pro-inflammatory activity, which contributes to the persistence of high blood pressure in patients with RH. A thorough analysis of the immune signature might be of advantage to improve patient selection for RDN.

4. Materials and Methods

4.1. Study Population

A total of 30 patients with resistant hypertension, which was defined according to ESH guidelines, were included in this prospective observational study [8]. The other inclusion criteria were age older than 18 years, no presence of acute infection and no presence of advanced chronic renal failure with an estimated glomerular filtration rate (eGFR) below 30 mL/min/1.73m^2, and the ability to give informed consent for participation in the study. All the patients with hypertension were on stable antihypertensive treatment for at least 4 weeks and had undergone a diagnostic evaluation for secondary hypertension, which was excluded in all the patients included in the study. In addition to this, 17 healthy controls were included in the study. These subjects had no history of any cardiovascular disease and demonstrated a systolic blood pressure below 130 mmHg without any antihypertensive drugs. This study was approved by the local ethics committee of the Medical Faculty at the Heinrich-Heine University, Düsseldorf, Germany (study number 3848 and 5365R), and it is in line with the Declaration of Helsinki, as revised in 2013.

4.2. Renal Denervation Procedure

Bilateral renal denervation (RDN) was performed for the patients with resistant hypertension using a radiofrequency ablation catheter (Symplicity Flex ® or Symplicity Spiral ®, Medtronic, Palo Alto, CA, USA) as described previously [51]. Briefly, the ablation catheter was placed via a. femoralis communis and positioned in the renal artery using X-ray guidance. After that, the electrode(s) were heated to about 55–65 degrees Celsius, a temperature high enough to induce nerve fiber damage within the artery wall. In the cases where a Symplicity Flex catheter was used, the catheter was withdrawn 5mm after each ablation to ensure circular ablation within the renal artery, as the catheter uses one radiofrequency electrode. The Symplicity Spiral catheter, however, has a helical shape, and the catheter consists of four electrodes enabling ablation at multiple points simultaneously. The procedure was performed in both renal arteries, and 4–6 points were ablated per renal artery.

4.3. Follow-Up of Patients with RH after Renal Denervation Procedure

The patients with resistant hypertension were followed-up for 6 months. Before RDN and 6 months after the procedure, a 24 h ambulatory blood pressure measurement (Mobil-O-Graph NG, Struck Medizintechnik GmbH, Enger Germany), as well as automated office blood pressure measurement (Boso Medicus Vollautomat, Bosch+Sohn GmbH & Co., Jungingen, Germany), was performed, and blood samples were drawn. According to the blood pressure measurements, the patients were divided into two groups: responders and non-responders. The patients with a reduction in systolic automated office blood pressure of at least 10 mmHg or more were defined as responders.

4.4. Measurement of Inflammation Markers

The concentrations of inflammatory markers in patients' serum were measured at two time points—before and 6 months after RDN. In addition, the same inflammation markers were measured in the 17 healthy controls. The samples were tested for interleukin 6 (IL-6), tumor necrosis factor-α (TNF-α) and high-sensitivity C-reactive protein (hsCRP) using an ELISA kit according to the manufacturer's instructions (Die Quantikine® Human CRP/TNF-α/IL-6 ELISA Kit, R&D Systems Inc., Minneapolis, MN, USA).

4.5. Measurement of T-Cell Signature

The subsets of T cytotoxic (CD8+) and T helper (CD4+) cells were analyzed in the peripheral blood mononuclear cells (PBMCs) of patients with RH before and 6 months after the treatment. Blood samples were drawn using Ficoll (CPT cell preparation tubes, Becton, Dickinson and Company, Franklin Lakes, NJ, USA) tubes. PBMC isolation and freezing were performed as described in the literature [52]. An analysis of the T-cell signatures was performed after all patient samples were collected. For the flowcytometric (FACS) measurements, the frozen PBMCs aliquots were thawed and stained according to recommendations [53]. An antibody cocktail consisting of anti-CD45, anti-CD3, anti-CD4, anti-CD8, anti-CD45RA and anti-CCR7 was added to the thawed PBMCs (Supplementary Table S1). The cell populations of the PBMCs were separated via FSC (forward-angle light scatter) and SSC (side-angle light scatter). Leukocytes were identified using CD45 staining, and T cells were identified using CD3 antibodies. Further gating was performed using CD4 and CD8 antibodies. According to the expressions of the surface markers CD45RA and CCR7, CD4+ and CD8+ cells were divided into T central memory cells (T_{CM} = CD45RA−, CCR7+), T effector memory cells (T_{EM} = CD45RA−, CCR7−), T effector memory cells (T_{EMRA} = CD45RA+, CCR7−) and T naïve cells ($T_{naïve}$ = CD45RA+, CCR7+), as presented in Supplementary Figure S2. According to previously published protocols, intracellular staining was performed [54]. In short, T cells were sorted using the antibodies listed in Supplementary Table S1 utilizing a high-speed digital cell sorter (Beckton Dickinson and Company) and re-stimulated for 4 h at 37 °C, 5% CO_2 in RPMI-1640 medium (Sigma Aldrich, St. Louis, MI, USA) supplemented with 10% fetal calf serum, 100 U/mL penicillin (Sigma Aldrich), 100 mg/mL streptomycin (Sigma Aldrich), 50 ng/mL PMA (Sigma Aldrich), 250 ng/mL ionomycin (Sigma Aldrich) and 0.65 µL/mL Golgistop (BD, Franklin Lakes, NJ, USA). The restimulated cells were fixed and permeabilized with a Foxp3/Transcription Factor Staining Buffer kit (eBioscience, San Diego, CA, USA) according to the recommendations of the manufacturer. Intracellular cytokines were detected using antibodies against human tumor necrosis factor alfa (TNF-α) and interferon gamma (IFN-γ), as presented in Supplementary Figure S3. An analysis of T-cell subsets was carried out using flow cytometry (CytoFlex®, Beckman Coulter, Brea, CA, USA) and the software Kaluza ®® (Beckmann Coulter).

4.6. Data Analysis

A statistical analysis was performed using SPSS version 23 (SPSS Inc., Chicago, IL, USA) and Graph Prism 8 (GraphPad Software, San Diego, CA, USA). The type of data distribution was assessed, and the continuous variables are expressed as mean ± standard deviation (SD) or median with the interquartile range expressed as two numbers, Q1–Q3. Categorical variables are expressed as numbers (percentages). The differences between the groups were assessed using unpaired or paired *t*-test, Mann–Whitney test or Wilcoxon rank test where appropriate. Univariate logistic regression was used to indicate variables associated with a positive response to renal denervation. p values less than 0.05 were considered statistically significant.

Supplementary Materials: The following supporting information can be downloaded at https://www.mdpi.com/article/10.3390/ijms24032493/s1, Figure S1: Comparison of T-cell subsets among responders with regard to sex. Table S1: Antibodies used for extracellular and intracellular lymphocyte staining and their specifications. Figure S2: Gating strategy on T-cell memory phenotypes. Figure S3: Gating strategy on T-cell effector memory residential cells and their intracellular cytokines.

Author Contributions: Conceptualization: O.V., D.N.M., L.C.R. and J.S.; methodology: M.Y., J.C.F. and A.M.; formal analysis: M.K., M.Y., P.H., A.M., B.D. and J.C.F.; investigation: M.Y., P.H., A.M., C.S., B.D., P.M., L.S. and G.A.; writing—original draft preparation: M.K. and J.S.; writing—review and editing: J.C.F., D.N.M., O.V., P.M., L.S., G.A., S.A.P., U.Z. and L.C.R.; supervision: L.C.R. and J.S. All authors have read and agreed to the published version of the manuscript.

Funding: This work was supported by the Erben Felder Foundation, Heinrich-Heine University Düsseldorf and the German Research Society (DFG STE2042/2-1).

Institutional Review Board Statement: This study was approved by the local ethics committee of the Medical Faculty at the Heinrich-Heine University, Düsseldorf, Germany (study number 3848 and 5365R), and it is in line with the Declaration of Helsinki, as revised in 2013.

Informed Consent Statement: Informed consent was obtained from all subjects involved in the study.

Data Availability Statement: Raw statistical data can be provided upon request. Please contact the corresponding author for more information.

Acknowledgments: We are thankful to all the members of the Department of Nephrology, University Hospital Düsseldorf for the in- and out-patient care and the organization of control visits.

Conflicts of Interest: The authors declare no conflict of interest.

References

1. Zhou, B.; Carrillo-Larco, R.M.; Danaei, G.; Riley, L.M.; Paciorek, C.J.; Stevens, G.A.; Gregg, E.W.; Bennett, J.E.; Solomon, B.; Singleton, R.K.; et al. Worldwide trends in hypertension prevalence and progress in treatment and control from 1990 to 2019: A pooled analysis of 1201 population-representative studies with 104 million participants. *Lancet* **2021**, *398*, 957–980. [CrossRef] [PubMed]
2. Kearney, P.M.; Whelton, M.; Reynolds, K.; Muntner, P.; Whelton, P.K.; He, J. Global burden of hypertension: Analysis of worldwide data. *Lancet* **2005**, *365*, 217–223. [CrossRef] [PubMed]
3. Zhou, D.; Xi, B.; Zhao, M.; Wang, L.; Veeranki, S.P. Uncontrolled hypertension increases risk of all-cause and cardiovascular disease mortality in US adults: The NHANES III Linked Mortality Study. *Sci. Rep.* **2018**, *8*, 9418. [CrossRef]
4. Dai, H.; Bragazzi, N.L.; Younis, A.; Zhong, W.; Liu, X.; Wu, J.; Grossman, E. Worldwide Trends in Prevalence, Mortality, and Disability-Adjusted Life Years for Hypertensive Heart Disease from 1990 to 2017. *Hypertension* **2021**, *77*, 1223–1233. [CrossRef]
5. Wei, J.; Galaviz, K.I.; Kowalski, A.J.; Magee, M.J.; Haw, J.S.; Narayan, K.M.V.; Ali, M.K. Comparison of Cardiovascular Events among Users of Different Classes of Antihypertension Medications: A Systematic Review and Network Meta-analysis. *JAMA Netw. Open* **2020**, *3*, e1921618. [CrossRef]
6. Hopstock, L.A.; Eggen, A.E.; Lochen, M.L.; Mathiesen, E.B.; Nilsen, A.; Njolstad, I.; Wilsgaard, T. Blood pressure target achievement and antihypertensive medication use in women and men after first-ever myocardial infarction: The Tromsø Study 1994–2016. *Open Hear.* **2018**, *5*, e000746. [CrossRef] [PubMed]
7. Banegas, J.R.; López-García, E.; Dallongeville, J.; Guallar, E.; Halcox, J.P.; Borghi, C.; Massó-González, E.L.; Jiménez, F.J.; Perk, J.; Steg, P.G.; et al. Achievement of treatment goals for primary prevention of cardiovascular disease in clinical practice across Europe: The EURIKA study. *Eur. Heart J.* **2011**, *32*, 2143–2152. [CrossRef]
8. Williams, B.; Mancia, G.; Spiering, W.; Rosei, E.A.; Azizi, M.; Burnier, M.; Clement, D.L.; Coca, A.; de Simone, G.; Dominiczak, A.; et al. 2018 ESC/ESH Guidelines for the management of arterial hypertension. *Eur. Heart J.* **2018**, *39*, 3021–3104. [CrossRef]
9. Cardoso, C.R.L.; Salles, G.F. Refractory hypertension and risks of adverse cardiovascular events and mortality in patients with resistant hypertension: A prospective cohort study. *J. Am. Heart Assoc.* **2020**, *9*, e017634. [CrossRef]
10. Kaczmarski, K.R.; Sozio, S.M.; Chen, J.; Sang, Y.; Shafi, T. Resistant hypertension and cardiovascular disease mortality in the US: Results from the National Health and Nutrition Examination Survey (NHANES). *BMC Nephrol.* **2019**, *20*, 138. [CrossRef]
11. Grassi, G. The Sympathetic Nervous System in Hypertension: Roadmap Update of a Long Journey. *Am. J. Hypertens.* **2021**, *34*, 1247–1254. [CrossRef] [PubMed]
12. Tsioufis, C.; Kordalis, A.; Flessas, D.; Anastasopoulos, I.; Tsiachris, D.; Papademetriou, V.; Stefanadis, C. Pathophysiology of resistant hypertension: The role of sympathetic nervous system. *Int. J. Hypertens.* **2011**, *2011*, 642416. [CrossRef]
13. Hering, L.; Rahman, M.; Potthoff, S.A.; Rump, L.C.; Stegbauer, J. Role of α2-Adrenoceptors in Hypertension: Focus on Renal Sympathetic Neurotransmitter Release, Inflammation, and Sodium Homeostasis. *Front. Physiol.* **2020**, *11*, 566871. [CrossRef] [PubMed]
14. Hering, L.; Rahman, M.; Hoch, H.; Markó, L.; Yang, G.; Reil, A.; Yakoub, M.; Gupta, V.; Potthoff, S.A.; Vonend, O.; et al. α2A-Adrenoceptors modulate renal sympathetic neurotransmission and protect against hypertensive kidney disease. *J. Am. Soc. Nephrol.* **2020**, *31*, 783–798. [CrossRef]
15. Saxena, M.; Schmieder, R.E.; Kirtane, A.J.; Mahfoud, F.; Daemen, J.; Basile, J.; Lurz, P.; Gosse, P.; Sanghvi, K.; Fisher, N.D.L.; et al. Predictors of blood pressure response to ultrasound renal denervation in the RADIANCE-HTN SOLO study. *J. Hum. Hypertens.* **2022**, *36*, 629–639. [CrossRef]
16. Fengler, K.; Reimann, P.; Rommel, K.P.; Kresoja, K.P.; Blazek, S.; Unterhuber, M.; Besler, C.; von Roeder, M.; Böhm, M.; Desch, S.; et al. Comparison of long-term outcomes for responders versus non-responders following renal denervation in resistant hypertension. *J. Am. Heart Assoc.* **2021**, *10*, e022429. [CrossRef]

17. Azizi, M.; Sanghvi, K.; Saxena, M.; Gosse, P.; Reilly, J.P.; Levy, T.; Rump, L.C.; Persu, A.; Basile, J.; Bloch, M.J.; et al. Ultrasound renal denervation for hypertension resistant to a triple medication pill (RADIANCE-HTN TRIO): A randomised, multicentre, single-blind, sham-controlled trial. *Lancet* **2021**, *397*, 2476–2486. [CrossRef] [PubMed]
18. Burlacu, A.; Brinza, C.; Floria, M.; Stefan, A.E.; Covic, A.; Covic, A. Predicting Renal Denervation Response in Resistant High Blood Pressure by Arterial Stiffness Assessment: A Systematic Review. *J. Clin. Med.* **2022**, *11*, 4837. [CrossRef]
19. Carnagarin, R.; Matthews, V.; Zaldivia, M.T.K.; Peter, K.; Schlaich, M.P. The bidirectional interaction between the sympathetic nervous system and immune mechanisms in the pathogenesis of hypertension. *Br. J. Pharmacol.* **2019**, *176*, 1839–1852. [CrossRef]
20. Ziegler, K.A.; Ahles, A.; Wille, T.; Kerler, J.; Ramanujam, D.; Engelhardt, S. Local sympathetic denervation attenuates myocardial inflammation and improves cardiac function after myocardial infarction in mice. *Cardiovasc. Res.* **2018**, *114*, 291–299. [CrossRef]
21. Xiao, L.; Kirabo, A.; Wu, J.; Saleh, M.A.; Zhu, L.; Wang, F.; Takahashi, T.; Loperena, R.; Foss, J.D.; Mernaugh, R.L.; et al. Renal denervation prevents immune cell activation and renal inflammation in Angiotensin II-induced hypertension. *Circ. Res.* **2015**, *117*, 547–557. [CrossRef] [PubMed]
22. Marvar, P.J.; Thabet, S.R.; Guzik, T.J.; Lob, H.E.; McCann, L.A.; Weyand, C.; Gordon, F.J.; Harrison, D.G. Central and peripheral mechanisms of T-lymphocyte activation and vascular inflammation produced by angiotensin II-induced hypertension. *Circ. Res.* **2010**, *107*, 263–270. [CrossRef] [PubMed]
23. Werdan, K. The activated immune system in congestive heart failure-from dropsy to the cytokine paradigm. *J. Intern. Med.* **1998**, *243*, 87–92.
24. Slota, C.; Shi, A.; Chen, G.; Bevans, M.; Weng N ping. Norepinephrine preferentially modulates memory CD8 T cell function inducing inflammatory cytokine production and reducing proliferation in response to activation. *Brain Behav. Immun.* **2015**, *46*, 168–179. [CrossRef]
25. Xiao, L.; do Carmo, L.S.; Foss, J.D.; Chen, W.; Harrison, D.G. Sympathetic enhancement of memory T-cell homing and hypertension sensitization. *Circ. Res.* **2020**, *126*, 708–721. [CrossRef] [PubMed]
26. Trott, D.W.; Thabet, S.R.; Kirabo, A.; Saleh, M.A.; Itani, H.; Norlander, A.E.; Wu, J.; Goldstein, A.; Arendshorst, W.J.; Madhur, M.S.; et al. Oligoclonal CD8+ T cells play a critical role in the development of hypertension. *Hypertension* **2014**, *64*, 1108–1115. [CrossRef] [PubMed]
27. Chen, J.; Bundy, J.D.; Hamm, L.L.; Hsu, C.Y.; Lash, J.; Miller, E.R.; Thomas, G.; Cohen, D.L.; Weir, M.R.; Raj, D.S.; et al. Inflammation and apparent treatment-resistant hypertension in patients with chronic kidney disease: The results from the CRIC study. *Hypertension* **2019**, *73*, 785–793. [CrossRef]
28. Chamarthi, B.; Williams, G.H.; Ricchiuti, V.; Srikumar, N.; Hopkins, P.N.; Luther, J.M.; Jeunemaitre, X.; Thomas, A. Inflammation and hypertension: The interplay of interleukin-6, dietary sodium, and the renin-angiotensin system in humans. *Am. J. Hypertens.* **2011**, *24*, 1143–1148. [CrossRef]
29. Delgado-Silva, J.; Rodrigues-Santos, P.; Almeida, J.S.; Santos-Rosa, M.; Gonçalves, L. Dynamics of Soluble Factors and Double-Negative T Cells Associated with Response to Renal Denervation in Resistant Hypertension Patients. *J. Pers. Med.* **2022**, *12*, 343. [CrossRef]
30. Rodriguez, I.J.; Lalinde Ruiz, N.; Llano León, M.; Martínez Enríquez, L.; Montilla Velásquez, M.D.P.; Ortiz Aguirre, J.P.; Rodríguez Bohórquez, O.M.; Velandia Vargas, E.A.; Hernández, E.D.; Parra López, C.A. Immunosenescence Study of T Cells: A Systematic Review. *Front. Immunol.* **2021**, *11*, 3460. [CrossRef]
31. Olson, N.C.; Doyle, M.F.; Jenny, N.S.; Huber, S.A.; Psaty, B.M.; Kronmal, R.A.; Tracy, R.P. Decreased Naive and Increased Memory CD4+ T Cells Are Associated with Subclinical Atherosclerosis: The Multi-Ethnic Study of Atherosclerosis. *PLoS ONE* **2013**, *8*, e71498. [CrossRef]
32. De Ciuceis, C.; Rossini, C.; Airò, P.; Scarsi, M.; Tincani, A.; Tiberio, G.A.M.; Piantoni, S.; Porteri, E.; Solaini, L.; Duse, S.; et al. Relationship between different subpopulations of circulating CD4+ T-lymphocytes and microvascular structural alterations in humans. *Am. J. Hypertens.* **2017**, *30*, 51–60. [CrossRef] [PubMed]
33. Rattik, S.; Engelbertsen, D.; Wigren, M.; Ljungcrantz, I.; Östling, G.; Persson, M.; Fredrikson, G.N.; Bengtsson, E.; Nilsson, J.; Björkbacka, H. Elevated circulating effector memory T cells but similar levels of regulatory T cells in patients with type 2 diabetes mellitus and cardiovascular disease. *Diabetes Vasc. Dis. Res.* **2019**, *16*, 270–280. [CrossRef]
34. Kundu, S.; Freiberg, M.S.; Tracy, R.P.; So-Armah, K.A.; Koethe, J.R.; Duncan, M.S.; Tindle, H.A.; Beckman, J.A.; Feinstein, M.J.; McDonnell, W.J.; et al. Circulating T Cells and Cardiovascular Risk in People with and Without HIV Infection. *J. Am. Coll. Cardiol.* **2022**, *80*, 1633–1644. [CrossRef] [PubMed]
35. Padgett, L.E.; Dinh, H.Q.; Wu, R.; Gaddis, D.E.; Araujo, D.J.; Winkels, H.; Nguyen, A.; McNamara, C.A.; Hedrick, C.C. Naive CD8+T Cells Expressing CD95 Increase Human Cardiovascular Disease Severity. *Arterioscler. Thromb. Vasc. Biol.* **2020**, *40*, 2845–2859. [CrossRef] [PubMed]
36. Lang, D.; Nahler, A.; Lambert, T.; Grund, M.; Kammler, J.; Kellermair, J.; Blessberger, H.; Kypta, A.; Steinwender, C.; Auer, J. Anti-Inflammatory Effects and Prediction of Blood Pressure Response by Baseline Inflammatory State in Catheter-Based Renal Denervation. *J. Clin. Hypertens.* **2016**, *18*, 1173–1179. [CrossRef] [PubMed]
37. Fengler, K.; Rommel, K.P.; Blazek, S.; Von Roeder, M.; Besler, C.; Hartung, P.; Desch, S.; Thiele, H.; Lurz, P. Predictors for profound blood pressure response in patients undergoing renal sympathetic denervation. *J. Hypertens.* **2018**, *36*, 1578–1584. [CrossRef]
38. Balasubramanian, P.; Hall, D.; Subramanian, M. Sympathetic nervous system as a target for aging and obesity-related cardiovascular diseases. *GeroScience* **2019**, *41*, 13–24. [CrossRef]

39. Zaldivia, M.T.K.; Rivera, J.; Hering, D.; Marusic, P.; Sata, Y.; Lim, B.; Eikelis, N.; Lee, R.; Lambert, G.W.; Esler, M.D.; et al. Renal Denervation Reduces Monocyte Activation and Monocyte-Platelet Aggregate Formation: An Anti-Inflammatory Effect Relevant for Cardiovascular Risk. *Hypertension* **2017**, *69*, 323–331. [CrossRef]
40. Dörr, O.; Liebetrau, C.; Möllmann, H.; Mahfoud, F.; Ewen, S.; Gaede, L.; Troidl, C.; Hoffmann, J.; Busch, N.; Laux, G.; et al. Beneficial effects of renal sympathetic denervation on cardiovascular inflammation and remodeling in essential hypertension. *Clin. Res. Cardiol.* **2015**, *104*, 175–184. [CrossRef]
41. Fliser, D.; Buchholz, K.; Haller, H. Antiinflammatory effects of angiotensin II subtype 1 receptor blockade in hypertensive patients with microinflammation. *Circulation* **2004**, *110*, 1103–1107. [CrossRef] [PubMed]
42. Awad, K.; Zaki, M.M.; Mohammed, M.; Banach, M. Effect of the renin angiotensin system inhibitors on inflammatory markers: A systematic review and meta-analysis of 32 randomized controlled trials. *Eur. Heart J.* **2021**, *42* (Suppl. S1), ehab724.2383. [CrossRef]
43. Caccamo, N.; Joosten, S.A.; Ottenhoff, T.H.M.; Dieli, F. Atypical Human Effector/Memory CD4+ T Cells with a Naive-Like Phenotype. *Front. Immunol.* **2018**, *9*, 2832. [CrossRef] [PubMed]
44. Liu, Y.; Rafferty, T.M.; Rhee, S.W.; Webber, J.S.; Song, L.; Ko, B.; Hoover, R.S.; He, B.; Mu, S. CD8+ T cells stimulate Na-Cl co-transporter NCC in distal convoluted tubules leading to salt-sensitive hypertension. *Nat. Commun.* **2017**, *8*, 14037. [CrossRef]
45. Carnevale, D.; Carnevale, L.; Perrotta, S.; Pallante, F.; Migliaccio, A.; Iodice, D.; Perrotta, M.; Lembo, G. Chronic 3D Vascular-Immune Interface Established by Coculturing Pressurized Resistance Arteries and Immune Cells. *Hypertension* **2021**, *78*, 1648–1661. [CrossRef]
46. Kirabo, A.; Fontana, V.; De Faria, A.P.C.; Loperena, R.; Galindo, C.L.; Wu, J.; Bikineyeva, A.T.; Dikalov, S.; Xiao, L.; Chen, W.; et al. DC isoketal-modified proteins activate T cells and promote hypertension. *J. Clin. Investig.* **2014**, *124*, 4642–4656. [CrossRef] [PubMed]
47. Henegar, J.R.; Zhang, Y.; Hata, C.; Narciso, I.; Hall, M.E.; Hall, J.E. Catheter-based radiofrequency renal denervation: Location effects on renal norepinephrine. *Am. J. Hypertens.* **2015**, *28*, 909–914. [CrossRef]
48. Tzafriri, A.R.; Mahfoud, F.; Keating, J.H.; Spognardi, A.M.; Markham, P.M.; Wong, G.; Highsmith, D.; O'Fallon, P.; Fuimaono, K.; Edelman, E.R. Procedural and anatomical determinants of multielectrode renal denervation efficacy: Insights from preclinical models. *Hypertension* **2019**, *74*, 546–554. [CrossRef]
49. Cohen-Mazor, M.; Mathur, P.; Stanley, J.R.L.; Mendelsohn, F.O.; Lee, H.; Baird, R.; Zani, B.G.; Markham, P.M.; Rocha-Singh, K. Evaluation of renal nerve morphological changes and norepinephrine levels following treatment with novel bipolar radiofrequency delivery systems in a porcine model. *J. Hypertens.* **2014**, *32*, 1678–1692. [CrossRef]
50. Madhur, M.S.; Harrison, D.G. Senescent T cells and hypertension: New ideas about old cells. *Hypertension* **2013**, *62*, 13–15. [CrossRef]
51. Mahfoud, F.; Schmieder, R.; Davies, J.; Kandzari, D.E.; Weil, J.; Whitbourn, R. Catheter-based renal sympathetic denervation—Long-term symplicity^{TM} renal denervation clinical evidence, new data and future perspectives. *Interv. Cardiol. Rev. Res. Resour.* **2013**, *8*, 118–123.
52. Puleo, A.; Carroll, C.; Maecker, H.; Gupta, R. Isolation of PBMCs Using Vacutainer®Cellular Preparation Tubes (CPTTM). *Bio-Protocol* **2017**, *7*, e2103. [CrossRef]
53. Barcelo, H.; Faul, J.; Crimmins, E.; Thyagarajan, B. A Practical Cryopreservation and Staining Protocol for Immunophenotyping in Population Studies. *Curr. Protoc. Cytom.* **2018**, *84*, 35. [CrossRef] [PubMed]
54. Gupta, S.; Maecker, H. Intracellular Cytokine Staining (ICS) on Human Lymphocytes or Peripheral Blood Mononuclear Cells (PBMCs). *Bio-Protocol* **2015**, *5*, e1442. [CrossRef] [PubMed]

International Journal of
Molecular Sciences

MDPI

Article

Moxonidine Increases Uptake of Oxidised Low-Density Lipoprotein in Cultured Vascular Smooth Muscle Cells and Inhibits Atherosclerosis in Apolipoprotein E-Deficient Mice

Yutang Wang [1,*,†], Dinh Tam Nguyen [1,†], Jack Anesi [1], Ahmed Alramahi [1], Paul K. Witting [2], Zhonglin Chai [3], Abdul Waheed Khan [3], Jason Kelly [4], Kate M. Denton [5,6] and Jonathan Golledge [7,8]

1 Discipline of Life Science, Institute of Innovation, Science and Sustainability, Federation University Australia, Ballarat, VIC 3350, Australia
2 Molecular Biomedicine Theme, School of Medical Sciences, Faculty of Medicine and Health, Charles Perkins Centre, The University of Sydney, Sydney, NSW 2006, Australia
3 Department of Diabetes, Central Clinical School, Monash University, Melbourne, VIC 3004, Australia
4 Fiona Elsey Cancer Research Institute, Ballarat, VIC 3350, Australia
5 Department of Physiology, Monash University, Melbourne, VIC 3800, Australia
6 Cardiovascular Disease Program, Monash Biomedicine Discovery Institute, Monash University, Melbourne, VIC 3800, Australia
7 Queensland Research Centre for Peripheral Vascular Disease, College of Medicine and Dentistry, James Cook University, Townsville, QLD 4811, Australia
8 Department of Vascular and Endovascular Surgery, The Townsville University Hospital, Townsville, QLD 4814, Australia
* Correspondence: yutang.wang@federation.edu.au
† These authors contributed equally to this work.

Citation: Wang, Y.; Nguyen, D.T.; Anesi, J.; Alramahi, A.; Witting, P.K.; Chai, Z.; Khan, A.W.; Kelly, J.; Denton, K.M.; Golledge, J. Moxonidine Increases Uptake of Oxidised Low-Density Lipoprotein in Cultured Vascular Smooth Muscle Cells and Inhibits Atherosclerosis in Apolipoprotein E-Deficient Mice. *Int. J. Mol. Sci.* **2023**, *24*, 3857. https://doi.org/10.3390/ijms24043857

Academic Editor: David Della-Morte

Received: 29 November 2022
Revised: 7 February 2023
Accepted: 13 February 2023
Published: 14 February 2023

Abstract: This study aimed to investigate the effect of the sympatholytic drug moxonidine on atherosclerosis. The effects of moxonidine on oxidised low-density lipoprotein (LDL) uptake, inflammatory gene expression and cellular migration were investigated in vitro in cultured vascular smooth muscle cells (VSMCs). The effect of moxonidine on atherosclerosis was measured by examining aortic arch Sudan IV staining and quantifying the intima-to-media ratio of the left common carotid artery in apolipoprotein E-deficient ($ApoE^{-/-}$) mice infused with angiotensin II. The levels of circulating lipid hydroperoxides in mouse plasma were measured by ferrous oxidation-xylenol orange assay. Moxonidine administration increased oxidised LDL uptake by VSMCs via activation of α2 adrenoceptors. Moxonidine increased the expression of LDL receptors and the lipid efflux transporter ABCG1. Moxonidine inhibited mRNA expression of inflammatory genes and increased VSMC migration. Moxonidine administration to $ApoE^{-/-}$ mice (18 mg/kg/day) decreased atherosclerosis formation in the aortic arch and left common carotid artery, associated with increased plasma lipid hydroperoxide levels. In conclusion, moxonidine inhibited atherosclerosis in $ApoE^{-/-}$ mice, which was accompanied by an increase in oxidised LDL uptake by VSMCs, VSMC migration, ABCG1 expression in VSMCs and lipid hydroperoxide levels in the plasma.

Keywords: moxonidine; atherosclerosis; inflammation; cell migration

1. Introduction

Atherosclerosis is a progressive disease of large- and medium-sized arteries in which dyslipidaemia plays an instrumental role in the disease pathogenesis [1–3]. When an atherosclerotic plaque develops an unstable phenotype, it is prone to rupture, which is thought to be the mechanism of acute cardiovascular events such as myocardial infarction and stroke [1]. Despite tremendous advances in the medical management of atherosclerosis through lifestyle changes and pharmacological control of low-density lipoprotein (LDL) cholesterol and triglyceride levels, blood pressure, diabetes and thrombosis, atherosclerosis-associated morbidity and mortality remains the number one health threat in most countries [1]. This highlights the need to develop innovative therapeutic strategies to prevent atherosclerotic plaque formation and rupture.

The sympathetic nervous system governs the "fight-or-flight" response [4]. The main overall end effect of the system is to prepare the body for physical activity, while also affecting many organs (e.g., kidney [5]) and physiological functions (e.g., metabolism [6]). The sympathetic nervous system may affect atherosclerosis. For example, stress, which leads to the activation of the sympathetic nervous system [7,8], is a critical risk factor for atherogenesis [9]. The class of β blocker drugs inhibit sympathetic nervous activity, but their adverse effects on lipoprotein levels and glucose metabolism prohibit their use as bespoke inhibitors of atherosclerosis [10]; some new-generation β blockers have some favourable effects on lipoprotein and glucose metabolism, although their effect on atherosclerosis prevention remains unclear [10]. Renal sympathetic denervation has been reported to inhibit atherosclerosis in mice [11], although this is not without controversy [12]. Furthermore, renal sympathetic denervation may permanently damage renal nerves and could result in renal artery stenosis in some patients [13]. Therefore, there is a need to investigate the effect of the sympathetic nerve system on atherosclerosis with other non-invasive approaches such as the use of sympatholytic drugs.

It is unknown whether moxonidine (a sympatholytic drug) inhibits atherosclerosis, although the drug has some activities that may be linked to an anti-atherogenic role, such as the inhibition of inflammation [14,15]. This study aimed to investigate the effect of moxonidine on atherosclerosis.

Angiotensin II can promote atherosclerosis formation [16]. On the other hand, blocking angiotensin II formation by angiotensin-converting enzyme (ACE) inhibitors can inhibit atherogenesis [17] and prevent cardiovascular events (e.g., myocardial infarction, stroke and cardiovascular mortality) [18]. Therefore, the angiotensin-infusion-induced atherosclerosis model is a clinically relevant model to study atherosclerosis. In addition, angiotensin stimulates the sympathetic nervous system [19], and overactivation of the sympathetic nervous system is seen in many conditions associated with atherosclerosis such as myocardial infarction [20] and stroke [21]. Therefore, this study chose angiotensin II infusion as the model of atherosclerosis to investigate the effect of moxonidine (a sympatholytic drug) on atherosclerosis.

Oxidised LDL plays a role in foam cell formation [22], a hallmark of atherosclerosis. This study also aimed to investigate the effect of moxonidine on oxidised LDL uptake by cultured vascular smooth muscle cells (VSMCs), as recent evidence suggests that VSMCs may be a major pathway for foam cell formation [23]. For example, 50% of foam cells within human coronary artery lesions were derived from VSMCs, as indicated by the expression of the smooth muscle cell-specific marker SM α-actin [24]. In addition, in apolipoprotein E-deficient ($ApoE^{-/-}$) mice supplemented with a Western diet for 6 weeks, 70% of foam cells in aortic arch lesions were derived from VSMCs as indicated by the expression of smooth muscle cell-specific fluorescent proteins [25].

2. Results

2.1. *Moxonidine Increased Oxidised LDL Uptake via α2 Adrenoceptors by VSMCs In Vitro*

Incubation of cultured VSMCs with the sympathetic activator norepinephrine (0.1 μM) decreased the uptake of oxidised LDL (Figure 1). Consistently, incubation of the cells with the sympatholytic drug moxonidine increased the uptake of oxidised LDL as indicated by the microscopic analysis (Figure 2) and total cholesterol levels (Figure S1). Moxonidine is an agonist for both α_2 adrenoceptors and imidazoline I_1 receptors [26–28]. To study this moxonidine-mediated increase in oxidised LDL uptake further, a series of inhibitor and activator studies were performed. Notably, oxidised LDL uptake was reversed by co-incubation with RX821002 (α_2 adrenoceptor inhibitor, 10 μM) or efaroxan (inhibitor for both α_2 adrenoceptor and I_1 receptor, 10 μM) with the drugs employed at final doses known to specifically target their respective receptor [29,30]. By contrast, the uptake of oxidised LDL was not enhanced by activation of the I_1 receptor-specific activator AGN192403 [31] (100 μM) (Figure 2). Together, these observations suggest that moxonidine increased the uptake of oxidised LDL through the activation of α_2 adrenoceptors. Gene expression studies showed that in the presence of oxidised LDL, moxonidine (10 μM) increased gene expression of the LDL receptor but not the scavenger receptor (Figure 3A,B). In addition, moxonidine (10 μM) increased the expression of the lipid efflux gene ABCG1, but not ABCA1 (Figure 3C,D).

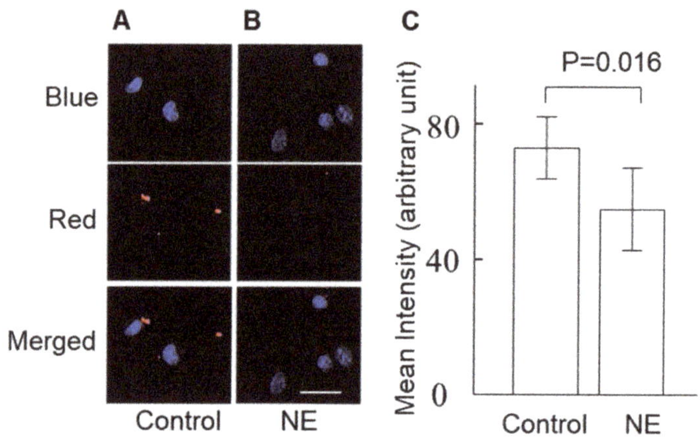

Figure 1. Norepinephrine (NE) increased oxidised low-density lipoprotein (LDL) uptake in vascular smooth muscle cells (VSMCs). VSMCs were incubated with NE (0.1 μM) or phosphate-buffered saline (PBS, control) for 30 min, and then the cells were treated with Dil-labelled oxidised LDL (25 μg/L) for 4 h. After staining the nucleus with Hoechst, the intracellular red fluorescence (Dil-labelled oxidised LDL) and blue fluorescence (nucleus) were imaged via a confocal microscope. (**A**,**B**) Representative images of oxidised LDL (red channel), nucleus (blue channel) and merged channels of cells in PBS-incubated (**A**) and NE-incubated cells (**B**). Scale bar = 40 μm. (**C**) The mean intensity of intracellular Dil-labelled oxidised LDL fluorescence. The mean intensity of Dil-labelled oxidised LDL fluorescence (red) in the cells was analysed using ImageJ. The difference between the groups was analysed using the Mann–Whitney U test. Data represent mean ± SD; N = 8.

2.2. *Moxonidine Inhibited Inflammatory Gene Expression in Lipopolysaccharide-Treated VSMCs and Endothelial Cells In Vitro*

Incubation of VSMCs with lipopolysaccharide increased inflammatory gene expression (Figure 4). Moxonidine (0.2 μM) significantly decreased mRNA expression of interleukin 1 (IL-1), monocyte chemoattractant protein-1 (MCP-1) and tumour necrosis factor-α (TNF-α) (Figure 4). However, moxonidine did not affect the mRNA expression of inflammatory markers in cultured endothelial cells (Figure S2).

2.3. *Moxonidine Enhanced VSMC Migration Without Affecting VSMC Proliferation In Vitro*

Incubation with moxonidine (10 μM) stimulated the migration of VSMCs by 42% (Figure 5). However, incubation with moxonidine at concentrations of 0.015, 0.15, 1.5, 15 and 150 μM for 24 h did not affect cell proliferation as assessed by the MTS assay (Figure S3). Corroborating results were obtained when VSMC proliferation was assessed by the trypan blue method (Figure S4).

2.4. *Moxonidine Decreased Atherosclerosis in Angiotensin-Infused ApoE$^{-/-}$ Mice*

To investigate the effect of moxonidine on atherosclerosis directly, we used angiotensin II infusion for 4 weeks to induce atherosclerosis formation in ApoE$^{-/-}$ mice. Compared with the control (without moxonidine), moxonidine administration (18 mg/kg/day) decreased atherosclerosis formation in the aortic arch as assessed by Sudan IV staining (Figure 6) as well as in the left common carotid artery as assessed by morphometry analysis (intimal/medial area ratio, Figure 6). Surprisingly, moxonidine administration increased lipid peroxide levels in the plasma of the mice (Figure 7). Association studies showed that higher lipid peroxide levels were associated with the presence of atherosclerosis [32], suggesting that lipid peroxide may be proatherogenic. Therefore, the plasma lipid peroxide data appeared in disagreement with the in vivo finding that moxonidine administration decreased atherosclerosis.

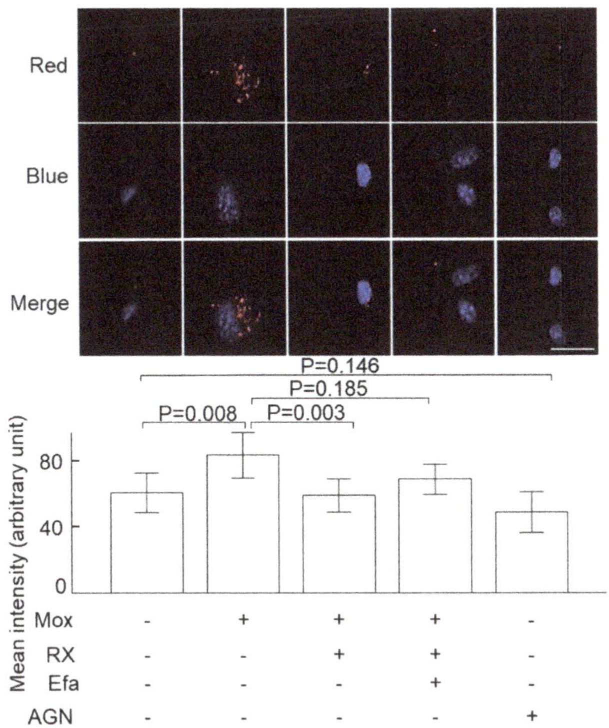

Figure 2. The effect of moxonidine on oxidised LDL uptake by VSMCs. VSMCs were incubated with RX821002 (10 μM, α2 adrenoceptor inhibitor), efaroxan (10 μM, α2 adrenoceptor and I1 receptor inhibitor) or AGN192403 (100 μM, I1 receptor activator) for 30 min and moxonidine (10 μM) or phosphate-buffered saline (PBS, control) was then added to the cells. After another 2 h, Dil-labelled oxidised LDL (25 μg/L) was added to all the cells and the red fluorescence (engulfed Dil-labelled oxidised LDL) inside of each cell was visualised by a confocal microscope after 4 h. The mean fluorescence intensity of the cells in the dish was calculated using ImageJ. The differences among groups were analysed using Kruskal–Wallis one-way ANOVA followed by Bonferroni's post hoc tests. Data represent mean ± SD; N = 8. Scale bar = 40 μm. AGN, AGN192403; Efa, efaroxan; Mox, moxonidine; RX, RX821002.

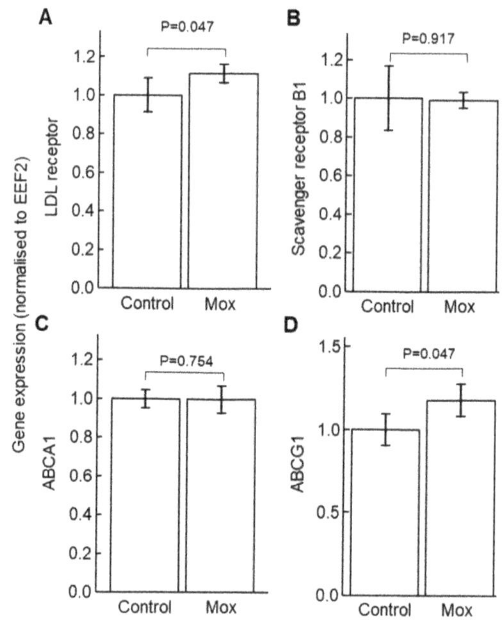

Figure 3. The effect of moxonidine on mRNA expression of genes related to lipid uptake and efflux. VSMCs were incubated with moxonidine (10 μM) or phosphate-buffered saline (PBS, control) for 2 h and the cells were then incubated with oxidised LDL (10 μg/L) for an additional 4 h. mRNA was extracted and the expressions of the LDL receptor (**A**), scavenger receptor B1 (**B**), ABCA1 (**C**) and ABCG1 (**D**) were quantified via quantitative PCR. The difference was analysed by Mann–Whitney U test. Data represent mean fold-change (vs. control) ± SD; N = 5. ABCA1, ATP binding cassette subfamily A member 1; ABCG1, ATP binding cassette subfamily G member 1; EEF2, eukaryotic elongation factor; LDL, low-density lipoprotein; Mox, moxonidine; VSMC, vascular smooth muscle cells.

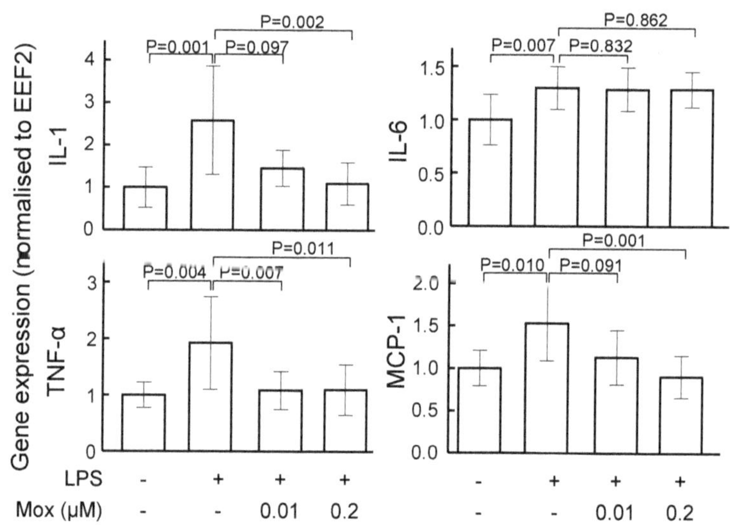

Figure 4. The effect of moxonidine on mRNA expression of inflammatory markers in VSMCs. Cells were incubated with lipopolysaccharide for 2 h in the absence or presence of moxonidine (0.01 or 0.2 μM). Then,

mRNA was isolated and the expressions of IL1, IL-6, MCP-1 and TNF-α were quantified via quantitative PCR. The difference was analysed by Kruskal–Wallis one-way ANOVA followed by Bonferroni's post hoc test. Data represent mean fold-change (vs. control) ± SD; N = 7–12. EEF2, eukaryotic elongation factor; IL, interleukin; LPS, lipopolysaccharide; MCP-1, monocyte chemoattractant protein-1; Mox, moxonidine; TNF-α, tumour necrosis factor-α; VSMC, vascular smooth muscle cells.

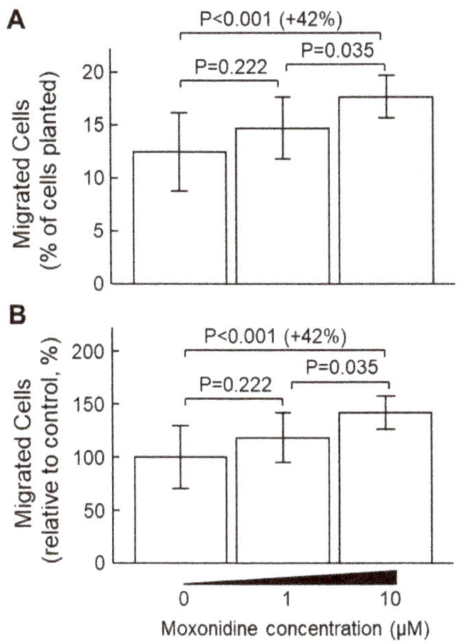

Figure 5. Effect of moxonidine on cellular migration of vascular smooth muscle cells (VSMCs). VSMCs in serum-free DMEM media were placed in the upper chamber of cell migration chambers, with or without moxonidine at the indicated concentrations. The lower chamber contained 150 μL of DMEM medium containing 20% foetal bovine serum. After 48 h incubation, the migrated cells in the lower chamber were stained and the fluorescence (an indicator of cell numbers) was measured (excitation/emission = 530/590 nm). (**A**) Migratory cells were represented as a percentage of the total number of cells planted in the upper chamber. (**B**) Migratory cell numbers relative to the untreated controls (0 μM moxonidine). The difference among the group was analysed using Kruskal–Wallis one-way ANOVA followed by Bonferroni's post hoc tests. Data represent mean ± SD; N = 11–12 per group.

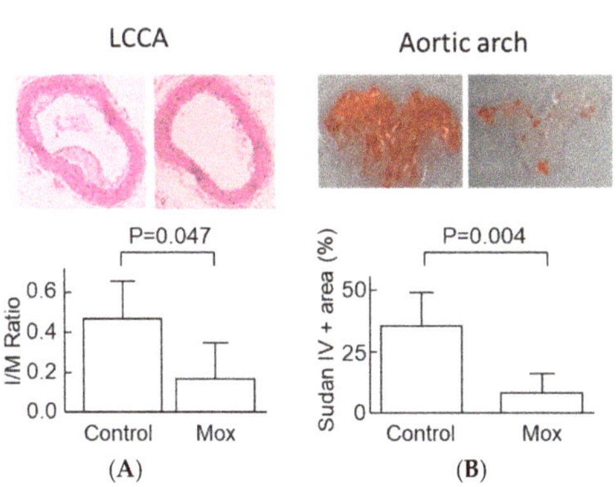

Figure 6. Effect of moxonidine on atherosclerosis in apolipoprotein E-deficient mice. (**A**) The left common carotid artery was stained with the H&E method. The ratio of the lesion area over the medial

area (I/M ratio) was calculated. Magnification = 10×. (**B**) The aortic arch was stained with Sudan IV to visualise the lipid in the lumen surface of the aortic arch. The difference between the two groups was analysed by the Mann–Whitney U test. Error bars represent SD; N = 6 per group. H&E, hematoxylin and eosin; LCCA, left common carotid artery; Mox, moxonidine.

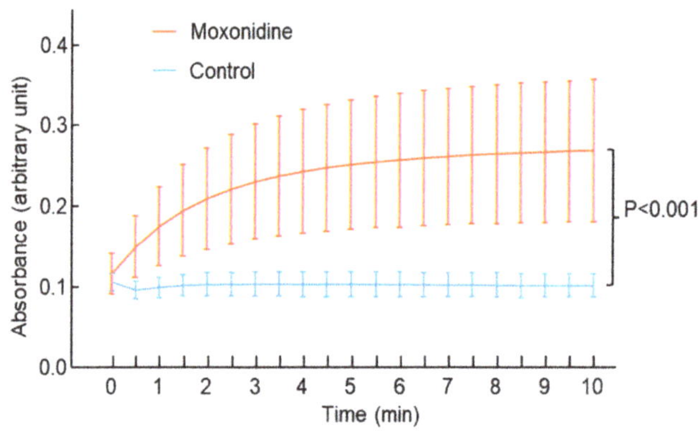

Figure 7. Effect of moxonidine administration on plasma lipid peroxide levels. Mice were treated with moxonidine (18 mg/kg body weight/day) via drinking water until the end of the experiment, while the control mice received normal drinking water only. Three days after the initiation of moxonidine administration, angiotensin II was subcutaneously infused into all the mice for 28 days to induce atherosclerosis. Mice were culled at the end of the angiotensin II infusion and plasma was collected. Lipid peroxide levels in the plasma were measured every 30 s. Higher absorbance = higher lipid peroxide levels. The difference in plasma lipid peroxide levels between the two groups was analysed using multiple linear regression: dependent variable = absorbance and independent variables = groups and time. Data represent mean ± standard error; N = 5 per group.

3. Discussion

This study found that the sympatholytic drug moxonidine increased the uptake of oxidised LDL, stimulated mRNA expression of the LDL receptor and the ABCG1 transporter, enhanced cell migration (but not proliferation) and inhibited inflammatory gene expression in vitro in cultured VSMCs. In addition, moxonidine administration in ApoE$^{-/-}$ mice inhibited atherosclerosis formation induced by angiotensin II infusion.

Moxonidine is a sympatholytic drug with blood pressure lowering properties [26]. Biological activity for moxonidine involves the deactivation of the sympathetic nervous system with parallel decreases in the plasma norepinephrine level [28,33]. Herein, this study found, for the first time, that moxonidine decreased atherosclerosis formation in a mouse model of atherosclerosis that was induced by subcutaneous infusion of angiotensin II. This mouse model of atherosclerosis is clinically relevant [16], as it mimics increased angiotensin signalling in patients with atherosclerosis and cardiovascular disease [18,20,21].

At four weeks (the end of the experiment), the body weight of the mice was similar between the moxonidine and control groups (Figure S5), and so was the body weight change during the experiment (Figure S6). Administration of moxonidine for 4 weeks did not change organ weight, including of the kidney, spleen and heart (Figure S7). In addition, the mice treated with moxonidine did not show apparent signs of illness during the experiment (e.g., diarrhoea, abnormal gait and abnormal breathing), suggesting that 4-week administration of moxonidine may not cause apparent toxicity. However, the toxicity associated with longer-term (>4 weeks) administration of moxonidine is unknown, as is the effect of moxonidine at the cerebral level. These questions need to be investigated in the future.

VSMCs in the intima have been considered beneficial for plaque stability, as VSMCs constitute the main cellular component of the protective fibrous cap within lesions and are responsible for synthesising extracellular matrix components that stabilise the cap [23,34]. Therefore, an increase in VSMC migration may be beneficial. The current study, for the first time, showed that incubation with moxonidine stimulated VSMC migration that potentially contributes to plaque stabilisation.

Atherosclerosis is an inflammatory disease [35,36]. Pro-inflammatory cytokines and chemokines (e.g., MCP-1 and TNF-α) play a key role in the initiation and progression of atherosclerosis [37]. The current study showed that moxonidine inhibited the lipopolysaccharide-induced increase in TNF-α expression. This is consistent with a previous report which demonstrated that moxonidine treatment in hypertensive postmenopausal women decreased circulating TNF-α levels [14]. We also showed for the first time that moxonidine decreased MCP-1 mRNA expression. Our results and those from the literature [14,15] support the notion that moxonidine has an anti-inflammatory effect, which might play an important role in mediating the anti-atherosclerotic effect of moxonidine.

Endothelial dysfunction is a major contributor to atherogenesis, and enhanced inflammation is a key mechanism underlying endothelial dysfunction [38,39]. Our results showed that treatment of endothelial cells with moxonidine did not affect the gene expression of inflammatory markers, including IL-1, IL-6, MCP-1 and TNF-α. This suggests that the effect of moxonidine may be cell type specific, i.e., moxonidine may target VSMCs rather than endothelial cells. This seems in agreement with the sympathetic innervation pattern of the vasculature. It is well known that arteries are innervated with sympathetic nerves [40] and the nerve endings are distributed in the smooth muscle layer but not the endothelial layer [41]. Therefore, VSMCs, rather than endothelial cells, may be the key target of moxonidine in blood vessels. However, the current study did not investigate the effect of moxonidine on endothelial function and this needs to be investigated in the future.

High levels of LDL are a risk factor for atherosclerosis formation. In localised intimal microenvironments, where antioxidant defences have been overwhelmed, LDL can be oxidised to initiate atherogenesis [42]. Although moxonidine does not affect plasma LDL levels [27,43] or subclass pattern nor oxidation susceptibility [43], it may inhibit oxidation of LDL in the inflammatory subendothelial space, as moxonidine inhibited inflammatory gene expression in VSMCs.

Oxidised LDL plays a significant role in atherogenesis. In the inflammatory intimal microenvironments, LDL can be oxidised and taken up by macrophages, leading to foam cell formation [44,45]. VSMCs are recently recognised as an important source of foam cells in atherosclerosis lesions [23]. For example, VSMC-derived foam cells accounted for 50% of foam cells within advanced human coronary artery lesions [24] and accounted for 70% of foams cells in the aortic arch lesions in apolipoprotein E-deficient ($ApoE^{-/-}$) mice [25].

Our results showed that moxonidine increased oxidised LDL uptake by activation of α2-adrenoceptors, which was associated with an increase in the mRNA expression of the LDL receptor. In addition, plasma lipid peroxide levels in the mice were increased after moxonidine treatment. These results suggest that moxonidine may be proatherogenic. However, this interpretation contradicted the in vivo finding that moxonidine decreased atherosclerosis, which is supported by the anti-inflammatory effect of moxonidine. The contradicting observations suggest that an alternative explanation may be needed.

The functions of the recently discovered VSMC-derived foam cells are poorly understood. These cells are believed to promote atherogenesis [24,25], because they showed a decrease in the lipid efflux transporter ABCA1 [24,25]. Moxonidine did not affect ABCA1 gene expression. Interestingly, it increased the expression of ABCG1, another key lipid efflux transporter. This is the first report that showed that gene expression of lipid efflux transporters could be increased in lipid-laden VSMCs. This suggests that moxonidine may change the phenotype of the VSMC-derived foam cells from accumulating oxidised LDL inside the cells to effluxing oxidised LDL out of the blood vessel to circulation for detoxification and elimination by the liver. This hypothesis fits the in vivo finding that moxonidine administration decreased atherosclerosis. In addition, this oxidised LDL efflux hypothesis is supported by the finding that moxonidine administration increased plasma lipid peroxide levels. Moreover, increased VSMC migration by moxonidine may also support this hypothesis, as VSMC migration to a favourable location could facilitate both uptake of oxidised LDL and its efflux back to circulation. This hypothesis seems to harmonise the observed contradicting results, but it is speculative in nature and needs to be investigated in the future.

Therefore, we propose the following mechanism to explain the atherosclerosis inhibitory effect of moxonidine: moxonidine inhibits the expression of inflammatory genes (e.g., IL-1, MCP-1 and TNF-α), which subsequently may inhibit the oxidation of LDL (Figure 8). In addition, moxonidine increases the uptake of oxidised LDL by VSMCs via activation of α_2-adrenoceptors. As moxonidine increases VSMC migration this may result in relocation of VSMCs to facilitate the uptake of oxidised LDL via the LDL receptor and efflux back to circulation via the ABCG1 transporter for detoxification and elimination by reverse cholesterol transport (Figure 8). This hypothesis warrants further investigation in the future.

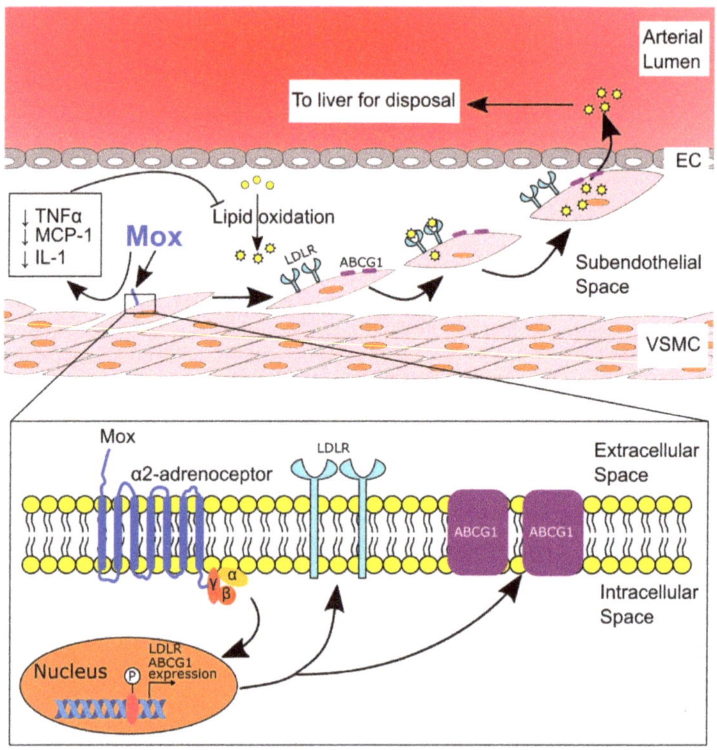

Figure 8. Hypothesis: proposed mechanism underlying moxonidine-induced inhibition of atherosclerosis. Moxonidine decreases the expression of inflammatory genes (e.g., TNF-α), which inhibit the oxidation of LDL. Moxonidine enhances VSMC migration. These VSMCs then migrate to a location that could facilitate both oxidised LDL uptake via the LDL receptor and its efflux back to circulation via the ABCG1 transporter for detoxification and elimination by the liver. Thus, moxonidine may inhibit atherosclerosis by inhibiting inflammation and promoting oxidised LDL clearance from the atherosclerotic plaque. ABCG1, ATP binding cassette subfamily G member 1; EC, endothelial cell; IL, interleukin; LDL, low-density lipoprotein; LDLR, low-density lipoprotein receptor; MCP-1, monocyte chemoattractant protein-1; Mox, moxonidine; TNF-α, tumour necrosis factor-α; VSMC, vascular smooth muscle cell.

4. Materials and Methods

4.1. Animals

Male ApoE$^{-/-}$ mice (3 months old) were purchased from the Animal Resources Centre, Perth, Australia. All experiments were conducted in a temperature-controlled animal house (21 ± 1 °C) under a 12:12 h light–dark cycle, and mice were given standard chow and water ad libitum.

4.2. Experimental Protocol

Twenty ApoE$^{-/-}$ mice were randomised into two groups according to their age and body weight (N = 10 per group): the control group and the moxonidine-treated group. The mice in the control group received plain drinking water, whereas mice in the moxonidine treatment group received moxonidine via drinking water (18 mg/kg/day) [46]. Three days after the initiation of the moxonidine administration, angiotensin II was administered subcutaneously to all the mice via a micro-osmotic pump (Model 2004, ALZET, Cupertino, CA, USA) at a rate of 1 µg/kg body weight/min for 28 days to induce atherosclerosis [12,16,47].

Four mice from each of the two groups died of aortic rupture, which is common in angiotensin II-infused mice [47,48]. The quality of the aortic tissue from these dead mice prevented them from inclusion in atherosclerosis assessment. Therefore, atherosclerosis was only assessed in the surviving animals (N = 6 per group).

4.3. Quantification of Atherosclerotic Lesion Area

Atherosclerosis in the aortic arch was quantified by en face staining as described previously [12]. Briefly, the aortic arch was opened longitudinally and pinned down on a wax-coated petri dish. Tissue samples were stained with 0.1% *w/v* Sudan IV for 10 min to identify areas of atherosclerosis. Sudan-IV-stained areas were quantified using ImageJ 1.53e and expressed as a percentage of the total aortic arch luminal surface area.

Atherosclerosis in the left common carotid artery was assessed using morphometry analysis [49]. In brief, formalin-fixed and paraffin-embedded left common carotid arteries were sectioned (5 µm thickness) starting from the labelled proximal end. The location where the first complete arterial structure (a circular structure) appeared was designated the location of 0 µm. Four serial arterial sections (at locations 0, 160, 320 and 480 µm) were subsequently obtained and stained with hematoxylin and eosin (H&E). Images of the stained sections were captured using a light microscope (Nikon, Tokyo, Japan). The area of the atherosclerotic lesion and medial area was quantified using Photoshop (Microsoft, version 22.0.0) and the corresponding intima-to-media area was calculated for each section. The average of the ratios from the 4 serial sections was assigned as the final measurement of atherosclerosis in that left common carotid artery.

4.4. Ferrous Oxidation-Xylenol Orange (FOX) Assay

A FOX assay was used to assess levels of lipid hydroperoxides in mouse plasma [50]. Briefly, 50 µL FOX solution (250 µM ammonium sulphate, 100 mM D-sorbitol and 125 µM xylenol orange) was added to 50 µL mouse plasma and the mixture was incubated at 20 °C. Absorbance was measured using a FLUOstar Omega reader (BMG LABTECH Pty. Ltd, Mornington, Australia) and a time-dependent change in output at 560 nm was determined at 30 s intervals with the linear rate of xylenol orange compared between different treatment groups.

4.5. Cell Culture

VSMCs were isolated from the mouse aorta [51] and cultured as previously described [52]. Briefly, VSMCs were cultured at 37 °C in Dulbecco's modified Eagle medium (DMEM) supplemented with 10% *v/v* foetal bovine serum, 100 U/mL penicillin and 100 µg/mL streptomycin in a cell culture incubator containing 5% CO_2. The cells were split with 0.05% *w/v* trypsin when they reached 80% confluency and subcultured for further passages.

Human aortic endothelial cells were purchased from Lonza Australia Pty Ltd. (Mount Waverley, Australia). These cells were cultured in EGM™-2 Endothelial Cell Growth Medium supplemented with growth factors required for culturing endothelial cells (Lonza Australia Pty Ltd.).

4.6. Confocal Microscopy for Oxidised LDL Uptake

VSMCs (5×10^5 cells) in DMEM + 1% foetal bovine serum were seeded in MatTek confocal dishes. Following 24 h of subculture, cells were then incubated separately with 10 μL of RX821002 (α_2 adrenoceptor inhibitor [29], final concentration = 10 μM), efaroxan (inhibitor for both α_2 adrenoceptor and I_1 receptor [30], final concentration = 10 μM) or AGN192403 (I_1 receptor activator [31], final concentration = 100 μM). The cells were equilibrated further for 30 min at 37 °C, and then 10 μL of moxonidine was added to each of the confocal dishes (final concentration = 10 μM) and the cells were further incubated for 2 h. Dil-labelled oxidised LDL (Thermo Fisher Scientific Australia Pty Ltd., Scoresby, Australia, final concentration = 25 μg/L) was added and the cells were incubated for another 4 h away from light. Next, 2 μL of Hoechst was added to the cell dishes for 10 min to stain the nuclei and the cells were washed with phosphate-buffered saline before the culture medium was replaced by a phenol-free DMEM. The red fluorescence (engulfed Dil-labelled oxidised LDL) was imaged by a confocal microscope and the mean fluorescence intensity of each cell on the image was measured using ImageJ. Eight images were taken for each dish at random locations across the whole dish. The mean intracellular fluorescence intensity from these 8 images was calculated as the final intracellular fluorescence intensity for that dish.

4.7. Gene Expression Analysis

Effect of moxonidine on inflammatory gene expression in VSMCs: VSMCs (6×10^5 cells per well) in 4 wells of 6-well plates were incubated with moxonidine (0, 0, 0.01 or 0.2 μM) for 12 h. Then, the cells were treated with lipopolysaccharide (0, 100, 100 or 100 ng/mL, respectively). After a further 2 h incubation, RNA was extracted using the TRI reagent (Merck, Bayswater, Australia).

Effect of moxonidine on the expression of genes related to lipid uptake and efflux: VSMCs (1×10^6 cells per well) in 6-well plates were cultured for 24 h and then incubated with moxonidine (10 μM) or phosphate-buffered saline (PBS, control) for 2 h. Oxidised LDL (10 μg/L) was then added to the cells which were cultured for an additional 4 h. Next, mRNA was extracted using the TRI-reagent (Merck).

Effect of moxonidine on inflammatory gene expression in endothelial cells: cells (2.5×10^5 cells per well) in 3 wells of 6-well plates were cultured for 48 h. Then, the cells were treated with moxonidine (0, 1 or 1 μM) for 2 h. Following this, the cells were treated with lipopolysaccharide (0, 100 or 100 ng/mL, respectively). After a further 2 h incubation, RNA was extracted using the TRI reagent (Merck).

The extracted RNA was reverse transcribed to cDNA using the High-Capacity Reverse Transcription Kit (Life Technologies, Carlsbad, CA, USA). Gene expression was assessed by quantitative PCR using SYBR reagents (Bioline Global Pty Ltd., Gregory Hills, Australia). Primer sets are outlined in Table S1. The cycling conditions were as follows: a hold at 95 °C for 2 min, followed by 40 cycles at 95 °C for 15 s, 58 °C for 20 s and 72 °C for 20 s. Relative gene expression was assessed using the $2^{-\Delta\Delta Ct}$ method [53]. Gene expression analysis was represented using relative gene expression compared with the control gene eukaryotic translation elongation factor 2 (EEF2) [12].

4.8. Migration Assay

Migration assay was conducted using a cell migration assay kit from Abcam (Cambridge, UK) according to the manufacturer's instructions. In brief, 50 μL of VSMCs (50,000 cells) in a serum-free DMEM were added to the top chamber in addition to 50 μL of serum-free DMEM containing 0, 2 or 20 μM moxonidine. The final concentration of moxonidine in the top chamber was 0, 1 or 10 μM. The bottom chamber contained 150 μL of DMEM + 20 % of foetal bovine serum per well. The cells were incubated in a CO_2 incubator at 37 °C for 48 h. The migrated cells in the lower chamber were washed and stained and the fluorescence (an indicator of cell numbers) was measured using a plate reader (excitation/emission = 530/590 nm).

4.9. Cell Proliferation

Cell proliferation was conducted using an MTS cell proliferation assay kit (Abcam) as previously described [51,54]. In brief, 200 μL of VSMCs (0.5×10^6 cells/mL) was added to each well of a 96-well flat-bottom plate and kept at 37 °C in an incubator overnight. Next, 2 μL of moxonidine at different concentrations was added to give a final concentration of 0, 0.015, 0.15, 1.5, 15 or 150 μM. After the cells were incubated for 24 h, 20 μL of MTS reagent was added to each well and the cells were incubated for another 2 h. Finally, the absorbance was recorded using a plate reader at 520 nm.

Cell proliferation was also assessed using the trypan blue method [51,54]. In brief, 2 mL of VSMCs (5×10^4 cells/mL) was placed in wells of 6-well plates and incubated in a 5% CO_2 incubator at 37 °C for 24 h. Then, various concentrations of moxonidine were added to give a final concentration of 0.01, 0.1, 1 or 10 μM. After 24 h of incubation, the cells were trypsinised, stained with trypan blue and then counted using Countess Automated Cell Counter (Invitrogen, Waltham, MA, USA).

4.10. Total Cholesterol

VSMCs (5×10^4 cells) were seeded in 96-well plates. Following 24 h of subculture, 2 μL of moxonidine (final concentration in the wells = 10 μM) or PBS was added, and the cells were further incubated for 2 h. Oxidised LDL (final concentration = 25 μg/L) was added to all the wells and the cells were incubated for another 4 h. The cells were washed and lysed and the supernatant was collected after centrifugation at 12,000 g for 10 min. Total cholesterol in the supernatant was then measured using a commercial kit from Abcam according to the manufacturer's instructions [55].

4.11. Statistical Analyses

The difference between two groups was analysed using Mann–Whitney U test [56] and the difference among multiple groups was analysed using Kruskal–Wallis one-way AVOVA. The difference in plasma lipid peroxide levels between two groups (with or without moxonidine) was analysed using multiple linear regression [57,58]: dependent variable = absorbance (i.e., lipid peroxide levels) and independent variables = groups (with or without moxonidine) and time. The null hypothesis was rejected for two-sided p values of <0.05. All analyses were performed using SPSS version 27.0 (IBM SPSS Statistics for Windows, Armonk, NY, USA, IBM Corporation).

Supplementary Materials: The following supporting information can be downloaded at: https://www.mdpi.com/article/10.3390/ijms24043857/s1.

Author Contributions: Conceptualization: Y.W., K.M.D. and J.G.; investigation: Y.W., D.T.N., J.A., A.A. and P.K.W.; formal analysis: Y.W., D.T.N., J.A., A.A. and P.K.W.; writing—original draft preparation, Y.W. and D.T.N.; writing—review and editing: Y.W., D.T.N., J.A., A.A., P.K.W., Z.C., A.W.K., J.K., K.M.D. and J.G.; Resources, Z.C., A.W.K. and J.K.; funding acquisition, Y.W. The following two authors contributed equally to this work: Y.W and D.T.N. All authors have read and agreed to the published version of the manuscript.

Funding: This research was funded by the National Health and Medical Research Council of Australia, grant number 1062671. Dinh Tam Nguyen was supported by a George Collins Memorial Scholarship and an Australian Government Research Training Program (RTP) Stipend and Fee-Offset Scholarship through Federation University Australia. Jack Anesi was supported by a Fee-Offset Scholarship through Federation University Australia.

Institutional Review Board Statement: The study was conducted in accordance with the Declaration of Helsinki and approved by the Institutional Ethics Committee of Federation University Australia (protocol code, 14-019; date of approval, 21 October 2014).

Informed Consent Statement: Not applicable.

Data Availability Statement: Not applicable.

Conflicts of Interest: The authors declare no conflict of interest.

References

1. Puylaert, P.; Zurek, M.; Rayner, K.J.; De Meyer, G.R.Y.; Martinet, W. Regulated Necrosis in Atherosclerosis. *Arterioscler. Thromb. Vasc. Biol.* **2022**, *42*, 1283–1306. [CrossRef] [PubMed]
2. Amarenco, P.; Labreuche, J.; Lavallée, P.; Touboul, P.-J. Statins in Stroke Prevention and Carotid Atherosclerosis. *Stroke* **2004**, *35*, 2902–2909. [CrossRef] [PubMed]
3. Gaggini, M.; Gorini, F.; Vassalle, C. Lipids in Atherosclerosis: Pathophysiology and the Role of Calculated Lipid Indices in Assessing Cardiovascular Risk in Patients with Hyperlipidemia. *Int. J. Mol. Sci.* **2022**, *24*, 75. [CrossRef]
4. Alshak, M.N.; Das, J.M. Neuroanatomy, Sympathetic Nervous System. In *StatPearls [Internet]*; StatPearls Publishing: Treasure Island, FL, USA, 2019. Available online: https://www.ncbi.nlm.nih.gov/books/NBK542195/ (accessed on 22 November 2022).
5. Wang, Y.; Seto, S.W.; Golledge, J. Therapeutic effects of renal denervation on renal failure. *Curr. Neurovascular Res.* **2013**, *10*, 172–184. [CrossRef] [PubMed]
6. Wang, Y.; Golledge, J. Neuronal nitric oxide synthase and sympathetic nerve activity in neurovascular and metabolic systems. *Curr. Neurovascular Res.* **2013**, *10*, 81–89. [CrossRef] [PubMed]
7. Won, E.; Kim, Y.K. Stress, the Autonomic Nervous System, and the Immune-kynurenine Pathway in the Etiology of Depression. *Curr. Neuropharmacol.* **2016**, *14*, 665–673. [CrossRef]
8. Weissman, D.G.; Mendes, W.B. Correlation of sympathetic and parasympathetic nervous system activity during rest and acute stress tasks. *Int. J. Psychophysiol.* **2021**, *162*, 60–68. [CrossRef]
9. Yao, B.C.; Meng, L.B.; Hao, M.L.; Zhang, Y.M.; Gong, T.; Guo, Z.G. Chronic stress: A critical risk factor for atherosclerosis. *J. Int. Med. Res.* **2019**, *47*, 1429–1440. [CrossRef]
10. Vrablik, M.; Corsini, A.; Tůmová, E. Beta-blockers for Atherosclerosis Prevention: A Missed Opportunity? *Curr. Atheroscler. Rep.* **2022**, *24*, 161–169. [CrossRef]
11. Wang, H.; Wang, J.; Guo, C.; Luo, W.; Kleiman, K.; Eitzman, D.T. Renal denervation attenuates progression of atherosclerosis in apolipoprotein E-deficient mice independent of blood pressure lowering. *Hypertension* **2015**, *65*, 758–765. [CrossRef]
12. Wang, Y.; Dinh, T.N.; Nield, A.; Krishna, S.M.; Denton, K.; Golledge, J. Renal Denervation Promotes Atherosclerosis in Hypertensive Apolipoprotein E-Deficient Mice Infused with Angiotensin II. *Front. Physiol.* **2017**, *8*, 215. [CrossRef] [PubMed]
13. Wang, Y. What is the true incidence of renal artery stenosis after sympathetic denervation? *Front. Physiol.* **2014**, *5*, 311. [CrossRef] [PubMed]
14. Pöyhönen-Alho, M.K.; Manhem, K.; Katzman, P.; Kibarskis, A.; Antikainen, R.L.; Erkkola, R.U.; Tuomilehto, J.O.; Ebeling, P.E.; Kaaja, R.J. Central sympatholytic therapy has anti-inflammatory properties in hypertensive postmenopausal women. *J. Hypertens.* **2008**, *26*, 2445–2449. [CrossRef] [PubMed]
15. Mukaddam-Daher, S.; Gutkowska, J. Imidazoline receptors in the heart: A novel target and a novel mechanism of action that involves atrial natriuretic peptides. *Braz. J. Med. Biol. Res.* **2004**, *37*, 1239–1245. [CrossRef]
16. Daugherty, A.; Manning, M.W.; Cassis, L.A. Angiotensin II promotes atherosclerotic lesions and aneurysms in apolipoprotein E-deficient mice. *J. Clin. Investig.* **2000**, *105*, 1605–1612. [CrossRef]
17. Candido, R.; Jandeleit-Dahm, K.A.; Cao, Z.; Nesteroff, S.P.; Burns, W.C.; Twigg, S.M.; Dilley, R.J.; Cooper, M.E.; Allen, T.J. Prevention of Accelerated Atherosclerosis by Angiotensin-Converting Enzyme Inhibition in Diabetic Apolipoprotein E–Deficient Mice. *Circulation* **2002**, *106*, 246–253. [CrossRef]
18. Danchin, N.; Cucherat, M.; Thuillez, C.; Durand, E.; Kadri, Z.; Steg, P.G. Angiotensin-Converting Enzyme Inhibitors in Patients With Coronary Artery Disease and Absence of Heart Failure or Left Ventricular Systolic Dysfunction: An Overview of Long-term Randomized Controlled Trials. *Arch. Intern. Med.* **2006**, *166*, 787–796. [CrossRef]
19. Reid, I.A. Interactions between ANG II, sympathetic nervous system, and baroreceptor reflexes in regulation of blood pressure. *Am. J. Physiol.* **1992**, *262*, E763–E778. [CrossRef]
20. Charkoudian, N.; Rabbitts, J.A. Sympathetic neural mechanisms in human cardiovascular health and disease. *Mayo Clin. Proc.* **2009**, *84*, 822–830. [CrossRef]
21. Nakagawa, T.; Hasegawa, Y.; Uekawa, K.; Ma, M.; Katayama, T.; Sueta, D.; Toyama, K.; Kataoka, K.; Koibuchi, N.; Maeda, M.; et al. Renal denervation prevents stroke and brain injury via attenuation of oxidative stress in hypertensive rats. *J. Am. Heart Assoc.* **2013**, *2*, e000375. [CrossRef]
22. Ganesan, R.; Henkels, K.M.; Wrenshall, L.E.; Kanaho, Y.; Di Paolo, G.; Frohman, M.A.; Gomez-Cambronero, J. Oxidized LDL phagocytosis during foam cell formation in atherosclerotic plaques relies on a PLD2–CD36 functional interdependence. *J. Leukoc. Biol.* **2018**, *103*, 867–883. [CrossRef]

23. Bennett, M.R.; Sinha, S.; Owens, G.K. Vascular Smooth Muscle Cells in Atherosclerosis. *Circ. Res.* **2016**, *118*, 692–702. [CrossRef]
24. Allahverdian, S.; Chehroudi, A.C.; McManus, B.M.; Abraham, T.; Francis, G.A. Contribution of intimal smooth muscle cells to cholesterol accumulation and macrophage-like cells in human atherosclerosis. *Circulation* **2014**, *129*, 1551–1559. [CrossRef]
25. Wang, Y.; Dubland, J.A.; Allahverdian, S.; Asonye, E.; Sahin, B.; Jaw, J.E.; Sin, D.D.; Seidman, M.A.; Leeper, N.J.; Francis, G.A. Smooth Muscle Cells Contribute the Majority of Foam Cells in ApoE (Apolipoprotein E)-Deficient Mouse Atherosclerosis. *Arterioscler. Thromb. Vasc. Biol.* **2019**, *39*, 876–887. [CrossRef]
26. Prichard, B.N.; Graham, B.R. The use of moxonidine in the treatment of hypertension. *J. Hypertens. Suppl.* **1997**, *15*, S47–S55. [CrossRef] [PubMed]
27. Wenzel, R.R.; Spieker, L.; Qui, S.; Shaw, S.; Luscher, T.F.; Noll, G. I1-imidazoline agonist moxonidine decreases sympathetic nerve activity and blood pressure in hypertensives. *Hypertension* **1998**, *32*, 1022–1027. [CrossRef] [PubMed]
28. Kirch, W.; Hutt, H.J.; Planitz, V. Pharmacodynamic action and pharmacokinetics of moxonidine after single oral administration in hypertension patients. *J. Clin. Pharmacol.* **1990**, *30*, 1088–1095. [CrossRef]
29. Clarke, R.W.; Harris, J. RX 821002 as a tool for physiological investigation of alpha(2)-adrenoceptors. *CNS Drug Rev.* **2002**, *8*, 177–192. [CrossRef] [PubMed]
30. Tolentino-Silva, F.P.; Haxhiu, M.A.; Waldbaum, S.; Dreshaj, I.A.; Ernsberger, P. alpha(2)-adrenergic receptors are not required for central anti-hypertensive action of moxonidine in mice. *Brain Res.* **2000**, *862*, 26–35. [CrossRef] [PubMed]
31. Munk, S.A.; Lai, R.K.; Burke, J.E.; Arasasingham, P.N.; Kharlamb, A.B.; Manlapaz, C.A.; Padillo, E.U.; Wijono, M.K.; Hasson, D.W.; Wheeler, L.A.; et al. Synthesis and pharmacologic evaluation of 2-*endo*-amino-3-*exo*- isopropylbicyclo[2.2.1]heptane: A potent imidazoline1 receptor specific agent. *J. Med. Chem.* **1996**, *39*, 1193–1195. [CrossRef]
32. Stringer, M.D.; Görög, P.G.; Freeman, A.; Kakkar, V.V. Lipid peroxides and atherosclerosis. *Br. Med. J.* **1989**, *298*, 281–284. [CrossRef]
33. Mitrovic, V.; Patyna, W.; Huting, J.; Schlepper, M. Hemodynamic and neurohumoral effects of moxonidine in patients with essential hypertension. *Cardiovasc. Drugs Ther.* **1991**, *5*, 967–972. [CrossRef] [PubMed]
34. Harman, J.L.; Jørgensen, H.F. The role of smooth muscle cells in plaque stability: Therapeutic targeting potential. *Br. J. Pharmacol.* **2019**, *176*, 3741–3753. [CrossRef] [PubMed]
35. Hansson, G.K.; Robertson, A.K.; Söderberg-Nauclér, C. Inflammation and atherosclerosis. *Annu. Rev. Pathol.* **2006**, *1*, 297–329. [CrossRef] [PubMed]
36. González, L.; Rivera, K.; Andia, M.E.; Martínez Rodriguez, G. The IL-1 Family and Its Role in Atherosclerosis. *Int. J. Mol. Sci.* **2022**, *24*, 17. [CrossRef]
37. Tedgui, A.; Mallat, Z. Cytokines in atherosclerosis: Pathogenic and regulatory pathways. *Physiol. Rev.* **2006**, *86*, 515–581. [CrossRef]
38. Pober, J.S.; Min, W.; Bradley, J.R. Mechanisms of endothelial dysfunction, injury, and death. *Annu. Rev. Pathol.* **2009**, *4*, 71–95. [CrossRef]
39. Xiao, J.; Li, N.; Xiao, S.; Wu, Y.; Liu, H. Comparison of Selenium Nanoparticles and Sodium Selenite on the Alleviation of Early Atherosclerosis by Inhibiting Endothelial Dysfunction and Inflammation in Apolipoprotein E-Deficient Mice. *Int. J. Mol. Sci.* **2021**, *22*, 11612. [CrossRef]
40. Ruffolo, R.R., Jr.; Nichols, A.J.; Stadel, J.M.; Hieble, J.P. Structure and function of alpha-adrenoceptors. *Pharmacol. Rev.* **1991**, *43*, 475–505.
41. Marieb, E.N.; Hoehn, K. The Peripheral Nervous System and Reflex Activity. *Hum. Anat. Physiol.* **2018**, *13*, 521–562.
42. Carr, A.C.; McCall, M.R.; Frei, B. Oxidation of LDL by Myeloperoxidase and Reactive Nitrogen Species. *Arterioscler. Thromb. Vasc. Biol.* **2000**, *20*, 1716–1723. [CrossRef] [PubMed]
43. Elisaf, M.S.; Petris, C.; Bairaktari, E.; Karabina, S.A.; Tzallas, C.; Tselepis, A.; Siamopoulos, K.C. The effect of moxonidine on plasma lipid profile and on LDL subclass distribution. *J. Hum. Hypertens.* **1999**, *13*, 781–785. [CrossRef] [PubMed]
44. Mironova, M.A.; Klein, R.L.; Virella, G.T.; Lopes-Virella, M.F. Anti-modified LDL antibodies, LDL-containing immune complexes, and susceptibility of LDL to in vitro oxidation in patients with type 2 diabetes. *Diabetes* **2000**, *49*, 1033–1041. [CrossRef] [PubMed]
45. Schneider, W.J. The low density lipoprotein receptor. *Biochim. Biophys. Acta* **1989**, *988*, 303–317. [CrossRef]
46. Rupp, H.; Maisch, B.; Brilla, C.G. Drug withdrawal and rebound hypertension: Differential action of the central antihypertensive drugs moxonidine and clonidine. *Cardiovasc. Drugs Ther.* **1996**, *10* (Suppl. 1), 251–262. [CrossRef]
47. Krishna, S.M.; Li, J.; Wang, Y.; Moran, C.S.; Trollope, A.; Huynh, P.; Jose, R.; Biros, E.; Ma, J.; Golledge, J. Kallistatin limits abdominal aortic aneurysm by attenuating generation of reactive oxygen species and apoptosis. *Sci. Rep.* **2021**, *11*, 17451. [CrossRef]
48. Moran, C.S.; Seto, S.W.; Krishna, S.M.; Sharma, S.; Jose, R.J.; Biros, E.; Wang, Y.; Morton, S.K.; Golledge, J. Parenteral administration of factor Xa/IIa inhibitors limits experimental aortic aneurysm and atherosclerosis. *Sci. Rep.* **2017**, *7*, 43079. [CrossRef]
49. Beck, K.; Wu, B.J.; Ni, J.; Santiago, F.S.; Malabanan, K.P.; Li, C.; Wang, Y.; Khachigian, L.M.; Stocker, R. Interplay Between Heme Oxygenase-1 and the Multifunctional Transcription Factor Yin Yang 1 in the Inhibition of Intimal Hyperplasia. *Circ. Res.* **2010**, *107*, 1490–1497. [CrossRef]
50. Youssef, P.; Chami, B.; Lim, J.; Middleton, T.; Sutherland, G.T.; Witting, P.K. Evidence supporting oxidative stress in a moderately affected area of the brain in Alzheimer's disease. *Sci. Rep.* **2018**, *8*, 11553. [CrossRef]

51. Wang, Y.; Nguyen, D.T.; Yang, G.; Anesi, J.; Chai, Z.; Charchar, F.; Golledge, J. An Improved 3-(4,5-Dimethylthiazol-2-yl)-5-(3-Carboxymethoxyphenyl)-2-(4-Sulfophenyl)-2H-Tetrazolium Proliferation Assay to Overcome the Interference of Hydralazine. *Assay Drug Dev. Technol.* **2020**, *18*, 379–384. [CrossRef]
52. Wang, Y.; Liu, H.; McKenzie, G.; Witting, P.K.; Stasch, J.P.; Hahn, M.; Changsirivathanathamrong, D.; Wu, B.J.; Ball, H.J.; Thomas, S.R.; et al. Kynurenine is an endothelium-derived relaxing factor produced during inflammation. *Nat. Med.* **2010**, *16*, 279–285. [CrossRef] [PubMed]
53. Livak, K.J.; Schmittgen, T.D. Analysis of relative gene expression data using real-time quantitative PCR and the $2^{-\Delta\Delta CT}$ method. *Methods* **2001**, *25*, 402–408. [CrossRef] [PubMed]
54. Wang, Y.; Nguyen, D.T.; Yang, G.; Anesi, J.; Kelly, J.; Chai, Z.; Ahmady, F.; Charchar, F.; Golledge, J. A Modified MTS Proliferation Assay for Suspended Cells to Avoid the Interference by Hydralazine and β-Mercaptoethanol. *Assay Drug Dev Technol* **2021**, *19*, 184–190. [CrossRef] [PubMed]
55. Liu, C.; Wu, J.; Jia, H.; Lu, C.; Liu, J.; Li, Y.; Guo, M. Oncostatin M promotes the ox-LDL-induced activation of NLRP3 inflammasomes via the NF-κB pathway in THP-1 macrophages and promotes the progression of atherosclerosis. *Ann. Transl. Med.* **2022**, *10*, 456. [CrossRef] [PubMed]
56. Wang, Y. Definition, prevalence, and risk factors of low sex hormone-binding globulin in US adults. *J. Clin. Endocrinol. Metab.* **2021**, *106*, e3946–e3956. [CrossRef]
57. Wang, Y.; Charchar, F.J. Establishment of sex difference in circulating uric acid is associated with higher testosterone and lower sex hormone-binding globulin in adolescent boys. *Sci. Rep.* **2021**, *11*, 17323. [CrossRef] [PubMed]
58. Cheng, W.; Wen, S.; Wang, Y.; Qian, Z.; Tan, Y.; Li, H.; Hou, Y.; Hu, H.; Golledge, J.; Yang, G. The association between serum uric acid and blood pressure in different age groups in a healthy Chinese cohort. *Medicine* **2017**, *96*, e8953. [CrossRef] [PubMed]

MDPI
St. Alban-Anlage 66
4052 Basel
Switzerland
www.mdpi.com

International Journal of Molecular Sciences Editorial Office
E-mail: ijms@mdpi.com
www.mdpi.com/journal/ijms

www.ingramcontent.com/pod-product-compliance
Lightning Source LLC
Chambersburg PA
CBHW061552210926
51850CB00009B/2084